Data-Driven Approaches in Transportation Engineering

DATA-DRIVEN APPROACHES IN TRANSPORTATION ENGINEERING

Dr. Bechoo Lal

ARCLER
PRESS

www.arclerpress.com

Data-Driven Approaches in Transportation Engineering

Dr. Bechoo Lal

Arcler Press
4164 Lakeshore Road
Burlington, ON L7L 1A4
Canada
www.arclerpress.com
Email: orders@arclereducation.com

ISBN: 978-1-77956-978-3 (Hardcover)

ABOUT THE AUTHOR

Bechoo Lal, PhD. became a Member (M) of IAENG: International Association of Engineers, USA with membership (108820) in 2010, a Senior Member (SM) in 2019. I am doctorate PhD in Computer Science, PhD- Information System from University of Mumbai, M.Tech-CSE, Master of Computer Application (MCA) - Banaras Hindu University (BHU), India, and PGP- Data Science from Purdue University, USA. Currently I am working as an Associate Professor in Department of Computer Science & Engineering, Koneru Lakshmaiah Education Foundation (KLEF) - KL University Vijayawada Campus Andhra Pradesh, India. In addition to this I am supervising PhD research scholars from SJJT University, Rajasthan, India. My research areas are data science, big data analytics and Machine Learning.

TABLE OF CONTENTS

LIST OF FIGURES

PREFACE

Transport networks are at the core of modern economies, shaping the movement of people, freight, and services into and out of cities, nations, and continents. The performance, safety, and sustainability of transport networks are no longer as much a function of traditional engineering imperatives but more of the power of data. Every sensor reading along roads, every GPS ping from an auto, and every swipe by a passenger on a smart card generates streams of valuable information. Unleashing this information requires a shift toward approaches that put data-driven insights first to plan, design, and manage transportation systems with greater precision and efficiency. In a time of rapid urbanization, climate stresses, and technological advancement, information is turning out to be the most significant asset for transforming mobility.

The transition from intuition and experience-driven decision-making to data-driven methods has transformed the profession. No longer do planners and engineers have to wait for post-hoc analysis; instead, they can now forecast congestion patterns, predict accident hazards, and monitor infrastructure performance in real time. Machine learning, intelligent transportation systems, and high-performance analytics are transforming the way professionals learn about mobility needs and respond to them. This is not just a technical revolution—it is social, as better decisions in transportation engineering directly affect the public's safety, accessibility, and quality of life. By connecting information across numerous layers of transportation planning, researchers and practitioners are building wiser, more resilient, and more adaptive systems that are capable of facing the challenges of today and tomorrow.

Data-Driven Approaches in Transportation Engineering has been produced to offer an integrated guide to understanding data's role in shaping the development of transport practice. It balances theoretical concepts with empirical methods, pointing out how novel tools and techniques can transform the planning, operation, and upkeep of transport networks. The chapters presented here highlight not just the principles of data collection and management but also the practical application of analytics, machine learning, and smart technologies to develop safer and more efficient mobility solutions. The book seeks to bridge the gap between academic knowledge and practitioner use so that readers not only get conceptual insights but also useful information.

Chapter 1 gives an introduction to data-driven transport systems and an overview of why data is central to engineering practice today. It outlines how the discipline emerged, what form transport data takes, and the advantages of moving towards evidence-based decision making.

Chapter 2 also covers data collection instruments and techniques, both traditional methods such as surveys and traffic counts and innovative techniques such as sensors, GPS tracking, and remote sensing. It examines the accuracy, reliability, and limitations of each technique while emphasizing the value of correlating various sources of data to expose the complexity of transport systems.

Chapter 3 discusses data processing and management, highlighting the need for organized means of storing, cleaning, and preparing data for use. Database systems, big data platforms, and preprocessing techniques that ensure data quality are all discussed in this chapter. It also outlines how nicely preprocessed data can form the foundation for more advanced analytical techniques.

Chapter 4 covers statistical and analytical methods existing for inference and interpretation in transportation engineering. Important topics include regression models, hypothesis testing, time series, and spatial statistics. The chapter places emphasis on how quantitative methods help identify concealed patterns, facilitate evidence-based conclusions, and validate decision-making.

Chapter 5 examines the application of machine learning to transportation and the way algorithms identify patterns, predict results, and enable sophisticated decisions automatically. From demand prediction to traffic prediction, the chapter shows how supervised learning and unsupervised learning can solve entrenched issues in transportation engineering.

Chapter 6 is dedicated to smart transport systems (ITS) where technology and data unite to achieve optimum mobility. The topic discusses issues like adaptive traffic control, connected vehicles, and real-time monitoring platforms. The chapter highlights how ITS applications provide improved safety, reduce congestion, and aid efficient system management.

Chapter 7 covers data-driven public transport planning, explaining the way data supports service design, timetabling, route planning, and passenger demand modeling. With growing pressure to deliver a more efficient and sustainable transport, this chapter demonstrates how data can enable authorities to create people-friendly and effective systems.

Chapter 8 discusses safety and risk analysis with data, illustrating the ways in which predictive models can identify risky locations, assess the driving behavior of individuals, and design safer roads. It is stressed how data plays a role in reducing accidents as well as increasing resilience, so that safety factors are always given priority when making engineering decisions.

Chapter 9 discusses sustainable and smart mobility solutions. It delves into how data can be used to reduce emissions, facilitate electric vehicle integration, and make mobility-as-a-service platforms. The chapter connects sustainability goals with data-driven methodologies, reflecting the broader societal effort towards environmentally friendly alternatives.

Chapter 10 concludes with a look at policy, ethics, and the future of data-driven transportation. It addresses the challenges of privacy, data stewardship, and equity looking to the promise on the horizon presented by artificial intelligence, autonomous systems, and digital twins. The chapter leaves readers with respect for possibilities and duties in the evolving field.

This book has been authored for a general audience that cuts across students, researchers, engineers, and policy makers. For undergraduate and master's students of civil and transport engineering, the book presents an overt example of the blending of data science and core engineering principles. Researchers can look forward to a thorough review of contemporary techniques that will shape future innovation and inter-disciplinary research. Professional planners and engineers can use the content as a guide to applying cutting-edge tools and techniques in real-world projects. Policymakers and administrators will gain insights into how data-led methods can influence transportation policy and governance. In summary, this book is for anyone who wants to revolutionize the transportation system by tapping the power of data and enable them to make informed and effective decisions.

CHAPTER 1

INTRODUCTION TO DATA-DRIVEN TRANSPORTATION SYSTEMS

Contents

INTRODUCTION

The growth of transport networks has arrived at a revolutionary phase that is data-driven, and the ongoing enhancement in vehicle technology, communications networks, and analytics has resulted in the stakeholders making greater use of these technologies to optimize the transport of goods and people. Such technologies as ride-sharing platforms and real-time traffic management illustrate how data transforms mobility services with demand-supply matching. The workshop, "Integrating Electric Mobility Systems with the Grid Infrastructure," stressed the need for a collaborative research agenda responding to the dynamic interface between new technologies and traditional transportation infrastructures. With the rapid pace of technological development, effective measures must be determined to harmonize the development of transport systems with clean energy initiatives in a way that information is utilized as a foundation for sustainable and effective modes of transport (Cassandras et al., 2019).

Figure 1.1. Data-driven transportation technology systems.

Source: https://miro.medium.com/v2/resize:fit:1358/1*OUZm4c4S5a1XGuhdEV-8Sw.jpeg.

The introduction of electric vehicles into transport systems also underscores the necessity to apply the use of a data-driven framework. As explained in recent reports, cities with good public transport systems benefit economically, particularly when gas prices are rising, as a sign of necessity for investment in other sources of mobility. Urban households that use public transport and appreciate it indicate enhanced access to cheap mobility, as well as a sign of necessity for aggregate data analysis in modeling these impacts (N/A, 2005). Installation of intelligent charging points and vehicle-to-grid technologies is an example of how data systems converging with electric mobility leads to reduced congestion and better energy management. By adopting a data-driven approach to planning and implementing such systems, cities not only get maximum efficiency from their transport networks but also contribute to broader environmental goals. With data-centric transport dynamics, policymakers are faced with a variety of challenges and uncertainties. With the expansion of on-demand mobility solutions, and the advent of autonomous vehicles, there is a need to have a highly evolved understanding of consumer behaviors and infrastructure demands. Powering electric cars and decongesting roads is an issue that requires thorough theoretical and empirical analysis. The workshop convened academic and industry specialists to debate these pressing issues and to predict potential future scenarios. By means of interdisciplinary collaboration and utilizing cutting-edge data analytics, stakeholders can better predict the multifaceted impacts of novel mobility solutions on regional economies as well as on urban development. With these systems on the move, the emphasis on data-driven strategies will be the key in determining transportation's future in a manner that it becomes aligned with social and environmental objectives (Cassandras et al., 2019) (Figure 1.1).

Definition of Data-Driven Transportation Systems

In the context of modern urban mobility, data-driven transport systems are those transport systems that rely on vast amounts of data to optimize the efficiency and effectiveness of transport solutions. These systems employ various technologies, such as sensors, GPS, and data analytics, to collect, process, and act on information related to travel behavior, vehicle performance, and infrastructure conditions. This method enables real-time decision-making, leading to optimized routes, less congestion, and enhanced delivery of service. By combining information from various sources such as public transport timetables, user preferences, and weather conditions, data-based systems empower operators to react ahead of changing dynamics. As city centers become more and more filled with congestion and air quality challenges, such systems rise to the occasion in developing dependable solutions to transportation that can accommodate the needs of the residents and commuters. Data-driven transport systems are not merely gathering data; they are an evolving network that continues to get better incrementally based on predictive analysis as well as user feedback. For instance, as seen in emerging logistics models like Delivery as a Service (DaaS), businesses are relying increasingly on real-time data to streamline operations and meet customer demands in an on-demand environment. Integrating cutting-edge IT solutions into urban logistics operations enhances supply chain efficiency and produces a smooth transport process. In this case, timely and accurate examination of information is a landmark towards responsiveness in transport services to facilitate adaptive adjustments and improve delivery performance. This not only benefits the providers but significantly boosts the satisfaction of users and proves that data-driven measures form the foundation of future transport practice. One of the most significant things about data-driven transport systems is that they have the ability to support cooperative strategies in unifying disparate stakeholders like government organizations, private enterprises, and citizens. The implementation of new systems, as defined in visions like the Virtual Integration Platform (VIP), relies on the unification of inputs from various sectors in an attempt to achieve integrated transport strategies. This coordination is essential in handling the complexities of urban air transport concepts, for instance. By employing both human sense and technological capability in planning, transport systems can be oriented more towards future needs of safety, sustainability, and efficiency. The ability to derive meaning from trends as well as user input also assists in formulating guidelines and policies that not only optimize operation but also resonate with community values and needs. Essentially, data-based transport systems represent a tremendous step toward integrated, user-centered mobility solutions that are crucial to urban settings in the future.

Importance of Data in Modern Transportation

In the current transport era, the role of data cannot be overstated. With growing populations and cities expanding further, transport infrastructure has also become exceedingly complex, and hence the utilization of data in order to make it more efficient and sustainable. Transport networks typically bear the problems of congestion, fluctuating levels of service, and environmental demands, for which the utilization of data becomes essential in an attempt to address the same. With the application of big data, transport planners can study traffic flow, predict variations in demand, and modify routes accordingly. All this heavy reliance on information not only allows real-time decisions to be put into effect according to changing circumstances but also facilitates longer-term planning that improves overall transit system resilience. Data-driven approaches are also revolutionizing the regulation of transport safety. Use of real-time data analytics assists in monitoring traffic conditions, identification of dangerous areas, and preemption of accidents before they happen. Safety of vehicles through collection and analysis of data is one of the most prioritized issues in current transport systems as it minimizes risks to drivers, passengers, and pedestrians alike. Additional levels of data collection, such as behavioral analysis through smartphone apps and onboard vehicle monitoring sensors, are employed to augment a fuller understanding of safety practice. Not only does this process induce proactive responses in safety management but also regulatory compliance and safer practice across the transport sector. The effect of data extends far beyond safety and efficiency; they have a powerful impact on sustainability too. Accurate data collection has established a linear trend between transport systems and carbon emissions, suggesting the tracing of environmental footprints. Understanding demographic variations, such as socioeconomic factors and their influence on transport behavior, allows interventions to be accordingly targeted to facilitate green behavior. Specifically, the study has demonstrated that socioeconomic capital is associated with transport emission patterns, calling for particular strategies that address different community needs (Ronald et al., 2019). Further, when one is studying the inventory management systems of the pharmaceutical industry, it is apparent that transportation data-driven strategies also need to consider the intricate interdependencies between multiple supply chain drivers, pointing towards a multidimensional approach towards sustainability and efficiency (Andreu et al., 2017). In putting data utility as a top concern, the transport sector can enable intelligent decision-making that, in the end, equates to more sustainable and inclusive systems (Figure 1.2).

Figure 1.2. Importance of data in modern transportation.

Source: https://www.ptvgroup.com/sites/default/files/styles/1x1_2560/public/2023-07/autonomous-cars.jpeg.webp?h=8c8994dd&itok=hSYe8q3S.

1.1. HISTORICAL CONTEXT OF TRANSPORTATION SYSTEMS

The evolution of transport systems manifests a growth stimulated largely by technological advances, the needs of society, and economic imperatives. Previous transport modes, such as footpaths and animal-drawn vehicles, gave way to more sophisticated networks. The Industrial Revolution brought sweeping changes in the form of the steam engine usage giving way to the internal combustion engine, which transformed movement and freight transport radically. As the urban areas expanded, the cities developed with extensive rail and road infrastructure, resulting in greater connectivity and economic progress. These advances brought with them new challenges, like congestion, pollution, and unequal access to transport resources, that helped underscore the need for reform and integration of fresh transport sustainability and diversity of technology solutions. As constituents increasingly recognize the effects of such prior decisions, they must balance innovation with responsibility and ethical demands for the social and environmental effects of transportation (Cassen et al., 2008). The innovation of car and airplane technology revolutionized business and personal transport during the 20th century, further embedding cars into daily life. With the rise of automobile culture, cities evolved, inclining towards road infrastructure at the cost of mass transportation networks. This shift transformed not only urban patterns but also lifestyle trends, encouraging suburban growth and increasing dependence on fossil fuels. The middle century saw embryonic discussions on sustainability, taking into account the environmental cost of transport and the need for alternative energy sources. The growth of hydrogen and hybrid technology soon developed into future controversies over the direction of transport systems, emphasizing the need for multi-modal transport rather than a single technological solution. Stakeholders' changing attitude towards societal values regarding mobility reflects increasing sensitivity to the need for change in transportation modes (Buscaglia, 2025). Historical evolution in transport encapsulates the complex relationships of technology, society, and the environment that shape ongoing systems to this day. As changing transport problems in the modern era necessitate a response, data-driven approaches led by historical factors become essential. Data analytics enable transport planners to establish the efficiency of various modes of transport, optimize traffic flow, and enhance user experience and facilitate sustainability. A multi-modal transport system that incorporates public transport, cycling, walking, and new technologies can make efficient travel within urban areas possible, eventually countering issues created by earlier infrastructural decisions. While planning ahead towards data-driven systems, it is useful to reflect on the past context for both revealing earlier mistakes as well as designing strong, adaptive, and equitable transportation systems in the future (Cassen et al., 2008).

1.1.1. Evolution of Transportation Methods

From wheel invention in some of the ancient civilizations to discovering the steam engine, every technological advancement has totally transformed the way of traveling as well as trading. Development of transport technology has revolutionized societies all over the world; not only does it enable people mobility but also transport of goods and services. Horse-drawn boats and carriages were primitive modes of transport that enabled local trade and movement, but with limited speed and efficiency. With the industrial revolution, there were significant advances in terms of railways and steamships that enabled mass transport and trade over long distances. With each new innovation, the efficiency of transport increased, giving rise to today's complex, data-driven transport systems incorporating multiple modes of transportation and logistics, all part of a continuum of improvement and adaptation.

Figure 1.3. Evolution of transportation methods.

Source: https://st.adda247.com/https://adda247jobs-wp-assets-adda247.s3.ap-south-1.amazonaws.com/jobs/wp-content/uploads/sites/2/2022/12/21153229/modes-of-transportation.png.

As technology progressed, so did the ability and sophistication of transport systems. The 20th century was marked by significant milestones in the invention of the car and commercial air transport, which altered societies' response to distance. The growth of cars spawned suburbanization, changing the spatial interaction within cities and introducing with it the beginning of new traffic jams and ecological issues. Concurrently, advances in airline travel significantly reduced the duration required to travel between destinations across the globe, making economic linkage between nations possible. Yet this also resulted in inefficiencies and the tendency toward the slower is faster (SIF) effect, through which individual high-speed efforts in transport can unintentionally reduce overall system performance in traffic and logistics (Gershenson et al., 2015). Recognizing these dynamics is essential to creating future urban and intercity transport models that successfully leverage data-driven approaches to reduce such problems. The transportation modes are experiencing another change today with the amalgamation of data-centric technologies, which is rewriting the mobility strategy. Technologies like GPS, real-time traffic sensing, and autonomous vehicles demonstrate the transition to the data-driven transportation models. Such a transformation is dependent upon the data collected and analyzed regarding transportation, which helps in making better micro and macro-level decision-making. Also, data-informed solutions can optimize logistics through improved routes efficiencies, demonstrating the potential to eliminate congestion and reduce carbon footprints in an immensely effective way. With design and operational decisions being furthered by data, the transport industry is bound to continue changing in the future by integrating more intelligent systems with the ability to adapt to user needs and alterations in the environment while relying on previous changes that have brought us to this point (Agostinelli et al., 2005) (Figure 1.3).

1.1.2. Introduction of Technology in Transportation

The integration of technology into transport has altered the dynamics of how mobility solutions are interacted with by society, reconstituting new paradigms. Expansion in communication technologies, and most notably the proliferation of smartphones, has made possible the establishment of Transportation Network Companies (TNCs) such as Uber and Lyft. The services are the manifestation of data-driven technology's impact on consumer convenience, providing on-demand mobility that dynamically responds to consumer needs. This shift to on-demand services, facilitated by sophisticated algorithms and instant data analysis, reflects the phasing out of old-fashioned stationary transportation systems in favor of more liquid and responsive ones. The concomitants of these changes hold the promise not only of greater efficiency in individual transport but also of reduced congestion and improved urban planning as cities adjust to such innovations. Further, the application of electric mobility technologies is expanding as part of an all-around transportation commitment to sustainability. Recent workshops, such as one sponsored by the Sloan Foundation, illustrate how the convergence of electric utilities and transportation systems is essential to the development of a cleaner energy infrastructure. As electric vehicles (EVs) become increasingly practical, the need for simultaneous improvement in charging infrastructure becomes apparent. The rapid development of these technologies demands reconsideration of existing transportation models and brings new challenges and opportunities for urban mobility. The context demands accelerating technology development coupled with policy-making with the aim of addressing the complexities in electric charging systems while fostering stability in transport and energy sectors, thereby enabling development towards sustainable transport models (Cassandras et al., 2019). In addition, the complete ramifications of such developments extend into aviation, where state-of-the-art methods are being explored for future air transportation concepts. The Virtual Integration Platform (VIP) concept is a synergistic approach using IT tools and collaborative methods to manage the challenge of air transportation systems complexity. With its focus on constructing future concepts that are Quiet and Clean for short hops, Comfortable and Clean for long-range travel, and Fast and Flexible for personal travel, the aviation sector mirrors ground transportations embracing data-driven approaches. This intersection of technological innovation across transportation modes not only encourages interconnectivity but also indicates the importance of a single, integrated approach to transportation planning. As these sectors continue to develop, the recognition of their interconnectedness will be critical in guiding effective, sustainable mobility systems (Gollnick et al., 2011) (Figure 1.4).

Figure 1.4. Introduction of technology in transportation.

Source: https://track2training.com/wp-content/uploads/2024/05/dallc2b7e-2024-05-14-13.24.51-a-futuristic-cityscape-showcasing-various-ai-applications-in-public-transport.-the-scene-includes-autonomous-buses-and-trains-smart-traffic-lights-p.webp.

1.1.3. Early Data Collection Methods in Transportation

The evolution of the initial data collection methods within transportation has played a crucial role in shaping the development of data-based systems. Data collection was conducted primarily manually, with observations on the levels of traffic and the volume of vehicles undertaken through observational surveys. Researchers employed practices such as manual counts, where transport specialists counted the number of vehicles at a single point within a specific time frame. This early simple approach, as limited in scope as it was, set the stage for more sophisticated subsequent data collection processes.

The simplicity with which these early approaches functioned belied their importance, as they accessed fundamental data on flow rates, travel time, and split modal, which was utilized to inform transport planning and policy development. These methods were, nonetheless, slow and prone to the vagaries of human influence, prompting the quest for higher levels of accuracy and efficiency in data gathering tools. With advancements in the transport sector, access to automated counting systems was a great step in terms of data collection procedures. Automated counting systems utilized technologies like inductive loops, infrared sensors, and cameras to measure traffic volume and vehicle type in real-time. Enhanced accuracy by automated systems reduced the requirement for manual input data so that more detailed data analysis could be done for multiple temporal and spatial settings. Development in these technologies not only improved the efficiency of data collection but also increased the rate and complexity of information, so that transport planning could be better informed. But that transition towards automated methods also introduced a necessity for thinking about calibration and maintenance processes so that integrity in data could be ensured. With an increase in the complexity of transportation infrastructures, the problem of data quality and interoperability between systems arose, necessitating regular upgrading of data gathering processes that, in the long run, would facilitate smart transport systems. Blending early data gathering strategies with robust transport policies has had long-lasting implications on environmental impact studies as well as mitigation. Importantly, fine-scale transportation emissions inventories are critical to understanding and reducing greenhouse gas emissions from car travel. The methods applied at these early stages set the trend for developing guidelines that could systematically assess emissions data to inform climate policies for sustainability. As research evolved, it became clear that the reconciliation of sound data collection and robust analytical methods was required to measure transportation systems' social and environmental effects. Subsequently, challenges continue to persist to bring together collected emissions data and policy frameworks to create utilisable information. As emphasized by linked research, methods for transportation emissions inventory remain vital in guiding strategic reduction measures, enabling the maintenance of continued improvement and data-informed methodology convergence in contemporary transport systems (Bishins et al., 2010).

1.2. KEY TECHNOLOGIES IN DATA-DRIVEN TRANSPORTATION

The integration of advanced technology in data-driven transport systems is transforming the way transport networks operate and react to users. One such technology is the Virtual Integration Platform (VIP) that employs an IT toolchain with collaborative human methodologies to tackle the complexities that form part of the nature of transport systems. With the facilitation of a synergistic development of air transport ideas, VIP platform promotes innovative design, resolving diversified requirements such as short-range efficiency, long-range comfort, and personal transportation flexibility. These developments play a key role in addressing ecological concerns and enhancing user satisfaction and point towards the determinative function of technology in shaping the transport system of the future whilst promoting sustainability and usability during the design phase (Gollnick et al., 2011). Another key technological advancement in transport via data is the application of state-of-the-art localization techniques, particularly for driver assistance applications and autonomous driving. GPS-based conventional systems have low road-level resolution that is inadequate for precise lane-level decision-making required in navigation. To overcome this deficiency, Light Detection and Ranging (LiDAR) sensors are an advanced technology used to offset global localization inaccuracies to deliver high accuracy. However, issues with computational burdens and costs still exist for LiDAR technology. Vision-based Lane change detection systems have consequently been in demand as a cheaper alternative, using deep learning models to analyze images captured using on-board cameras on vehicles. The approach has been highly effective in practical implementation, recording accuracy rates of around 87%, thus improving significantly the efficiency and safety of self-driving transportation systems (Barth et al., 2019). The convergence of these kinds of technologies highlights the importance of real-time data processing and adaptability in modern transportation planning. Data-driven methods not only maximize the performance of individual vehicles but also support more complex traffic management systems and are easily optimized for city mobility. With the implementation of real-time data and machine learning algorithms, transport systems become capable of dynamically adjusting to changing conditions, thus optimizing overall efficiency, and reducing traffic congestion. Also, with the increasing use of self-driving vehicles, technologies like the VIP and luxury localization systems will form the basis of creating smooth links among all modes of transport. Ultimately, these technologies are pushing the way to smart urban environments, where transport networks can address burgeoning mobility demands while remaining oriented towards safety and sustainability (Barth et al., 2019; Gollnick et al., 2011) (Figure 1.5).

Figure 1.5. Key technologies in data-driven transportation systems.

Source: https://aredgroup.com/wp-content/uploads/2024/03/How_Smart_Technology_is_Enhancing_Transportation_2024_03_20_07_13_45_334784_00_00.jpeg.

1.2.1. Internet of Things (IoT) and Its Applications

The Internet of Things (IoT) is a paradigm-changing phenomenon in the transport systems sector, where better-informed decision-making and optimized operations are made possible. Through the use of a sophisticated web of sensors and networked devices, transport organizations can tap into real-time data on vehicle performance, traffic flows, and weather conditions. The connectivity provides valuable data to urban planners and logistics companies, and it is now possible to develop smarter transport solutions. With continued advancements in IoT technologies, cities are increasingly becoming highly capable of routing traffic and optimizing route dynamically. There is a need to simulate these interrelated environments in order to forecast performance under various conditions, as uncovered in D'Angelo et al. (2017). By simulation, stakeholders are able to study the feasibility of adopting innovative transport strategies that utilize real-time information towards improved performances. But still another significant aspect of IoT in transport is its capacity to facilitate hybrid simulations that are effective in simulating complicated environments. Hybrid simulation techniques allow one to merge various simulation models and thereby create adaptive systems that can capture the multi-dimensional character of IoT systems. Hybrid simulations decompose such intricate environments into manageable segments, and each segment can be simulated using individual methods, as outlined in D'Angelo et al. (2017). In this way, transport planners can imagine how smart infrastructure can connect with real-time data from devices and vehicles. Further, with the syncretic application of agent-based and discrete-event simulation methods, practitioners can achieve high fidelity in simulating varied situations that characterize modern urban transport. The learnings are especially important in creating smart systems that not only capture optimal resource allocation but also enhance user experiences in real time. The role of IoT in promoting sustainability within transport infrastructure cannot be overstated. Intelligent devices and networked infrastructure enable the monitoring and management of resources in an environmentally friendly way. IoT-powered traffic management systems can avoid congestion and lower emissions by optimizing traffic lights and providing real-time information to drivers on alternative routes. Furthermore, data collected from several IoT applications may inform policymakers on the environmental footprint of transportation modes to facilitate data-driven interventions. While cities struggle with increasing populations and concomitant challenges, green uses of IoT will be pivotal in establishing green urban transport systems. By employing such technologies as discussed in D'Angelo et al. (2017), planners are able to attain innovative designs that facilitate efficient and sustainable mobility in cities for future generations (Figure 1.6).

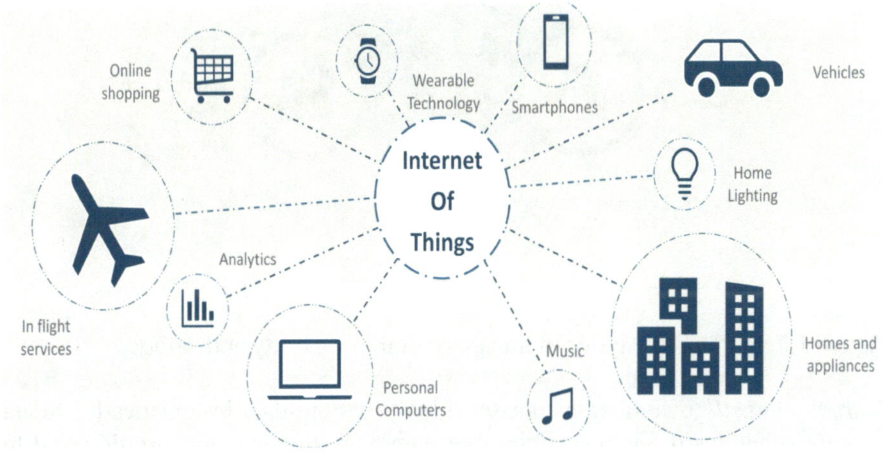

Figure 1.6. Internet of things (IoT) and its applications.

Source: https://www.edureka.co/blog/wp-content/uploads/2018/09/iot1.png.

1.2.2. Big Data Analytics in Transportation

The transport big data analytics revolutionary impact is clearer now as the industry works on enhancing various mobility and safety aspects. With large quantities of data provided by automobiles, traffic management systems, and infrastructure, decision-making can be informed by generating timely and potent insights. The advent of intelligent transportation systems (ITS) embodies the spirit of the shift, where driving experience, traffic flow management, and accident analysis are enhanced. Information technology developments, combined with the pervasiveness of the Internet of Things (IoT), provide real-time data collection and processing capacities and form a basis for more intelligent urban development and driving assistance systems. Ultimately, the strategic use of big data analytics not only fosters operational effectiveness but also provides a framework in which data-driven strategies can engage actively to solve problems of modern transportation (Arabnia et al., 2020).

Figure 1.7. Big data analytics in transportation of mobility and safety.

Source: https://gowombat.team/assets/blog-posts/big-data-in-logistics-how-data-analytics-changes-the-industry/how-big-data-analytics-in-transportation-and-logistics-is-used.webp.

The use of big data analytics for transportation relies heavily on interoperability between several forms of data sets, such as traffic flow, weather conditions, and vehicle telemetry. State transportation agencies have recognized the potential that their data stores contain, but the complexity and volume of these data sets call for sophisticated analytical capabilities. For example, the TITAN (Transportation data InTegration and ANalytics) platform demonstrates a cutting-edge approach integrating big databases to facilitate insights on mobility, safety, and transit performance. By offering a user-friendly interface along with an elastic back end, TITAN demonstrates how advanced analytics facilitates transportation data to improve the functionality, ultimately resulting in facilitating agencies to make educated decisions that minimize risks and optimize resources (Shu et al., 2025). The ongoing expansion of such platforms underscores the necessity to marry technology and data analysis in creating robust, data-informed transportation systems. Future expansion in big data analytics in transportation will not only affect operational efficiency but also environmental sustainability. Through the potential to analyze data to inform functions like traffic flow optimization, there can be revolutionary reductions in carbon emissions that address pressing ecological concerns. With advancements in technology in big data, applications can be expanded to predictive analytics for the detection and avoidance of accidents and for public safety directly. More collaboration among automobile companies, city planners, and traffic agencies will be needed to leverage the complete potential of these data systems. Therefore, building a culture of data sharing and collective intelligence is able to make stakeholders develop more effective transport solutions that are not only beneficial to users but also to the environment. The long-term emphasis on big data analytics guarantees that transport systems remain agile and responsive to the complex realities of urban mobility (Arabnia et al., 2020) (Figure 1.7).

1.2.3. Role of Artificial Intelligence in Optimizing Transport Systems

The union of artificial intelligence (AI) and transportation networks is a revolutionary step toward optimizing metropolitan mobility and efficiency. AI technologies harness enormous datasets to analyze traffic patterns, enabling dynamic routing and improved logistics planning. This data-driven approach not only minimizes travel time but also reduces environmental impact via optimal fuel consumption and emission reduction. By integrating machine learning technology into traffic management systems, cities can anticipate congestion points and adjust signal cycle lengths and public transport timetables dynamically. Such innovations have led to the development of intelligent transportation systems (ITS) that apply AI in making urban transportation solutions more streamlined, safe, and environmentally friendly, representing a broader move toward smart cities as referenced in Doctor et al. (2020). In addition, the implementation of AI overflows into vehicle autonomy, whereby it enables the development of autonomous technology. Utilizing advanced machine learning algorithms, vehicles would be able to scan their outer environment, thus making instant decisions. The advent of 6G wireless technology is likely to amplify these capabilities by providing seamless connectivity and data transfer among vehicles, infrastructure, and users. These advancements are crucial in establishing a collaborative transport system where cars and infrastructure can talk to each other efficiently, enhancing security and ease of use within transport systems. As cited in (Abbas et al., 2020), these advancements not only represent a step in the autonomy of vehicles but a revolution in the way transport systems operate and integrate with cityscapes. Furthermore, the role of AI in optimizing transport systems hinges on its ability to drive predictive analytics that inform policy and infrastructure investments. Through the analysis of big data generated from various transport modalities, stakeholders can identify trends and potential issues before they escalate. This proactive approach allows for informed decision-making regarding infrastructure development, ensuring resources are allocated effectively to meet future demands. The integration of AI with big data also encourages harmonization in multi-modal transport systems because diversified transport modes become connected through data-sharing systems. The growth of urban populations is increasing, and the integration of AI into transport systems will remain central to addressing congestion, connectivity, and sustainability challenges, promoting the shift to intelligent data-based transportation systems as noted in Doctor et al. (2020) (Figure 1.8).

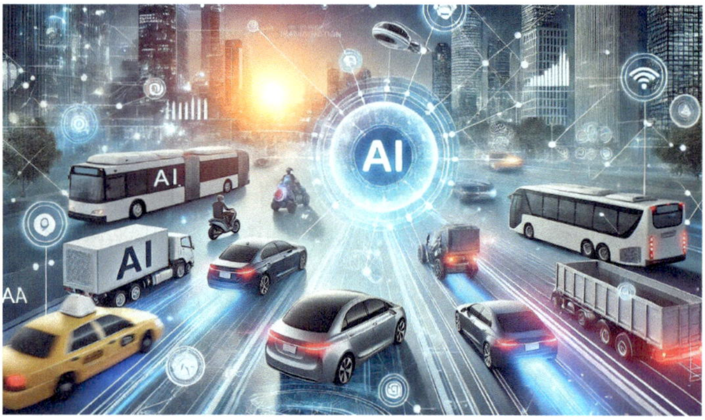

Figure 1.8. Role of artificial intelligence in optimizing transport systems.

Source: https://miro.medium.com/v2/resize:fit:1200/1*RONCugY9Kw67fsfyoEhfew.jpeg.

1.3. BENEFITS OF DATA-DRIVEN TRANSPORTATION SYSTEMS

Data-driven transport systems are a big benefit to urban mobility, primarily in terms of improved efficiency and safety. Sophisticated data analysis built into such systems enable real-time monitoring of traffic conditions and traffic patterns and provision of dynamic feedback for congestion. Real-time data, for instance, can be utilized in adjusting traffic signal changes, thereby reducing waiting times at intersections as well as maximizing traffic flow. Such adaptability is vital in reducing the adverse effects of traffic jams, as timely adaptive responses will lead to more effective travelling for passengers. Such systems also enable the management of varied traffic environments that comprise both human-driven vehicles and connected and autonomous vehicles (CAVs).

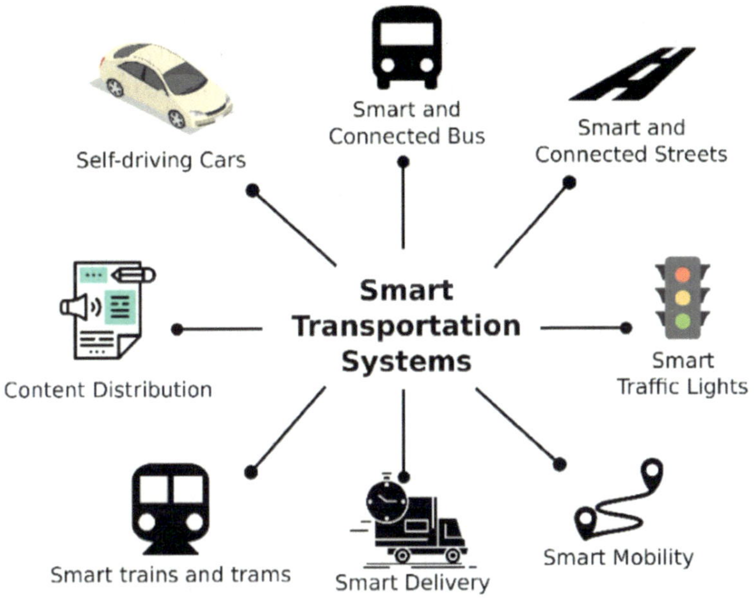

Figure 1.9. Benefits of data-driven transportation systems.

Source: https://uploads-ssl.webflow.com/62ebf26875eba5d0775548fe/63206e566f82a0b536f21ea0_hFKYQgIcyiT1rwKhjXp8bc2qwL4-rEL8VDJGSBHBANg-WOqxjqUm3GjWfFcXBTwDrxrhgUA_H7Sq3shALugvB5nmHLsz1Xdpjy8OC-TQvMW4eILIEL1GeURMpJdvqggsJ1iwaYEJgucAWtG5tgJZtEaZqh6bZW7IB-0g9QEhkLT9tXSGUhusHmrlFFZyg.png.

The realization of traffic state estimation and coordinated ramp metering can lead to safer merging maneuvers by cars, hence minimizing crashes on congested roads (Barth et al., 2019). In addition to its contribution to improved flow and safety, data-driven transportation systems also facilitate sustainability goals by lessening emissions. Optimized traffic control strategies and routing algorithms minimize unnecessary acceleration and idling, which are significant sources of vehicle emissions. With transport networks utilizing data to optimize the most efficient routes for cars, not only is fuel conserved but the carbon footprint of the entire urban transport system is minimized. With the application of green technologies, including electric vehicles, being more optimally deployed in a data paradigm, energy management can be enhanced and cleaner ways of commuting can be enhanced. By enabling the smart allocation of resources—i.e., electric car charging stations in traffic areas—such systems play a central role in promoting environmental sustainability in cities. Last but not least, there is the potential for heightened user happiness that comes center stage in data-driven transport systems. The aggregation of data from diverse sources, such as user opinions and real-time transit information, allows for a personalized journey experience. Passengers can be offered real-time updates of delays or route diversions by mobile apps in order to enable them through such information for making the appropriate travel decisions. Platforms can also offer multimodal modes of transport such as rideshares, public transport, and cycling within a seamless travel experience. The emphasis on accurate lane localization and lane change detection is of significant importance in augmenting decision-making processes in autonomous systems as well as human drivers (Barth et al., 2019). All these, together, point to the transformative role of data-driven technology in overall efficiency and satisfaction in transport systems (Figure 1.9).

1.3.1. Improved Efficiency and Reduced Congestion

Data-driven transport systems are needed to converge so that there can be enhanced efficiency and reduced congestion within city centers. Since urban locations are subject to increased mobility demands, car-centered policies have never been enough, leading to more traffic congestion and inefficiencies. Data-driven solutions can facilitate multi-modal transport in which all the available options such as public transport, cycling, and walking are utilized to the fullest. Through the utilization of intelligent traffic management systems which track real-time information, cities can eliminate travel time and delay due to congestion by a significant amount. Aside from these benefits, such developments also promote the utilization of alternative modes of transportation, ultimately leading to a more sustainable city transport system (N/A, 2008). Consciousness of the expense of congestion as well as inefficiencies within metropolitan areas is also required for efficient transportation planning. The Chicago Metropolitan Agency for Planning (CMAP) envisions the worth of long-term strategic planning in ensuring economic growth and maintaining quality of life. In this, data systems facilitate decision-making to impact growth and productivity directly. For instance, better traffic flow and reduced congestion can lead to improved accessibility for residents and businesses and develop a healthier economic environment. In consideration of CMAPs GO TO 2040 plan, it is imperative that the interlinked issues of transportation, land use, and economic development be addressed to establish a resilient urban system, making data-driven solutions important to implement (Kosarko et al., 2011). Finally, the relief of congestion through data-driven innovations not only serves individual commuters but also results in numerous societal benefits, including environmental sustainability. With cities becoming increasingly populous, the resultant rise in greenhouse gas emissions and air pollution presents dangerous health impacts. Transport systems that are backed by data can help solve such issues through smoother flows of traffic and encouraging the use of public transport, thus lowering per-capita emissions. Moreover, an increase in active transport types such as walking and cycling can enhance public health initiatives while decreasing congestion. By investing in alternate modes of transport, cities are able to achieve improved efficiency in the use of resources, improved livability for the residents, and ultimately an achievable city of the future (N/A, 2008).

1.3.2. Enhanced Safety and Risk Management

The presence of data-driven transport systems significantly enhances safety and the management of risks in all aspects of mobility. Applying the capabilities of sophisticated data analysis, the systems can identify patterns and predict potential dangers before they happen on the roads. For instance, real-time data collected from various sources such as vehicle sensors and mobile applications can help authorities predict traffic congestion or road accidents. This proactive approach facilitates timely intervention, thus reducing the occurrence of safety incidents. Further, analysis of historical data facilitates easy identification of high-risk areas, which can be prioritized by transport planners for resource deployment and application of safety measures to prevent identified risk exposures. This evidence-based decision-making, in turn, leads to improved safety travel conditions for all stakeholders on the road. Also, the implementation of vehicle-to-vehicle (V2V) and vehicle-to-infrastructure (V2I) communications is a significant move towards enhancing overall road safety. These assistive technologies enable automobiles to exchange essential information regarding their speed, direction, and ambient environmental conditions, encouraging an aggressive cooperative approach to risk management. By employing V2V and V2I communications, cars are able to warn one another of impending hazards, like unanticipated stops or inclement weather, thus improving general situational awareness on the roadway. Interconnectedness benefits reduced response times during emergency situations and allows for the development of automated safety functions. As argued in recent studies, the convergence of these communication methods can advance revolutionary advancements in road traffic management, hence promoting the European vision of safer and more efficient traffic solutions (Cartolano et al., 2008). Despite the immense potential offered by data-driven technologies, effective implementation of these systems necessitates extensive analysis through Field Operational Tests (FOTs). FOTs allow investigators and stakeholders to attain empirical response on the performance of wireless vehicular cooperative systems in operation, which are vital in guiding their impact on traffic safety and management (Bush et al., 2011). The testing helps in establishing future potential challenges and optimizing the technologies employed in a bid to ensure their efficiency in actual environments. Additionally, long-term results of FOTs provide useful data that can inform policymakers and transport organizations on actual gains from such systems in enhancing safety and managing risks effectively. Thus, through subsidizing research in rolling out such tests, stakeholders can develop a more efficient and safer transport system that tomorrow's data-driven systems will definitely fit.

1.3.3. Environmental Impact and Sustainability

The interconnection of transport systems and environmental sustainability keeps increasing in importance as urban centers continue to expand in terms of size. Evidence-based policies make it possible to monitor various transportation modes, hence being able to fully understand its environmental impacts. Car fleet management, for instance, is essential in evaluating the sustainability of a transport policy. An appropriate set of indicators can robustly assess the dual roles of car users as decision-makers and affected stakeholders in the general transport system. As data collection and analysis become more advanced, policymakers are better and better positioned to value the entire set of impacts on environmental, economic, and social dimensions. This interconnected perspective supports informed decision-making that is aligned with sustainability goals, which subsequently translates into efficient transport systems that yield fewer negative environmental effects while enhancing the quality of life in communities (Akinyemi et al., 2013). Hydrogen alternative fuel technology technologies yield profound implications for sustainable transport systems. Stakeholder conversation has unveiled the complex realities of hydrogens viability in transport systems. Unlike the simplification of hydrogen as one solution to sustainability, there is recognition of a range of responses in terms of modal shifts and attitudes towards commuting patterns. While hydrogen technologies are promising for limiting emissions and energy insecurity, these must be validated relative to other available technology options such as biofuels and hybrid cars. This combined concept of sustainability necessitates that stakeholders redefine purposes beyond traditional car driving, rather focusing on community values and sustainable mobility practices (Cassen et al., 2008).

Data-based transport systems also emphasize the need for accountability and responsiveness in adopting sustainable measures. Through data analysis in real time, cities can react to fluctuating traffic and public transport usage patterns, enabling more efficient resource allocation. Such sensitivity is not only desirable for operational efficiency but also for minimizing the environmental footprint of transport systems. Data-based sustainability indicators allow tracking of progress, evaluation of program effectiveness, and focusing of strategies on promoting cleaner behavior. As the systems get better over the years, they are also instruments for continuous improvement in climate change mitigation and ecological stewardship. The use of strong data analytics in transport planning thus becomes key in propelling urban areas towards sustainability while enhancing resilience to contemporary environmental stresses (Akinyemi et al., 2013) (Figure 1.10).

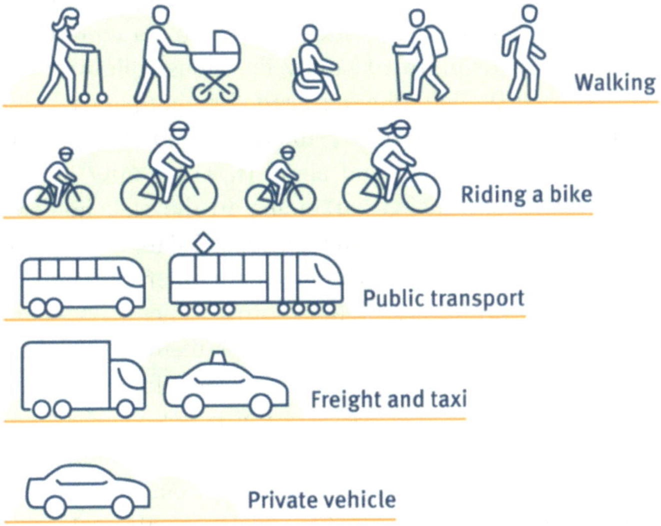

Figure 1.10. Environmental impact and sustainability of transportation systems.

Source: https://www.kindpng.com/picc/m/331-3312866_an-infographic-showing-the-sustainable-transport-hierarchy-car.png.

1.4. CHALLENGES IN IMPLEMENTING DATA-DRIVEN SYSTEMS

Implementing data-based systems in transportation is riddled with numerous problems that impede the potential functionality and effectiveness of these systems. One significant obstacle is the integration of varied data sources, which are in various forms, quality, and availability. Transporting agencies usually deal with systems and databases that have been developed in isolation over a period of time, leading to information silos. This lack of interoperability prevents the creation of a shared data architecture required for decision-making and optimization of services. Additionally, institutional resistance may impede development as stakeholders oppose switching to new technologies or methodologies. Successful implementation therefore requires not only technological advancement, but also organizational cultural change to embrace innovation and collaboration. Another significant challenge is to offer the precision and reliability of the information collected for transport systems. If the data sets are incomplete or contaminated, the consequences would be high, perhaps making the case of traffic congestion or safety concerns harder to cope with. The majority of the time, the collection processes of the data can be biased that undermines the decision-making process. For example, human behavior in traffic can be over reliant on older models, which do not represent the existing conditions accurately. Misinformed policy decisions follow as a result. This is among the issues that highlight the need for continuous monitoring and evaluation of the data quality and investment in technologies and algorithms that can improve data collection and analysis. The ability to maintain high degrees of data integrity remains paramount to the stakeholders who wish to have the capability to harness the power of data-driven transportation systems. Provision and intensity of resources are other barriers to data-driven system integration. The majority of transportation agencies operate and are confronted by limited budgets, which handicaps them from spending money on necessary technology or training personnel. Scarce funds will impede timely upgrades to existing systems or introduction of novel, sophisticated tools that could enhance functionality. Further, as raised in many studies, transitioning into data-driven solutions requires significant upfront investments in human capital and infrastructure (N/A, 2016; Russell et al., 2016). As such, securing consistent and reliable funding streams is essential to long-term sustainability. Partnerships between the public and private sectors can introduce innovative steps to fill fiscal gaps, enabling agencies to take full advantage of data-driven systems for transportation.

1.4.1. Data Privacy and Security Concerns

With data-driven transport systems evolving, they inevitably raise significant data security and privacy concerns that deserve careful scrutiny. The proliferation of smart devices and automobiles produces huge volumes of data, much of which is sensitive and personal in nature. The data is easy prey for cybercriminals to exploit, violating users' privacy. The extremely networked nature of sophisticated transport systems also creates the requirement for robust data protection mechanisms to guard against breaches. Failures in security can have catastrophic implications on not only individual privacy, but public safety as well, making it imperative that complete regulatory frameworks are established forthwith with safeguarding sensitive information as their utmost priority. Ultimately, as the technology behind these systems continues to improve, so too must the security that maintains the privacy of the individuals that are relying on them. The potential risks in bad data protection of transport systems are compounded by a changing view on privacy by the public. Under new emerging technologies like drones and autonomous vehicles, concerns have increased on ongoing surveillance and gathering of citizens' behavioral data. The public is paying closer attention and raising concerns about how their data is being collected, shared, and utilized. As is evident in recent discourses of civil drones and transportation, there exists a polar tension between desire for the economic benefits of these innovations and concern over privacy breaches by the publics. This strife highlights an essential feature of the discussion on transportation based on data: establishing a platform between policymakers, technology designers, and citizens to innovate systems without compromising user privacy rights (Biehle et al., 2020). Furthermore, as the Internet of Things (IoT) continues to evolve in the transportation sector, vulnerabilities in interconnected systems increase. The interconnectivity of devices and reliance on large networks of information create bountiful opportunities for cyberattacks that could manipulate not just personal privacy, but also the integrity of transport networks overall. Analysts anticipate that while there is a threat of such an attack, innovations in security protocols will most likely enable development in IoT for the transport industry (Anderson et al., 2017). Therefore, it is necessary that policymakers and developers work together to install firm security controls safeguarding against evolving threats such that the benefits of data-driven systems are not overshadowed by the threat of privacy misuses and data breaches. The intersection of innovation and security calls for ongoing attention and proactive efforts in ensuring a secure transport future.

1.4.2. Integration with Existing Infrastructure

The integration of data-driven transportation networks into existing infrastructure is necessary in order to optimize mobility, enhance operations, and reduce environmental imprints. While cities try to take on increased quantities, infrastructure has to be remodeled in order to take on new services and technologies. This calls for a general assessment of the existing systems, from road networks to transit hubs, as a basis for future upgrades. For instance, the Virtual Integration Platform (VIP) concept offers a solution method for synchronizing air transport systems with the current infrastructures via IT means and collaborative human methods in developing solutions for complex challenges (Gollnick et al., 2011). By encouraging harmonious transport mode communication, stakeholders are able to develop integrative solutions that consider altering demands to provide for easy transition with fewer disruptions and more usefulness. Integrating data-driven technology into current transportation networks also necessitates the investment in intelligent infrastructure like sensors, data analysis platforms, and networked systems. These technologies allow for real-time collection and analysis of data, allowing transportation agencies to base decisions on data to optimize system performance. For example, changing services according to passenger needs can lead to less congestion and more fluid traffic flows. Moreover, such integration allows for adaptive management of resources to adapt to shifting use patterns and environmental conditions. These adaptive measures are particularly highlighted by globally shifting shipping routes and shipping patterns due to climate change, representing a turning point between technology advancements and physical preparedness of infrastructure (Ashcraft et al., 2018). By combining advanced technology with efficient infrastructure, cities are able to develop robust transportation networks that will withstand future demands. Furthermore, proper integration also relies on teamwork among diverse stakeholders like the government agencies, private organizations, and the public. Incorporating these multiple stakeholders ensures the development of an integrated strategic framework in balancing the deployment of technology with local needs and legislative requirements. Cooperation is most critical in the case of potential traffic safety effects, environmental sustainability, and economic development. Furthermore, developing partnerships can open up new funding possibilities and drive innovation, leading to holistic solutions that integrate traditional infrastructure with new data-based ones. The ongoing evolution and integration of these building blocks are a giant leap toward the creation of transportation networks that not only work efficiently but also to the constantly evolving urban mobility environment. In this regard, stakeholder engagement and market foresight strategies will be essential in the realization of fully integrated data-driven transport systems.

1.4.3. Cost Implications and Funding Issues

The financial climate that surrounds data-based transportation networks is that of big cost considerations that could impact the viability of a project and its overall sustainability in the long run. In adopting such systems, it is important to understand the dual aspect of capital investment and ongoing operational expenditures. A substantial capital expenditure is generally pertinent in implementing complicated technologies, building overhauls, and acquiring data analytics capabilities. To cities that are attempting to be smart transportation systems, such capital expenditures impose budgetary strains that necessitate astute planning and fiscal stewardship. The economic environment, as in the broader metropolitan contexts, necessitates coordinated funding frameworks that leverage both public and private finance to alleviate such fiscal strains and finance the intended purpose of optimizing economic growth through effective public expenditure (Kosarko et al., 2011). Moreover, the funding mechanisms for data-based transport systems often require innovative solutions, especially amid uncertain economic circumstances. As various cities face fiscal constraints, the traditional sources of investment such as government grants and municipal taxes often fail to address the multi-faceted expenses of these technologies. In such situations, interdependency among stakeholders is imperative to align investments towards sustainable development in the long term. By building partnerships between government, private sector, and community organizations, cities can pursue alternative sources of financing including public-private partnerships (PPPs) or performance-based incentives in order to obtain the required financing. Through coordination, stakeholders can come up with effective financing mechanisms that promote economic growth and ensure that transportation innovations reach across different communities (Bierbaum et al., 2011). Last but not least, the sustainability of transport systems that are data-intensive relies not just on technological innovation but also on effective harmonization of cost elements and financial instruments. Strategic planning in response to extensive assessments of economic impact and local development dynamics is essential in navigating this daunting fiscal terrain. As demonstrated through initiatives like Chicago's GO TO 2040, organizing transportation investment in conjunction with other economic goals has the potential to foster a thriving urban economy while simultaneously addressing funding uncertainty. To this end, ongoing monitoring of transportation investments should focus on encouraging cost-effectiveness to the maximum extent and equal access so that they provide an environment that supports addressing near-term operational needs and long-term objectives of regional mobility and prosperity (Kosarko et al., 2011).

1.5. CASE STUDIES OF SUCCESSFUL DATA-DRIVEN TRANSPORTATION

The application of data-based transport systems has been successfully demonstrated in numerous case studies that showcase the revolutionary impact of technology on improving traffic management and safety. An example of the successful implementation of this is the deployment of vehicle-to-vehicle (V2V) communication-enabled connected and automated vehicle (CAV) technology, which uses advanced wireless communications tactics to enhance vehicle-to-vehicle engagements. A recent study suggested a Vissim-based simulation environment for traffic with an analytical dedicated short-range communication (DSRC) module. Not only did this environment achieve a phenomenal packet reception rate of over 93% in different scenarios, but it also provided insights into how the reliability of V2V communication varies with different deployment strategies. These disclosures underscore the potential of data-based approaches in facilitating real-time decision-making within traffic infrastructures to facilitate smoother traffic and safer roads in the face of increased urban congestion (Lee et al., 2018). Another compelling argument underscores the value proposition of data analytics in designing emergency response strategies for transport infrastructure against the occurrence of threats.

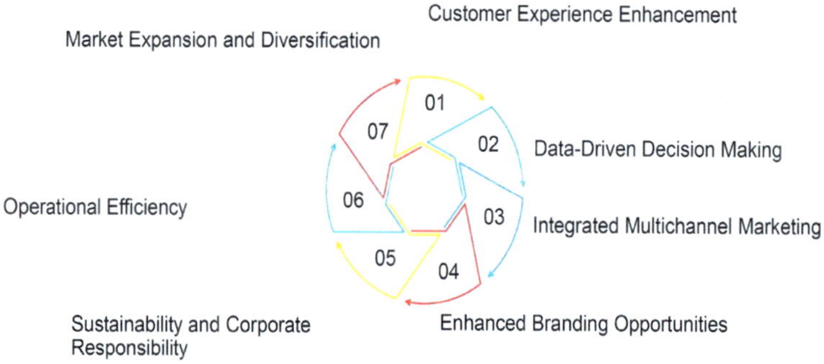

Figure 1.11. Case studies of successful data-driven transportation.

Source: https://fastercapital.co/i/Transport-Case-Studies--Marketing-Success-in-the-Transport-Sector--Case-Studies-and-Best-Practices--The-Role-of-Digital-Transformation-in-Transport-Marketing.webp.

The employment of vehicle ramming attacks by terrorists symbolizes a paradigm shift in safety concerns on roads. By analyzing these attack methods, researchers have given vital information to law enforcement to develop effective mitigation strategies. By analyzing many case studies, the study explains not just the attack methods of potential attackers but also effective countermeasures and response strategies that can save lives. This combination of data-driven analysis with emergency planning increases the importance of leveraging thorough analytics to improve security across transportation systems and thereby enable populations to react in a timely and efficient manner to unforeseen circumstances (Edwards et al., 2020). The importance of case studies in identifying the importance of data-driven transport systems also has its role in urban mobility and public transport systems. For instance, those municipalities using smart traffic signals and real-time transit information have realized tangible improvements in commuters' travel experience. Transport agencies are capable of smoothing traffic and route planning by analyzing the movement patterns and bottlenecks using data collected. Utilizing platforms that aggregate data from multiple sources offers an integrated view of transport dynamics, allowing proactive rather than reactive intervention. These successful changes reflect the bold integration of transport planning and data systems, thus indicating the imperatives for using data to craft resilient urban mobility systems. As cities keep evolving, these data-driven methods will be pivotal in facilitating active and sustainable modes of transport for all citizens (Figure 1.11).

1.5.1. Smart City Initiatives and Their Outcomes

Smart city initiatives are a cutting-edge method of urban living that employs innovative technologies to enhance the quality of life of citizens. These initiatives would most likely involve innovation such as the Internet of Things (IoT) and data analytics to design integrated systems that optimize transportation, energy usage, and public safety. Application of these technologies provides an environment that not only facilitates good real-time decision-making but also addresses the specific issues of urban people. For instance, cities can monitor the flow of traffic and energy consumption well, and ultimately this results in more green practices and improved urban infrastructure. By data-driven methods, such systems can be regularly assessed and improved by the stakeholders to enhance their effectiveness and responsiveness within dynamic city contexts, better positioning the smart city models to support sustainable urban growth. Further, integrating innovative technologies into smart city initiatives prompts interagency coordination among the stakeholders, which includes public agencies, private organizations, and academic institutions. As reiterated in public procurement studies, the implementation of innovative public procurement instruments such as PDTI (Public end Users Driven Technological Innovation) is able to drive technological innovation in urban settings. This collaborative model not only initiates research and development but also ensures that solutions are developed to meet citizens' needs in an adequate manner. For example, it has been possible for European cities to identify robotic urban challenges through this process and develop a reference for other smart cities to emulate in an effort to address such challenges. This collaboration enables the achievement of more efficient resource and knowledge deployment, resulting in more effective and viable outcomes in smart city projects, particularly in the category of transportation systems. Smart city program outcomes usually become the first determinants of success, particularly for transportation systems of cities. Using information and communication technologies (ICTs), cities are able to collect and utilize massive datasets, enabling cities to make highly informed decisions that respond to matters of common interest such as traffic and air pollution. The analytical solutions through such data not only enhance efficiency in operations but also empower citizens by providing them with real-time feedback on their city living. Moreover, the establishment of organizational frameworks that promote cooperation further builds upon such initiatives, leading to innovative solutions through data-driven insights. The constant evolution of smart city concepts demonstrates that they can develop responsive, adaptive urban systems that not only address current issues but also anticipate and avoid future issues, thereby improving general mobility and urban sustainability (Monte et al., 2017; Stahl et al., 2017).

1.5.2. Public Transportation Systems Leveraging Data

The convergence of public transit with data management has redefined the planning for mobility in cities. With the application of sophisticated data analytics, agencies can optimize operations and deliver better customer experiences on transit systems. Transit agencies, by analyzing historical and real-time data, can optimize routes and schedules so that service aligns with rider patterns. For instance, advanced geographic information systems (GIS) allow for the identification of hot spots of high demand, leading to pinpoint resource deployment. More than mere efficiency, data-informed systems can improve communication with travelers by providing them with real-time updates on service change or delay. Such responsiveness not only makes public systems more credible but also spurs higher ridership, ultimately resulting in a more sustainable urban transport system. Besides, the presence of electric mobility systems and public transport presents an opportunity for synergy, which is well facilitated by data analytics. The above workshop outlined in the electric mobility insights emphasizes the need for innovative frameworks that connect various types of modes of transport like electric vehicles and public transport systems (Cassandras et al., 2019). Through data, agencies are able to craft holistic charging plans that correlate with public transit schedules as well as private electric vehicle use, thus keeping the electric grid decongested and making commute patterns less complicated. This creates a twofold advantage of supporting a clean energy transition as well as ensuring optimal mobility access across urban planning. In understanding how these systems work interactively, innovation in new business models is achievable that can reconcile both environmental purity and customer satisfaction. Beyond that, the economic benefit of data-driven public transport is too large to be overlooked. Cities with more investment in their public transportation infrastructure are likely to bring serious economic dividends to their citizens, particularly as the price of gasoline fluctuates. A current study highlighted the fact that communities with better-developed public transport systems experience less economic strain as a result of rising gas prices, as residents enjoy affordable and reliable mobility (N/A, 2005). This highlights the important role of information in not only enhancing service delivery but also influencing broader economic policy. By considering transportation spending alongside household consumption, policymakers can better assess public transit investment social returns and, in the process, better enable transportation access equity and policy assistance for regional economic resilience. With such data integration, transit planning now emerges as both a facilitator of short-term operation improvement and long-term economic sustainability (Figure 1.12).

Figure 1.12. Public transportation systems leveraging data.

Source: https://img.freepik.com/premium-photo/efficient-public-transportation-system-with-buses-54xjpg_835197-13569.jpg?w=2000.

1.5.3. Freight and Logistics Optimization Through Data

Freight and logistics environment is now experiencing a big change with data-driven decision-making. The reason behind the change is the capacity to read and analyze huge amounts of data so that supply chain operations can be optimized by the logistics managers. For instance, the outsourcing of stand-alone production procedures has a great influence on freight volumes. As more and more business organizations concentrate on their core competences, they rely more on varied suppliers, which may result in higher freight traffic. That kind of complexity can be observed in the automotive industry, wherein fewer in-house productions create greater deliveries in fewer lots, thereby augmenting transportation needs. It is essential that firms grasp the dynamics of cutting operations while efficiently handling freight volumes, and thus that indicates the multi-faceted interconnectedness among corporate behavior, logistics policy, and data analysis (Thierstein et al., 2025). Added to that, advancements in data analysis enable organizations with the ability to handle just-in-time production and combine telematics solutions, which are able to reduce unwarranted freight traffic. By using data to make accurate demand forecasts, businesses can reduce overproduction and coordinate deliveries more tightly with genuine consumer demand. By enabling a clearer understanding of supply chain performance through data-driven insight, businesses can identify areas of inefficiency and take remedial action in a timely fashion. To illustrate, optimized truck routing and consolidation of consignments have the potential to cut the carbon emissions associated with freight operations, increasingly relevant in the context of international commitments, like those outlined in the Paris Agreement. In adopting these new methods, businesses not only optimize the performance of their operations but also positively contribute to broader green targets, illustrating the double benefit of data integration within logistics (Agenbroad et al., 2019). As companies grow into such data-driven tactics, they also have to deal with the issues and challenges posed by such high-level utilization of data. Foremost among these is ensuring that the data are accurate and dependable, which are highly critical in creating sound logistical decisions. Inaccurate data can lead to poor forecasting, ranging from excess inventories to stock shortages, both of which can produce detrimental impacts on freight optimization. Apart from that, investments in technology and training must be made to equip employees with the capabilities to utilize and act upon data insights effectively. By establishing a culture that values data-driven decision-making, organizations can bring about responsiveness and innovation within their logistics operations. As the field of study keeps evolving, the understanding of implications of data on freight optimization will grow increasingly critical in influencing the way commercial enterprise handles the complexity of modern transportation networks (Thierstein et al., 2025).

1.6. FUTURE TRENDS IN DATA-DRIVEN TRANSPORTATION

The evolution of data-based transport systems signals an overhaul of the way urban logistics operates. As cities become increasingly congested and customers require quicker deliveries, it is essential to embrace new models such as Delivery as a Service (DaaS). It embodies the essence of efficient, customer-focused logistics that cater to the requirements of the world of on-demand. The integration of Information Technology solutions into urban logistics operations not only improves the efficiency of supply chains but also aids in overall environmental objectives through promotion of clean means of transport. As can be observed in several European pilot projects, such developments express a vision for the urban logistics of the future drawn from strategic transport optimization and Internet of Things priorities, with long-term consequences of urban sustainability and efficiency (Huschebeck et al., 2020). Apart from logistics developments, the air transport sector is also confronted with data-driven revolution. Modern approaches, such as the Virtual Integration Platform (VIP), are an overall system method of air transportation design. The system uses an IT tool chain with a working collaboratively human method to handle air travel system complexity effectively.

Figure 1.13. Future trends in data-driven transportation.

Source: https://img.freepik.com/premium-photo/autonomous-vehicles-future-transportation_1271337-2833.jpg.

By enabling the possibility of designing future air concepts from Quiet and Clean short-range operations through to Comfortable and Clean long-range options, the VIP concept suggests significant improvements in operation effectiveness and passenger comfort. With cities expanding and the demand for travel increasing, using such methods ensures air transport to adapt to society's needs and concern for the sustainability of the environment (Gollnick et al., 2011). The convergence of innovative data analytics and transportation modes marks the advent of smart mobility solutions, which have the capacity to transform urban areas. The application of big data analytics enables transport authorities to make useful deductions for traffic flow optimization, improving security mechanisms, and reducing carbon footprint. Moreover, incorporating autonomous cars into the infrastructure calls for a robust data infrastructure that provides systems with immediate information exchange to improve efficiency and customer satisfaction. With predictive analysis and offering real-time routing alternatives, data-driven systems can relieve congestion and reduce delays. As communities become more and more interconnected in cities, continued development of these data-driven methods will be important for building responsive and resilient transportation systems that can cope with future challenges (Figure 1.13).

1.6.1. Autonomous Vehicles and Their Data Requirements

The introduction of autonomous vehicles has brought about a revolutionary shift in the transport sector, necessitating a sophisticated infrastructure for data acquisition and utilization. The vehicles rely on a robust array of sensors, from LIDAR, radar, and cameras, to collect real-time data surrounding their environment, which is vital for immediate driving choices. As automobiles get more and more embedded in digital environments, they are not just vehicles of transportation but also functioning nodes in data networks. As a result of such advanced data collection, traffic behavior and congestion can be comprehensively analyzed, thus enhancing urban mobility planning. Furthermore, the adoption of technologies compatible with concepts like Delivery as a Service (DaaS) places emphasis on a changing business model reliant upon maximum efficiency and meeting customer requirements, typical of the new urban logistics face (Huschebeck et al., 2020). The complexity of software architecture for autonomous vehicles further puts emphasis on the priority area of data management. Modern-day cars handle massive amounts of data, with sources stating a capability of around 750MB/s, which is similar to processing up to 180 concurrent 4K video streams (Berger et al., 2017). This bulk stream of data is required for autonomous driving, helping the vehicle scan its surroundings and respond effectively in real-time. Incorporation of advanced software systems not only contributes to improved vehicle performance but also provides a basis for Continuous Experimentation. Through the adoption of methods that support rapid feedback cycles, automobile engineers are in a position to enhance the software architecture continuously and make sure that automobiles evolve in line with technological advancements. Ongoing updating is paramount in maintaining safety and reliability in autonomous cars. Data requirements go past simply collecting; they involve storage, processing, and analysis to acquire usable insights. Autonomous vehicles must operate in an environment where data integrity and security are the priorities, particularly in the face of cyber threats. The creation of city logistics systems, based on platforms like DaaS, serves to highlight the connection between data usage and customer expectations (Huschebeck et al., 2020). Compelling data management strategies are called for to help ensure that autonomous vehicles can smoothly integrate with existing transport infrastructures. As we progress towards an increasingly autonomous future, strategic alignment of data governance policy and technological innovation will be essential in creating a transportation landscape that is not only efficient but also robust to expected emerging challenges.

1.6.2. The Role of Machine Learning in Transportation

The integration of machine learning into transport systems is a quantum step towards optimizing operational efficiency and security. This advancement relies considerably on the ability of machine learning algorithms to handle vast volumes of data generated by modern transport infrastructure. For instance, the use of graph-based machine learning approaches facilitates the simulation of complex transport infrastructure for improved decision-making in traffic management as well as route planning. The use of Intelligent Transportation Systems (ITS) illustrates such a scenario, where it consolidates data from various sources—roads, vehicles, and communication infrastructures—to optimize the movement of traffic and reduce congestion. Such systems are bound to produce significant enhancements in efficiency and safety performance since they provide real-time information that informs users and transport authorities about conditions prevailing on the road, hence making the transport environment more responsive (Liu et al., 2023). Secondly, the connection between big data analytics and transportation plays a fundamental role in the development of smart cities, where data-informed planning is essential to guide urban complexity. Through big data, transportation planners are able to unveil trends and patterns previously concealed, so that infrastructure investments and policy deployments can be better made. Machine learning algorithms are among the most effective analytic tools for this purpose since they can process and extract useful information from disparate data. For example, the application of computational intelligence in transportation not only optimizes performance in operations but also maximizes rider satisfaction and engagement by adjusting services according to consumers' needs more effectively. Big data's integration with transportation is particularly important in developing anticipatory and adaptive transportation systems that capture the dynamic characteristics of urban areas (Doctor et al., 2020). As machine learning continues to advance, its uses within the transportation system are set to rise exponentially. The upcoming technologies like driverless vehicles also rely extensively on machine learning algorithms to analyze sensory data, reach decisions in split seconds, and navigate through complex environments without incident. This not only reduces the potential for human error but also introduces new paradigms of mobility and accessibility for transportation. In addition, machine learning can optimize predictive maintenance of vehicles and infrastructure so that timely actions are taken before failure happens. These advances highlight the important role that machine learning will play in creating a sustainable, efficient, and safer transportation future. Their promise is also brought out through ongoing research into fresh methods and their implementation in real-world applications, providing the way for groundbreaking solutions that solve today's transport challenges (Liu et al., 2023) (Figure 1.14).

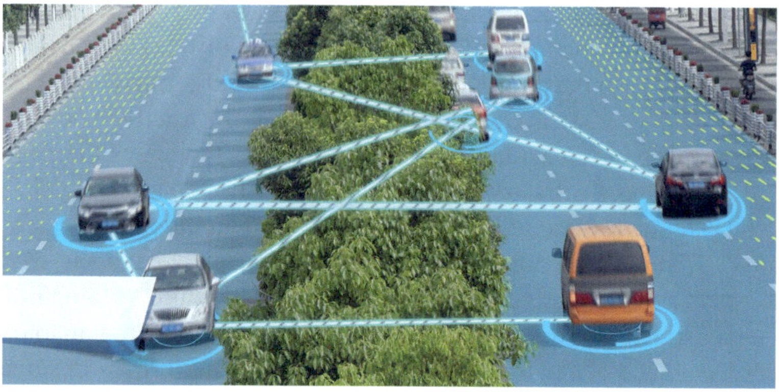

Figure 1.14. The role of machine learning in transportation.

Source: https://www.hitechwork.com/wp-content/uploads/2024/07/The-Role-of-AI-and-Machine-Learning-in-Transportation-Software.webp.

1.6.3. Predictions for the Next Decade in Transportation Technology

Turning to the future decade of transportation technology, a fundamental change will arise through developments in data-based systems. These systems will harness the power of big data, artificial intelligence, and machine learning to amplify decision-making potential across all modes of transport. As more and more data is exploited, predictability and analysis of traffic behavior, user behavior, and environmental impacts will probably grow tremendously. For instance, planning support systems (PSS) will emerge to provide dynamic solutions that exchange information with real-time data. These systems will integrate multiple sources of data to enable informed decision-making, capable of redefining city transport planning and enhancing stakeholder coordination. The sprawling character of modern PSS, highlighted by recent research, recognizes the necessity for harmonized instruments that will respond to the complex and diverse needs of modern cities (Batty et al., 2007). Beyond enhancing urban planning, the next decade will likely experience enormous advancements in autonomous and intelligent vehicle technology. Autonomous vehicles will increasingly be the norm, driven by technical leaps in sensors, artificial intelligence, and infrastructure development. As technology continues to evolve, integrating it into existing transport systems will require intense data analysis to ensure safety, efficacy, and reliability. The impact of weather disruption, for example, will be critical; data models can simulate the effect various conditions have on transport systems. Current studies have shown that agent-based models can effectively minimize delays caused by environmental factors, reflecting the importance of data in enhancing the resilience of air transport networks (Eguiluz et al., 2013). Hence, transport policies will persist in leveraging predictive analytics to create adaptive systems that can manage disruptions while maintaining operational integrity. Furthermore, the expansion of smart city projects will re-engineer mobility in urban cities over the next several years. As cities implement Internet of Things (IoT) technologies, connectivity of transportation infrastructure will facilitate greater collection and analysis of massive amounts of operational and environmental information. This shift will enable more efficient use of resources, reduced levels of congestion, and sustainable modes of commuting. Public-private partnerships will be essential to effectuate data to support better transit service and urban planning. For instance, integrated mobility platforms may become accessible with real-time information on modes of transport, weather, and forecasting based on historical trends. These technologies not only enhance the user experience but also enable a green mode to mobility in the city, so that the cities are set for future challenges and demands in transport technology (Figure 1.15).

Figure 1.15. Predictions for the next decade in transportation technology.

Source: https://i.ytimg.com/vi/YauHK0TIm3c/maxresdefault.jpg.

SUMMARY

In short, data-driven transportation systems have revolutionized city traffic and mobility management. As their strength, having real-time data analytics, they allow congestion trend anticipation and make traffic management measures more efficient. Road traffic analysis, particularly concerning big events like sports activities, is a case in point of the capability of data to optimize the transport flow while maintaining the least possible disruption. For instance, with the use of INRIX data to monitor traffic flows on game day, scientists have ascertained that there are persistent hotspots that affect travel patterns (Ahsani et al., 2019). Such application of data demonstrates how much analytical systems are crucial to understanding and optimizing urban mobility. Furthermore, the economic impact of effective transport systems cannot be overstated. Investment in public transportation infrastructure not only enhances mobility but also has economic gains for households. regions with good-developed public transport have reduced economic burdens from rising fuel prices since citizens adopt cheaper modes of transport. This effect fosters eco-friendly transport patterns and ensures an equitable transport environment, particularly in urban areas (N/A, 2005). The use of data analytics in transportation planning is a direct response to the challenges posed by fluctuating gas prices and urban traffic. In essence, the development of data-driven transportation systems is a revolutionary step towards more robust and efficient urban living. By understanding the dynamics of technology, infrastructure, and human behavior, cities can create robust transportation systems that respond to the problems of the day as well as the demands of the future. The knowledge gained from analyzing data enables city planners to set priorities for investments and design strategies that facilitate mobility and interconnection among the populace. With further maturation of such systems, their role in establishing sustainable, adaptable, and sensitive transport networks will become only more vital in answering the sophistication of city transport in the 21st century.

The use of data analytics to transport has emerged as a cornerstone innovation, enhancing the efficiency of operations and the user experience. Data analysis allows transit operators and logistics providers to tap into oceans of data from millions of sources, from sensors on the road to geographic information systems. By using predictive analytics, companies can lead to improved resource usage, best routes, and rapid response times to unexpected disruptions. Transportation networks can thus transition from reactive to proactive in their administration, ultimately leading to reduced congestion and operating costs. The shift to a data-based approach necessitates the adoption of superior technologies and software that can absorb and process real-time data, as offered in analyses that analyze existing trends within the industry. Under this approach, some case studies highlight the central role of data in understanding patterns of behavior addressing transportation dynamics. For instance, traffic impacts of major sporting events were of special interest in offering insights into the way temporary movement of population affects mobility. Traffic flow analysis for games of the Nebraska Cornhuskers illustrates the complexity involved with handling transport systems amidst orchestrated disruptions. With methods such as the Multi-EigenSpot in detecting hotspots, transport planners can identify dangerous zones where traffic congestion is likely to occur. These observations point towards the need for systematic strategies that address special conditions while having the capacity to adjust according to unplanned variations in demand, a recurring theme in the movement towards improved data-based transportation systems (Ahsani et al., 2019). Further, the importance of inventory management in supply chains and its relationship with transport systems cannot be overlooked. In drugs, for example, the complexity of inventory configurations often is defined by factors like safety stock levels and regulatory requirements. The quest for optimal inventory levels is directly associated with the effectiveness of transport logistics because addressing inventory problems reduces strain on networks of distribution. Transportation systems must adapt in a way to maintain levels of service while simultaneously recovering cost in storage and spoilage, capturing the intricate relationship between inventory control and the efficiency of transportation (Andreu et al., 2017). As data-driven solutions continue to become increasingly prevalent, they will allow for greater coordination between these areas, encouraging an integrated approach to the multi-faceted issues experienced in transportation systems today.

REFERENCES

1. Abbas, R., Ahmad, I., Ali, S., Amuru, S., et al. (2020). 6G white paper on machine learning in wireless communication networks. http://arxiv.org/abs/2004.13875 (accessed on 21st October 2025).
2. Agenbroad, J., Bai, Q., Chrisman, K., Creyts, J., et al. (2019). A roadmap for China to peak carbon dioxide emissions and achieve a 20% share of non-fossil fuels in primary energy by 2030. https://core.ac.uk/download/323308755.pdf (accessed on 21st October 2025).
3. Agostinelli, S., Allison, J., Amako, K., Apostolakis, J., et al. (2005). GEANT4: A simulation toolkit. https://core.ac.uk/download/14503261.pdf (accessed on 21st October 2025).
4. Ahsani, V., & Sharma, A. (2019). Assessing the impact of game day schedule and opponents on travel patterns and route choice using big data analytics. https://core.ac.uk/download/286729950.pdf (accessed on 21st October 2025).
5. Akinyemi, C. J., Banister, D., Banister, D., Baumann, H., Black, W., Bojkovic, D., Axon, C. J., et al. (2013). A methodology for measuring the sustainability of car transport systems. https://core.ac.uk/download/29140862.pdf (accessed on 21st October 2025).
6. Anderson, J., & Rainie, L. (2017). The Internet of Things connectivity binge: What are the implications? https://core.ac.uk/download/89472988.pdf (accessed on 21st October 2025).
7. Arabnia, H. R., Kim, T. H., Mohammed, S., Qu, G., et al. (2020). IEEE Access special section editorial: Big data technology and applications in intelligent transportation. https://core.ac.uk/download/389670059.pdf (accessed on 21st October 2025).
8. Ashcraft, C. M., Brewer, J. F., Burakowski, E., Coffin, M., et al. (2018). Preparing for a Northwest Passage: A workshop on the role of New England in navigating the new Arctic. https://core.ac.uk/download/215537367.pdf (accessed on 21st October 2025).
9. Barth, M. J., Hao, P., Wang, C., Wei, H., et al. (2019). Vision-based lane-changing behavior detection using deep residual neural network. https://core.ac.uk/download/305124840.pdf (accessed on 21st October 2025).
10. Barth, M. J., Wang, Z., Wu, G., Ye, X., et al. (2019). The state-of-the-art of coordinated ramp control with mixed traffic conditions. https://core.ac.uk/download/305124841.pdf (accessed on 21st October 2025).
11. Batty, M. (2007). Planning support systems: Progress, predictions, and speculations on the shape of things to come. https://core.ac.uk/download/1683968.pdf (accessed on 21st October 2025).

12. Berger, C., & Giaimo, F. (2017). Design criteria to architect continuous experimentation for self-driving vehicles. http://arxiv.org/abs/1705.05170 (accessed on 21st October 2025).

13. Biehle, T., Fischer, L., & Kellermann, R. (2020). Drones for parcel and passenger transportation: A literature review. https://core.ac.uk/download/323212148.pdf (accessed on 21st October 2025).

14. Bierbaum, A. H., McKoy, D. L., & Vincent, J. M. (2011). Opportunity-rich schools and sustainable communities: Seven steps to align high-quality education with innovations in city and metropolitan planning and development. https://core.ac.uk/download/71358710.pdf (accessed on 21st October 2025).

15. Bishins, A., & Duduta, N. (2010). Citywide transportation greenhouse gas emissions inventories: A review of selected methodologies. https://core.ac.uk/download/71350130.pdf (accessed on 21st October 2025).

16. Bolea Monte, Y., Casanovas Garcia, J., Grau Saldes, A., Puig-Pey Clavería, M., et al. (2017). Public entities driven robotic innovation in urban areas. https://core.ac.uk/download/224975022.pdf (accessed on 21st October 2025).

17. Buscaglia, E. (2025). The Economic Factors Behind Legal Integration: A Jurimetrics Analysis of the Latin American Experience. https://core.ac.uk/download/pdf/7154726.pdf (accessed on 21st October 2025).

18. Bush, S., Henning, T., McCarten, P., Omenzetter, P., et al. (2011). Risk based bridge data collection and asset management and the role of structural health monitoring. https://core.ac.uk/download/29465888.pdf (accessed on 21st October 2025).

19. Cassandras, C. G., Fox-Penner, P., & Kelly, H. (2019). Final report: Workshop on integrating electric mobility systems with the grid infrastructure. https://open.bu.edu/bitstream/2144/39613/1/SloanWshopFinalReport.pdf (accessed on 21st October 2025).

20. Cassen, C., Crassous, R., Dorin, B., Gitz, V., et al. (2008). A novel hybrid architecture for agriculture and land use in an integrated modeling framework: MATISSE (Methods and Tools for Integrated Sustainability Assessment). http://agritrop.cirad.fr/546835/1/document_546835.pdf (accessed on 21st October 2025).

21. D'Angelo, G., Ferretti, S., & Ghini, V. (2017). Distributed hybrid simulation of the Internet of Things and smart territories. http://arxiv.org/abs/1710.04252 (accessed on 21st October 2025).

22. D'Angelo, G., Ferretti, S., & Ghini, V. (2017). Modeling the Internet of Things: A simulation perspective. http://arxiv.org/abs/1707.00832 (accessed on 21st October 2025).

23. Doctor, F., Iqbal, R., Mahmud, S., More, A., et al. (2020). Big data analytics: Computational intelligence techniques and application areas. https://core.ac.uk/download/228156656.pdf (accessed on 21st October 2025).

24. Edwards, F. L., & Goodrich, D. C. (2020). Transportation, Terrorism and Crime: Deterrence, Disruption and Resilience. https://core.ac.uk/download/289097215.pdf (accessed on 21st October 2025).

25. Eguiluz, V. M., Fleurquin, P., & Ramasco, J. J. (2013). Data-driven modeling of systemic delay propagation under severe meteorological conditions. https://core.ac.uk/download/36198882.pdf (accessed on 21st October 2025).

26. Gershenson, C., & Helbing, D. (2015). When slower is faster. http://arxiv.org/abs/1506.06796 (accessed on 21st October 2025).

27. Gollnick, V., Lehner, S., Stumpf, E., Szodruch, J., et al. (2011). Virtual integration platforms (VIP)–A concept for integrated and interdisciplinary air transportation research and assessment. https://core.ac.uk/download/11155540.pdf (accessed on 21st October 2025).

28. Huschebeck, M., Huschebeck, M., Leonardi, J., Leonardi, J., et al. (2020). Approaching delivery as a service. https://core.ac.uk/download/305119210.pdf (accessed on 21st October 2025).

29. Kosarko, G., & Weissbourd, R. (2011). Economic impacts of GO TO 2040. https://core.ac.uk/download/71345969.pdf (accessed on 21st October 2025).

30. Lee, J., & Zhong, Z. (2018). Simulation framework for cooperative adaptive cruise control with empirical DSRC module. http://arxiv.org/abs/1810.06510 (accessed on 21st October 2025).

31. Liu, M., Wu, H., & Yan, S. (2023). Recent advances in graph-based machine learning for applications in smart urban transportation systems. http://arxiv.org/abs/2306.01282 (accessed on 21st October 2025).

32. Masvidal Andreu, C. (2017). Inventory drivers in a pharmaceutical supply chain. https://core.ac.uk/download/157811883.pdf (accessed on 21st October 2025).

33. N/A. (2005). Driven to spend: Pumping dollars out of our households and communities. https://core.ac.uk/download/71338874.pdf (accessed on 21st October 2025).

34. N/A. (2008). Active transportation for America. https://core.ac.uk/download/71343026.pdf (accessed on 21st October 2025).

35. N/A. (2016). A roadmap to reduce U.S. food waste by 20 percent. https://core.ac.uk/download/75784647.pdf (accessed on 21st October 2025).
36. Ronald, E. J. (2019). Creating low-carbon communities: Evaluating the role of individual agency and systemic inequality in San Jose, CA. https://core.ac.uk/download/215444299.pdf (accessed on 21st October 2025).
37. Russell, C. A., Hildreth, J. L., & Stevens, P. (2016). ExpandED Schools national demonstration: Lessons for scale and sustainability. https://core.ac.uk/download/75785743.pdf (accessed on 21st October 2025).
38. Shu, X. (2025). Transportation data integration and analytics. https://core.ac.uk/download/344664881.pdf (accessed on 21st October 2025).
39. Stahl, E., & Dupuis, N. (2017). Trends in smart city development. https://core.ac.uk/download/83111937.pdf (accessed on 21st October 2025).
40. Thierstein, A., & Schnell, K. D. (2025). Corporate strategies, freight transport and regional development. https://core.ac.uk/download/pdf/7046746.pdf (accessed on 21st October 2025).

CHAPTER 2

DATA COLLECTION METHODS AND TOOLS

Contents

INTRODUCTION

Within today's research context, the roles of data collection tools and methods become central in the pursuit of information across different disciplines in academia. With experts embarking on the task of deciphering complex occurrences, a well-set introduction becomes crucial in defining the research question and background. Data collection not only dictates the credibility of the research but also significantly contributes to the findings and conclusions derived.

This first phase sets the foundation for selecting appropriate methodologies that direct how researchers obtain, analyze, and interpret data. Sound understanding of data collection methods can highlight gaps in existing literature and inform follow-up inquiry, as observed in systematic reviews exploring diverse issues, like (Johnman et al., 2016).

Figure 2.1. Data collection methods and tools.

Source: https://cdn.brocoders.com/Data_Collection_Tool_1d7de818ee.png.

Such extensive preparation can open doors to informed discussion, hence adding to the body of knowledge of the fields. Furthermore, the availability of credible and accurate data collection instruments makes a big impact in improving research outcomes. Specific methodologies, be they qualitative or quantitative, contribute to the richness and wealth of the results, thus making it feasible for researchers to draw meaningful conclusions. For instance, in public health, data collection instruments have a primary contribution to appreciating the richness of community health problems. In studies on male partner engagement in prevention of mother-to-child transmission (PMTCT) services, the accuracy of data is responsible for the efficacy of intervention strategies. The systematicity of research applied in such studies underscores the significance of strong data collection systems that yield actionable data, considering the immediacy of urgent health issues in diverse global realities, as observed in (Johnman et al., 2016). Besides, a proper introduction stipulates the importance of ethicality in the utilization of data collection instruments and techniques. Researchers must walk carefully through the minefield of ethical considerations while gathering sensitive data to ensure participant welfare and anonymity. A case in point is the research on tuberculosis (TB) treatment outcomes, where ethical data collection is shown to be significant, given that researchers are confronted with issues such as scant resources and sociocultural impediments, as discussed in. An ethical frame not only justifies the research but also enhances public confidence in the research results. Thus, an introduction evoking both methodological precision and ethical integrity establishes a well-rounded view of the research process, ultimately guiding scholars in their quest to contribute positively to their respective field (Figure 2.1).

Definition of Data Collection

Data collection is the foundational column of research, which sustains the systematic gathering of information to be analyzed and interpreted. Data collection incorporates methods used to achieve relevant data that can be used by individuals or organizations to make well-informed decisions. Data collection as a process includes purposeful recruitment of subjects or participants, building instruments or modes of measurement, and procedure in maintaining precision and reliability. Effective data collection mechanisms enable researchers to record both qualitative and quantitative data, thereby enhancing the richness and extensiveness of their findings. Moreover, as evident in research on cross-border cooperation, the choice of data collection mechanisms significantly impacts the quality of conclusions derived from research, with systematic mechanisms playing a crucial role in spotting and developing relationships between various stakeholders (Adams et al., 2018). To successfully conduct data collection, researchers must explore various techniques that best suit their research question, target populations, and data type. These range from surveys and interviews to observation techniques and content analysis of public documents. Also, the role of utilizing digital platforms in data collection cannot be overestimated. As has been implied in current debates regarding research methods, the intentional use of digital tools can have the effect of making data collection efforts more efficient and effective (Kann et al., 2014). Integrating contemporary technology allows researchers to make the process more streamlined, cut down on mistakes, and get results that are closer to the representativeness of the larger population. Thus, the definition of data collection extends beyond mere information collection; it is the organized utilization of various methods under the constraints of the specific needs of research undertakings. Generally speaking, the definition of data collection is irrevocably intertwined with methods and tools applied by researchers to provide a stage on which data are gathered accurately, dependably, and meaningfully. With the nature of research being as rapid as it is in virtual space, there exists a perpetual necessity for having clearly articulated guidelines and procedures in data collection. There has to be critical consideration by researchers as to what implications a chosen approach would carry, such as the involvement of stakeholders and the use of technology. The resultant findings produced from soundly conducted data collection extend beyond information dissemination in scholarly debates but even provide the foundation for practical applications in practice. Thus, the evolution of data collection knowledge is essential in encouraging research-informed practice and policy in a wide range of fields (Adams et al., 2018).

Importance of Data Collection in Research

The collection of data is the backbone of research as it forms the foundation on which there are well-reasoned conclusions and analyses built. Not only does the process lead to acquiring empirical data, but it also dictates the choosing of steps in methodology by enlightening various sides of the research issue. While researchers try to understand complex social realities, the strategic choosing of instruments of data collection has the potential to make or destroy the validity of results. Specifically, new computational methods have broadened availability to diverse datasets, diversifying the research setting. Nonetheless, there is a need to keep maintaining a contemplative practice in terms of the data sources and types obtained, as highlighted by researchers that do not recommend the use of these new methods blindly, without considering their presuppositions and structures (Hukal et al., 2017). Defining the power of data collection enables researchers to construct contextual and informed discourses without falling into traps of misinterpretation of data. For sustainable development, reliable data is central in determining the environmental, economic, and social impacts of various practices, particularly on the construction end. As attention on sustainability grows, the amount of data pertaining to this goal allows for complete assessment of building performance at each stage of its life cycle. For instance, new environmental assessment tools are capable enough to be used in planning and designing sustainable buildings, provided they are supported by intensive data collection techniques. However, unless data collection is done in a strict fashion, such instruments will prove ineffective in addressing serious issues at hand since current practices are inclined to overlook the multifaceted interdependencies among environmental, economic, and societal aspects (Liu et al., 2014). Thus, good data collection not only helps improve research quality but also guarantees well-informed decision-making towards sustainable practice. Also, the establishment of data collection methods, or in this case, qualitative and quantitative methods, has given way to improved research designs. It allows researchers to triangulate the data so that the outcome is more reliable and valid. The combination of surveys and interviews, for instance, gives a better vision by showing the quantitative trend as well as the in-depth perspectives. This approach is also conducive to the integration of mixed data types, which adds strength to analysis as well as offering a deeper understanding of the research topic. As social science evolves, being critical of the processes of data collection is important to ensure that researchers value the ethical issues as well as possible biases that can arise (Hukal et al., 2017). Last but not least, effective data gathering techniques not only reinforce scientific strength but also provide the basis for successful, practical applications of research findings (Figure 2.2).

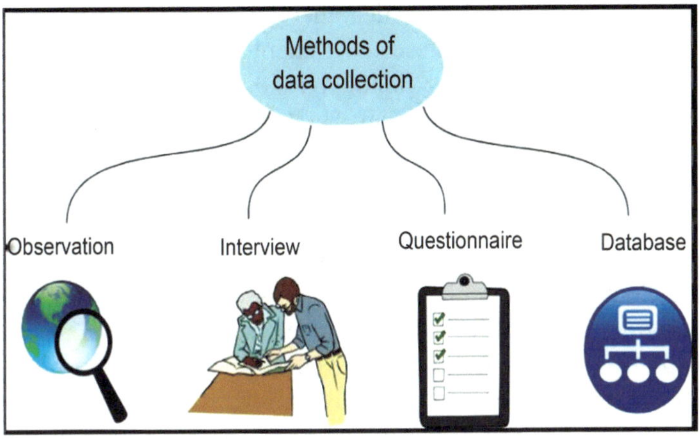

Figure 2.2. Importance of data collection in research.

Source: https://i0.wp.com/knowledgebase.centreforelites.com/wp-content/uploads/2022/04/Data-collection-instruments-in-Research.jpg.

Overview of Various Methods and Tools

Within the field of data acquisition, several methods and devices are available to meet a variety of research requirements and goals. Exploratory experimentation (EE) is one active procedure used across disciplines, optimally observed in disciplines such as nanotoxicology. Scholars utilizing EE typically engage with several methods, including the search for regularities, the development of new methods and apparatus, and the application of computer simulations to investigate large data sets. These strict methodologies not only enable patterns and relationships to be established but enable experimental controls to be manipulated in order to achieve more accurate results. As emphasized in current research, the development of a taxonomy which differentiates the objectives, theoretical underpinnings, and procedures employed in EE enhances our understanding and practice of exploratory research, justifying methodology's contribution to the data collection process (Elliott et al., 2008). The use of systematic assessment instruments further enhances the data gathering environment by providing provisions for systemic analysis. The European Organic Action Plan (EUOAP), for instance, has played a central role in shaping organic farming policy throughout the European Union. The ORGAP project is a best practice example of multi-stakeholder collaboration between government institutions and farmers' associations with the goal of developing an assessment toolbox to monitor the performance of action plans. This project introduces a sequential approach that guarantees comprehensive stakeholder consultation and capturing varied national priorities, thereby providing qualitative data for decision-making. The use of these assessment tools highlights the complex policy functionality-stakeholder input relationship, thereby imposing the depth and scope of data collection in Agri-structures (Dabbert et al., 2008). Moreover, the variety of qualitative and quantitative data-gathering methods corresponds with the necessity to tailor approaches to specific research settings. For instance, interviews and surveys are commonly employed to gather qualitative data, which provide depth and nuance that statistical estimates might lack. Conversely, quantitative measures—experiment and observation studies—provide quantifiable data amenable to resulting statistically significant outcomes applicable in broader settings. The use of participatory methods, with stakeholders being brought participatively into the research process itself, also contributes to data richness and validity. Such diversity in methodology not only reflects the nature of modern research settings, which are complex, but also emphasizes the potential for methodological triangulation, under which plurality of methodologies being brought together legitimizes findings and strengthens research integrity in general. Generally, an understanding of the various techniques and tools of data gathering is essential to researchers who aim for thorough and productive results.

2.1. QUALITATIVE DATA COLLECTION METHODS

Qualitative data collection methods are paramount in research since they allow researchers to achieve a deep understanding of phenomena that are complexly intricate. Such methods focus on demystifying meanings and lived experiences of individuals, providing rich and nuanced data that may not be discovered using quantitative approaches. Techniques such as interviews, focus groups, and observation studies allow such an analysis to be made possible since they document the perceptions of the participants within their natural settings. Flexibility of qualitative research enables researchers to adjust their method as the research process unfolds, rendering the gathered data more applicable. As can be witnessed in studies like that conducted by P2TP2A "Maharani," qualitative methods like document analysis and guided interviews have proven highly valuable in ascertaining program success when assisting victims of trafficking (Asriati et al., 2017). These approaches ensure that the experience and testimony of those affected are at the center of the research process. That multiple qualitative research methods of data collection allow for data triangulation means researchers can ensure high validity and reliability of findings.

Figure 2.3. Qualitative data collection methods.

Source: https://www.questionpro.com/blog/wp-content/uploads/2023/02/Qualitative-Data-Collection-Methods-1.jpg.

As an illustration, combining observations with interview data raises the general richness of understanding of a participant's context and behavior. As evidenced in inquiry-based learning educational research, explicit communication and observation techniques have the power to significantly increase knowledge related to student engagement in mathematics, energizing pedagogy changes (122 et al., 2013). Employing multiple sources of data not only enriches the findings but also enables proper analysis of social processes. Thus, qualitative methods can adapt to the needs of the research question, developing a good setup that entraps richer details of human interactions and experiences. Qualitative data collection methods depend on researchers being able to engage with participants empathetically and iteratively. The building of rapport is a central component because it enables participants to give truthful opinions and sentiments, thus creating quality data. Researchers mostly use open-ended questions to allow for discussion and establish an environment that stimulates narrative growth. Additionally, the qualitative inquiry process is cyclical, allowing ongoing refinement of research instruments, reactions to evolving themes arising during the gathering of data. Such adaptability not only enhances findings' depth but also allows researchers to address problems as they arise, thus allowing them to offer a more critical examination of the issue under study. Lastly, qualitative data collection methods enable researchers to tap into the experience of individuals themselves, and thus derive understanding that is critical to informing practice and policy (Figure 2.3).

2.1.1. Interviews

Interviews are a common method of qualitative data collection in most fields, particularly social science, market studies, and health. Interviews allow researchers to directly engage with participants and gain a robust comprehension of their ideas, experiences, and motivations. Conducting interviews can yield detailed contextual information that cannot be obtained with surveys since they allow researchers to probe deeper into participants' responses through follow-up questions. Thus, interviews can be useful to study complex issues that require nuanced understanding, as the conversational format of this method induces honesty and elaborate explanations between the respondents. However, interviews are highly reliant on interviewers' skills to elicit information and control the flow of talk—concerns that emphasize the necessity of thoughtful preparation and reflexivity during the research process. While interviews have their advantages, they do have limitations, and these can often include researcher bias and difficulties in qualitative data analysis. One major limitation is that interviews can be costly and time-consuming, both to conduct the interviews and to analyze the data created. Besides, they are likely to be affected by the interviewer's bias or the interviewer-interviewee relationship balance, which can, indirectly, reflect adversely on the validity of the information collected. Classical interviewing methods have also been criticized by researchers, particularly in market research, citing that excessive use of qualitative methods like interviews is bound to lead to weak customer needs comprehension. In these situations, other methodologies may be needed to collect in-depth information regarding consumer behavior and attitudes, and thereby enhance the importance of flexibility in data collection methods (Goffin et al., 2013). In ensuring more research rigor, pilot studies are typically carried out by researchers, which are extremely significant preparatory actions before the whole-scale research. A pilot study allows researchers to refine their research questions, methodology, and interview protocols upon reviewing preliminary findings. This guarantees that any challenges can be determined beforehand and that assumptions regarding the appropriateness of interview questions in drawing out information can be tested (Edwards et al., 2017). Researchers can make the main study more valid and reliable by resolving these issues beforehand. The value of the first phase can hardly be overstated, as it equips researchers with the knowledge necessary to conduct a structured interview process. Thus, integrating pilot studies in research design not only increases the quality of data collected through interviews but also the methodological strength in general, leading to more impactful and meaningful research findings.

2.1.2. Focus Groups

Focus groups are recognized as a qualitative data gathering method that takes advantage of group dynamics potential to gain an understanding of participants feelings and attitudes. The technique allows researchers to engage with various individuals in a straight way, generating broad discussions that have the ability to expose subtle feelings and views not uncovered by other methods such as surveys and interviews. However, the usefulness of focus groups has been questioned in academic literature. Critics have argued that traditional market research techniques like focus groups lag behind in measuring the complexity pertaining to break-through products, which call for in-depth customer needs and behaviors understanding (Goffin et al., 2013). Hence, researchers must weigh the benefits of deep qualitative insight against the threats of bias and limitation inherent in group situations, such as strong personalities overruling less assertive participants, distorting the data collected. The exploratory use of focus groups makes them particularly well-suited to the early stages of the research process, where having knowledge of participant opinion and attitudes is most on the agenda. By using an interactive environment, researchers can pose questions that will generate rich feedback, from product development ideas to consumer trends and tastes. The information gathered from such conversations can also be utilized as grounds for secondary quantitative research, informing hypotheses and guiding the creation of subsequent research tools. Yet, as noted in existing literature, not all market research methods are equally effective across all environments. For instance, focus groups may struggle to serve the needs of organizations that seek to innovate, particularly where traditional means are not adequate to capture the hopes and desires of customers (Gienapp et al., 2012). The researcher must then critically assess the appropriateness of focus groups for their own research purposes. Despite some of the criticism leveled against them, focus groups are a useful vehicle for gathering data, especially when used in conjunction with other methods. By blending the views gathered from these forums with quantitative data, researchers are in a better position to create an entire picture of the problem under investigation. Further, the incorporation of focus groups into a wider methodology allows triangulation—a method for checking findings from more than one source of data. Multiple perspectives can benefit organizations, as they guarantee that the customer voice is well represented across product development efforts. In the future, focus groups can continue to enhance through embracing new facilitation techniques and technology, enhancing participation, and overcoming constraints. Ultimately, developing a strategic plan on the use of focus groups in conjunction with other methods of information gathering will be able to enable organizations to make informed decisions which are customer needs and market conditions friendly (Goffin et al., 2013) (Figure 2.4).

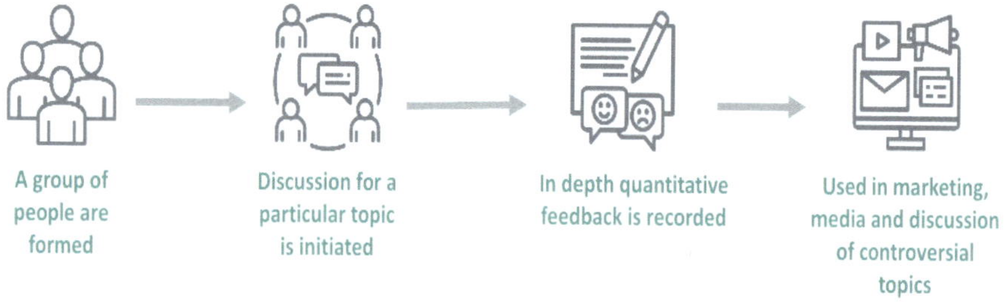

Figure 2.4. Focus groups are qualitative data collection.

Source: https://wallstreetmojo-files.s3.ap-south-1.amazonaws.com/2023/08/Focus-Group.png.

2.1.3. Observations

Observation as a data collection method has a core position in both quantitative and qualitative research, enabling researchers to gather first-hand information on behavior, events, or environments. The method is not only effective in capturing phenomena while they occur but is also employed in capturing important details about the environment in which events occur. Through direct observation, researchers can capture nuances that surveys or interviews may overlook. This kind of information is inevitable when making conclusions about the issue at hand. For instance, research in education commonly utilizes observation to examine teaching modes and student interactions in classrooms. Detailed records of interaction can be utilized to provide patterns that inform pedagogy and instruction intervention (Observatory et al., 2007). Thus, observation is about making real-life situation complex dynamics readable and comprehensible to researchers in order to make valid interpretations. Observation can also be differentiated into various forms such as participant observation and non-participant observation, each of which is applicable for a given purpose depending on research aims. For participant observation, researchers become part of social groups that are being studied and witness events from the insider's perspective. This method produces depth in understanding social practice, rituals, and interactions, as is evident in ethnographic studies. Non-participant observation, however, allows for the researcher to be somewhat detached while being able to capture behavior without contamination, which can further enhance objectivity. For example, research involving cultural rituals will tend to be largely observational so as to gain the full complexity of practices and beliefs without risking the introduction of biases through direct contact (Istiandini et al., 2014). Both modalities highlight the adaptability of observational methods in providing the rich qualitative information they do, thereby rendering them crucial assets in a researcher's toolbox. The primary limitation of observational studies, though, is the potential for biases and the challenge of achieving reliability and validity in the data collected. Observer bias would contort the interpretation of behavior because individual experience and knowledge can bias one's observation and recording of occurrences. In addition, subjects may modify their behavior as a result of awareness of observation, known as the Hawthorne effect, biasing results. To reduce these concerns, researchers must observe systematically in a manner that adheres to ethical practice while maintaining data collection rigor. This can include the use of structured observation schedules and triangulation with other data collection methods, for example, interviews or content analysis to cross-check findings. By this approach, while observation methods provide useful data, they must be keenly scrutinized so that the validity and integrity of the research findings are not compromised.

2.2. QUANTITATIVE DATA COLLECTION METHODS

The quantitative data collection field is characterized by systematic methodologies that aim to collect numerical data for analysis. Surveys, experiments, and observational studies are among them, each selected based on research question and context. Surveys, usually in the form of questionnaires, allow researchers to gather data from a large sample base at minimal cost, thus making statistical analysis of patterns and relationships possible. The use of experiments, though, involves altering variables to observe effects, thereby producing causations. Observational studies provide a further element by allowing researchers to collect data in natural settings, although they need to be well planned so that they are valid. Both of these methods form a core to quantitative research, as they yield data which can be quantified and computed statistically in an effort to learn about patterns in the phenomenon in question. The choice of using particular quantitative data collection techniques tends to rest on the context requirements and the nature of data needed. Studies in developing nations provide a testament to the intricacy of making such choices, highlighting that methodology as well as data output must be taken into account on a continuum of contextuality.

Figure 2.5. Quantitative data collection methods.

Source: https://cdn.brocoders.com/Quantitative_Data_Collection_Methods_ 8ce-a8f60fb.png.

For instance, while certain health utilization data might necessitate more contextual methods with more detailed findings, more typical non-contextual statistical methods might be preferable for more generalizable wider uses. The tension calls attention to the necessity of researchers critically studying their methodology in terms of their specific research questions. Moreover, as in contemporary research, the blend of qualitative insights and quantitative practices can contribute depth and applicability to insights, allowing scholars to approach multiple dimensions of the data collection process with respect (Hentschel et al., 2025). Moreover, the emerging world of population studies depicts the increasing collaborative partnership between quantitative and qualitative approaches. Quantitative expert researchers are now increasingly turning to qualitative tools, like semi-structured interviews, to supplement their knowledge of complex social processes. While this blending of data collection techniques is beneficial to multi-disciplinary research, it also poses challenges, particularly for researchers who utilize only quantitative data collection methods. Representative problems, respondent motivation, and interviewer attributes can complicate qualitative surveys and necessitate proper handling and interpretation of qualitative information. As evident in the comparative studies in nations like Zimbabwe and Senegal, proper application of these combined approaches can provide deeper insights that even simple quantitative information might not be able to provide. Thus, a peaceful and contemplative fusion of the two methodologies not only broadens the researcher's tool kit but also deepens the generated data analysis (Randall et al., 2025) (Figure 2.5).

2.2.1. Surveys and Questionnaires

Questionnaires and surveys are some of the data collection methods through which researchers are able to obtain quantitative and qualitative data from a given population. They enable the measurement of variables and attitudes through the responses received in forms that are either structured or semi-structured. Developed mostly for either hard copy or soft copy administration, surveys vary from complicated to simple depending on the research objective. For instance, good questionnaire design is crucial to minimizing biases and easing participants' responses to be an indicative sample of the general population. Proper question composition, like measures and open-ended questions, is crucial in the richness of the data acquired. Thus, careful attention to survey design enhances the validity and reliability of results, making surveys an integral instrument in various areas of research, from health research to social studies and beyond. Advances in technology have greatly influenced the history of surveys, particularly with the advent of digital platforms that can conduct online data gathering. Online surveys provide various advantages, such as fast dissemination, live collection of data, and lower operation cost. The versatility with digital forms means that researchers can easily shift survey instruments at a rapid pace, and hence introduce more flexibility in various research contexts. Nevertheless, as per existing literature, there are specific issues with embracing the digital form. These include participant attrition, response rates, and providing questionnaires to individuals with disabilities, among others, and which pose innovative methods in designing surveys. Particularly in mHealth research, specific frameworks are required to handle real-time data access and security concerns effectively, which translates to that the future of survey research may indeed depend on reconciling technology with sound methodological practice (Bilen et al., 2016). Researchers should also implement best practices regarding survey design and presentation to maximize the effectiveness of surveys and questionnaires. Foundational concepts of clarity, brevity, and reasonable ordering of questions are essential in guiding respondents through the survey process. Pilot testing of surveys can also identify potential question wording or format errors, so that the follow-up data collection yields high-quality data. Lastly, ethical considerations of informed consent and data confidentiality are essential to maintaining participant cooperation and trust. The uniform argument over menstrual hygiene management (MHM) confirms the importance of constructing the proper survey instruments fit to face present barriers and noted practice in multiple contexts (Vasquez et al., 2014). These focused endeavors not only enhance data quality but also allow researchers to create actionable results that guide effective interventions in numerous fields of social research (Figure 2.6).

	Survey	Questionnaire
Individual questions	✓	✓
Used for gathering data	✓	✗
Analyzes responses	✓	✗
Provides insights about an individual	✓	✓
Provides insights about a group of people	✓	✗

Figure 2.6. Surveys and questionnaires.

Source: https://cdn.prod.website-files.com/6718da5ecf694c9af0e8d5d7/6749b96c71627968f9d5b6cb_survey-vs-questionnaire.webp.

2.2.2. Experiments

Experiments are a pillar of support in the discipline of data collection methods and tools, forming a systematized process for hypothesis testing. Through variable changes under controlled conditions, researchers can distinguish the effect of single variables while eliminating confounding variables. This method not only yields greater cause-and-effect insights but also raises the validity of findings. The controlled experimental environment allows for replicability, a critical component of scientific inquiry. Furthermore, technology and advanced tools of data gathering have enhanced experimental method, with a precise measurement and simple analysis. As recommended in Bouhineau et al.

(2007), the establishment of guidelines for experimental procedures ensures that researchers are conducting best practices, therefore enhancing the overall quality and consistency of data gathered during the research process. Another equally critical aspect of experiment conduct is the design and analysis cycle that adapts based on initial results and hunches. This adaptability is crucial to refining research questions and framing methodologies to suit best the research goals. The pre-experiment phase identified in Bouhineau et al. (2007) creates a framework for addressing core research challenges. Such advance planning is crucial to directing the methodologies employed and data to be collected. In addition, the value of shared effort among researchers is also highlighted through ongoing sharing of input and progress during the experimental stage. The collaborative climate not only enriches the data collection process but also results in more refined explanations of outcomes since various angles make for a complete understanding of the data. Moreover, the role of simulation in experimental design has become more apparent as a means of enhancing teaching methods with the use of statistics. As, simulations provide students with an opportunity to experience real-world scenarios involving careful decision-making in data gathering techniques. By practicing the intricacies of spatial surveys or environmental monitoring, students can have firsthand experience of the intricacies of experiment design. Such a hands-on activity not only anchors theoretical ideas but also hones critical thinking abilities involved in interpreting data. The complementing of laboratory practice with traditional experimental methods emphasizes the dimensional nature of data collection in scientific research. Altogether, the evolution of experiments as a data collection method depicts a unifying interaction among theoretical foundations and new practices, which results in better understanding of complex phenomena across varied academic fields.

2.2.3. Secondary Data Analysis

Secondary analysis of data is a critical method in the grand framework of data collection tools and techniques. Secondary analysis allows researchers to use pre-existing data collected by another person, thus saving time and resources that could otherwise have been utilized for collecting primary data. Secondary analysis is particularly useful in exploratory research, where extensive datasets from previous research can expand the area of research. By examining existing data sets, researchers can identify emerging trends and areas of not knowing which direct more research endeavors. Additionally, such analysis emphasizes the importance of regarding context seriously; utilization of secondary data requires a sharp understanding of the methodology applied in the initial study to ensure the validity and reliability of findings obtained (Chorley et al., 2009).

Thus, secondary data analysis is not merely a supporting methodology but also an indispensable component in the generation of sound and comprehensive research results. Despite its advantages, secondary data analysis inherently possesses challenges to be addressed by researchers. To begin with, secondary data may be of varying quality depending on the effort and intention of the primary data collectors.

Figure 2.7. Secondary data analysis.

Source: https://www.phdbox.edu.in/img/secondry.png.

Researchers must critically assess the limitations and potential biases in such data sets before drawing conclusions. It is especially necessary in cases where the secondary data concern consumer behavior or market trends, where integrating the customer's voice has a heavy impact on product configuration (Goffin et al., 2013). Secondary data may also be not specific or detailed enough for the needs of this research, and thus, need to be interpreted cautiously. In such a situation, combining secondary data with other methodologies, say qualitative interviews, can bridge the gaps and give depth to analysis in order to achieve a more sensitive understanding of the subject. Furthermore, subsumption of secondary data analysis within a systematic research framework can enhance the general outcomes of a study. With triangulation of primary and secondary findings, researchers can take advantage of the strengths of both approaches in order to provide a more holistic picture. Integration helps to promote comprehensive knowledge that is in fact crucial in product development and marketing, where an understanding of customer taste needs an integrated approach for data collection (Chorley et al., 2009). Secondary data analysis should also consider using the rising practice of using the application of big data and analytics software to aid in secondary data analysis, which offers real-time analysis and predictive analytics. Eventually, when properly used, secondary data analysis does not only complement primary data collection but also improves the overall research process by offering differing perspectives and data-driven insights for informed decision-making (Figure 2.7).

2.3. MIXED METHODS APPROACH

The mixed methods approach is even more salient in the midst of the diversified range of data gathering procedures, since it combines qualitative and quantitative research methods to get a balanced insight into research matters. The eclecticism approach permits gathering and intersection of different types of data, hence adding varied dimensions to the analytical process. Notably, the integration process occurs on different levels, including design, methods, and interpretation. At the design level, researchers can use various designs such as exploratory sequential, explanatory sequential, or convergent designs to guide the methodology towards their specific investigation. By this flexibility, not only is the level of analysis enhanced but also responsive research strategies are encouraged to dynamically respond to emerging findings during the study, thus yielding stronger conclusions about complex health issues and phenomena (Creswell et al., 2013). In health services research, mixed methods is especially beneficial when assessing patient-reported outcomes (PROs) and their use in treatment decision-making. Different studies, including those that utilize electronic PRO questionnaires, have affirmed the feasibility of such combined method of obtaining rich patient experiences. For instance, Strength Through Insight app, applied in a study for prostate cancer patients, connected quantitative health-related quality of life measurements with qualitative interviews. The study demonstrated the power of mixed methods not only to engage with quantitative data but also to further have an interface with rich, qualitative data that informs clinical interventions and individualized care strategies. By combining various types of data, researchers expand their capability for exploring patient attitudes, and findings become more meaningful and actionable in real-life healthcare settings (Dicker et al., 2020). Furthermore, the strength of the mixed methods lies in the ability to break through the constraint related to single methodology orientations. Quantitative data alone may fail to capture the emotional and social dimensions of health experience, whereas qualitative methods are sometimes non-generalizable. Synthesizing these with a mix of methods enables researchers to cross-validate data and to give a fuller evaluation of health interventions. The interpretive process, in which researchers synthesize outcomes from qualitative and quantitative streams, enables more insight into the interaction between the factors influencing health outcomes. Therefore, mixed methods research not only maximizes strong data collection and analysis but also allows for a richer account that is crucial in informing practice and policy in health, since researchers blend the strengths of both methods to provide insightful interventions and suggestions (Creswell et al., 2013) (Figure 2.8).

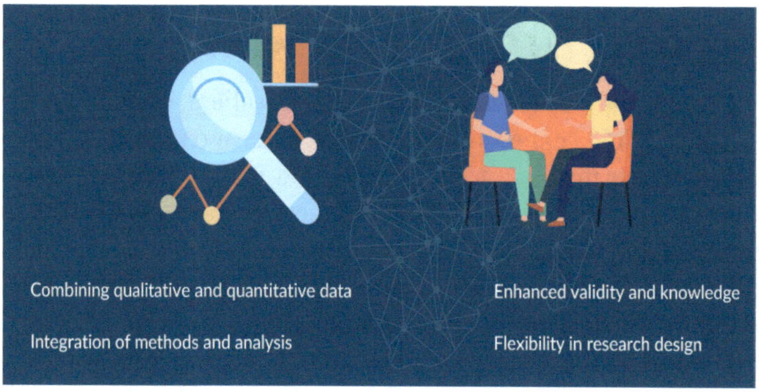

Figure 2.8. Mixed methods approach of data collection methods.

Source: https://www.inonafrica.com/wp-content/uploads/2023/11/Mixed-Method.png.

2.3.1. Definition and Significance

An understanding of data collection procedures and tools is essential for any reliable research project as the processes constitute the foundation and credibility of information gathered. Essentially, data collection is a systematic process of acquiring, quantifying, and analyzing information in an attempt to create answers, test hypotheses, or establish patterns. This very first step is crucial across fields, from scientific research to social sciences, and hence constitutes an essential part of the data analysis process. The reason behind employing proper data collection is the impact it has on the reliability as well as the accuracy of research findings. Without using properly chosen tools and techniques, researchers risk collecting inaccurate or biased information, thus lessening their results and conclusions. Also, understanding the nuances in data collection provides researchers with authority to select the best methods suited to their specific research objectives, thereby enhancing the depth and extent of their studies. Techniques in data collection range from surveys, interviews, observation, and data harvesting on online platforms, each playing an indispensable role in data collection. Each method has unique strengths and limitations, necessitating careful consideration at their selection. Surveys, for instance, can gather huge quantities of quantitative data quite quickly but could miss the fine points of people's opinions. Or alternatively, qualitative methods such as interviews can deliver rich details but are likely to have limited sample sizes, hindering generalizability. Recent research finds that the integration of mixed methodologies enhances research robustness and provides a richer understanding of complex issues. As reiterated by studies emphasizing the importance of data collection in various disciplines, choosing a method to use has a considerable effect on outcomes and thus an informed manner of conducting things is needed to achieve reliable and actionable results (Banfill et al., 2009). Also, the evolutionary character of data collection mechanisms, especially as shaped by technological developments, has a monumental effect on research approaches. Electronic sites and software enable not only the gathering of data but also ease the process of analysis, enabling researchers to process large datasets with greater ease. Utilization of electronic tools, such as data management systems and survey sites, enhances data collection processes by increasing accuracy and efficiency. Also, suggestions integrating the documentation and information management processes into research work point to the importance of systematic approaches in conservation and evaluation contexts (LeBlanc et al., 2008). With researchers applying these tools, they not only ensure comprehensive data management but also promote integrity of the findings. Lastly, as data collection techniques continue to innovate, they will play a more focal role in creating research developments in many disciplines.

2.3.2. Combining Qualitative and Quantitative Data

Combining qualitative and quantitative data provides a robust system of data collection and analysis that enables the researcher to benefit from the strengths of each approach. Quantitative data, with numbers and statistical significance, provides a solid basis for drawing broad conclusions, while qualitative data adds depth and context, explicating the inherent reason and conditions surrounding those numbers. The mixture of these two kinds of data is quite important in sophisticated research situations, particularly in health and social sciences, where recording subtleties of human activity and social patterns are vital. A well-considered mixed-methods design facilitates thorough understanding of research issues, enabling better insight and richer interpretations of data than mere correlations. The development of systems for concurrent collection and analysis of qualitative and quantitative data remains an area of innovation. From observing recent innovations, like the use of a cloud analytics platform for studying the built environment and its users, marrying sensor data from smart devices with qualitative observation is one way in which these methods can coexist and complement each other. In this case study, data obtained from older adults in care homes is one instance of how the integration of these data types allows researchers to identify trends and patterns that might not otherwise become apparent.

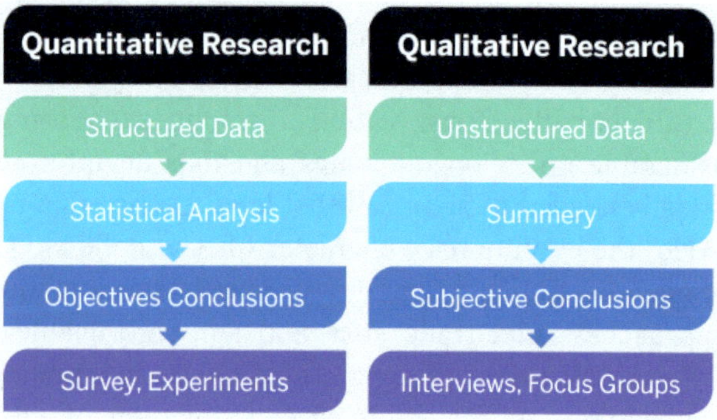

Figure 2.9. Combining qualitative and quantitative data of both methodologies.

Source: https://www.qualtrics.com/m/assets/wp-content/uploads/2022/01/Quantitative-Research1.png.

By placing quantitative data from sensors in close proximity to qualitative data from interviews or observation, researchers can construct an enhanced multimodal narrative that not only explains what is happening but also why it is happening, and therefore overall validity and usefulness of the research are enhanced (Dowsland et al., 2017). The philosophical debates regarding mixed-methods research also serve to bring to light the complexities involved in the integration of qualitative and quantitative effectively. From existing literature and even mixed-methods studies in Nepal, researchers have to work around different theoretical frameworks to decide on the most appropriate sequencing and blending of their data sources. This means looking into the underlying philosophical assumptions of each method, for instance, interpretivism for qualitative and positivism for quantitative tests. Meeting pragmatic requirements, including reconciliation of findings using different methodologies and consolidation of the aforementioned findings, is crucial to ensure a common interpretation of the research findings. Successful synthesis of opposing data sets not only increases analytical robustness but also builds an integrated picture of the research problem, thereby leading to the enhanced reliability and usability of the findings in real-world contexts (Bryers et al., 2014) (Figure 2.9).

2.3.3. Examples of Mixed Methods Studies

Mixed methods studies have gained prominence in research as it is able to utilize both qualitative and quantitative information to yield a full picture of complex problems. A good case in point is in the design and evaluation of interactive simulations for learning quantum mechanics, where authors utilized the utilization of mixed methods. This procedure allowed educators not only to quantify the effectiveness of QuVis simulations in student performance data but also to assess them qualitatively through observational data and student feedback. By determining how students interacted with these simulations, researchers were able to enhance education plans effectively. The incorporation of diverse methods of data collection enhanced the validity of the study and provided a rich tale of students' learning experiences, demonstrating the strength of mixed methods in education research (Kohnle et al., 2014). In another revealing example of manufacturing engineering, a mixed methods investigation examined combined New Product Introduction (NPI) practices. This research used both surveys and case studies to determine how current manufacturing practice compares with lean integration principles. The case studies provided detailed information on current processes, while the survey quantified the implementation of these processes on a series of component development programs in a major aerospace manufacturer. Identifying challenges and formulating guiding principles for better NPI practice were byproducts of this two-way methodology. Here, the mixed approach made it possible to explore contextual insight and general trends alike, demonstrating the versatility of combining the qualitative and quantitative measures in resolving real-world problems in organizational practices and engineering (Beadle et al., 2013). Proper application of mixed methods studies can result in higher insights, especially where there is a need for inter-disciplinary approaches. For instance, an integrated social science and healthcare study may employ surveys to gauge patient satisfaction and interviews to procure the personal experiences underlying such quantitative results. Employing both methods gives researchers validation of results between methods and therefore enhances the validity of the research results overall. By employing varying data-collection strategies, researchers can tackle the sensitivity and complexity of human action and society. These mixed strategies ensure that findings achieved are not only statistically valid but also contextually applicable, making mixed methods pivotal in achieving comprehensive conclusions in today's research studies.

2.4. DIGITAL DATA COLLECTION TOOLS

As the world becomes more digitalized, the selection of appropriate data gathering tools is the key to successful research. Digital data gathering tools are a broad software and hardware application, utilised to enhance the gathering and analysis of data through increased efficiency and accuracy. They are employed for a diverse array of functions, from simple surveys undertaken online to sophisticated data management systems using multiple sources of information. With the prevalence of mobile technology and cloud computing, data gathering has been transformed to allow for instant real-time data gathering and instant retrieval. With researchers adapting to these advancements, it is crucial that they critically assess each tool's capabilities and shortcomings in making a decision regarding its suitability for specific research objectives. The advent of digital technologies has likewise resulted in the creation of an ethics perspective on data privacy and security that emphasizes responsible data management and informed consent as the hallmarks of contemporary research practice. The application of mobile applications that support data gathering is particularly significant in this context, as they have revolutionized research methodologies in numerous fields. Particularly, the popularity and usability of Android devices keep increasing such that the need for a forensically viable practice of collecting evidence from these devices also increases. Android devices may carry a lot of evidentiary information that can be very important to a forensic practitioner while carrying out investigations. Collecting this type of information, however, requires the practitioner to understand and implement ways of analyzing data obtained from the device. The ability to collect information efficiently through mobile apps enhances the research process, where it becomes more rapid and possibly even more comprehensive compared to conventional methods. The shift also aligns with broader societal practices and preferences, as more individuals incorporate technology into their daily lives, something that researchers must consider when designing studies and selecting appropriate instruments. Apart from this, the application of digital data collection instruments has inherent challenges regarding the validity and reliability of collected data. Data collectors must remain vigilant in ensuring stringent criteria so that such instruments are able to give reliable results. Bias in data collection is always a risk, particularly with the use of automated devices or algorithms that may inadvertently lean towards particular groups or responses. As technology advances, there is an increased demand for sound methodology to verify and validate findings. Emphasizing the significance of ethical considerations during the period of gathering data is imperative, particularly because (Stephenson et al., 1999) underscores the importance of rigor in work submitted or published. Last but not least, the impact of computerized tools on data collection is profound since they necessitate up-to-date research practices to be examined in order to ensure that results remain valid and representative of human behavior in the digital age (Figure 2.10).

Figure 2.10. Digital data collection tools to effective research.

Source: https://www.evalcommunity.com/wp-content/uploads/2024/02/rsz_1data-collection-tools.png.

2.4.1. Online Survey Platforms

In the evolving context of data collection, internet-based survey instruments are an integral resource for researchers. These instruments enable efficient collection of quantitative and qualitative data across different populations, hence enhancing the representativeness of the databases. Researchers benefit from editable templates, variety in types of questions, and advanced analysis, which assist in developing surveys that can be standardized for specific research aims. In addition, these tools provide researchers with the ability to reach respondents through geographical spaces, which increases inclusivity in addition to diverse opinions. While researchers pursue research on the intricacies of web-based data collection, it is evident that web-based survey websites have the potential to ease the research process while expansively increasing the scope and depth of study. Online survey platforms also introduce new sets of methodological challenges and problems for researchers. For example, while these websites offer ease of use and convenience, they can inadvertently bring about respondent selection biases, in which individuals who do not have proper internet access are excluded from research. Additionally, the reliance on self-reported information raises concerns regarding the validity and honesty of the response. One of the fundamental foundations of using online survey technologies is that they will allow researchers to appropriately balance these biases and limitations. Critical scrutiny of the data collection process, including respondent engagement and retention plan, is required in an effort to enhance validity of results. Through acknowledgement of these challenges, researchers can revise their methods in an effort to enhance the analytical potential of data gathered through online surveys. Integration of online survey instruments into more complete qualitative study designs also presents new opportunities for innovation. As the case of Lusophone African multinational business companies shows, the employment of social media in identifying and qualifying research populations mirrors the diversified utilization of computer tools in data collection. Employing web survey sites can be supplemented directly by web scraping techniques reported in recent literature so that researchers can access real-time data and contextual details which may otherwise be difficult to accomplish (Back et al., 2013). Such synergy between web survey sites and web analysis lends strength to the belief that fine-grained data collection should be achieved to adapt to social complexity. With this changing relationship, scientists are prompted to remain fluid and creative in their methods, utilizing technology to expand the depth and relevance of the investigations (Smith et al., 2018) (Figure 2.11).

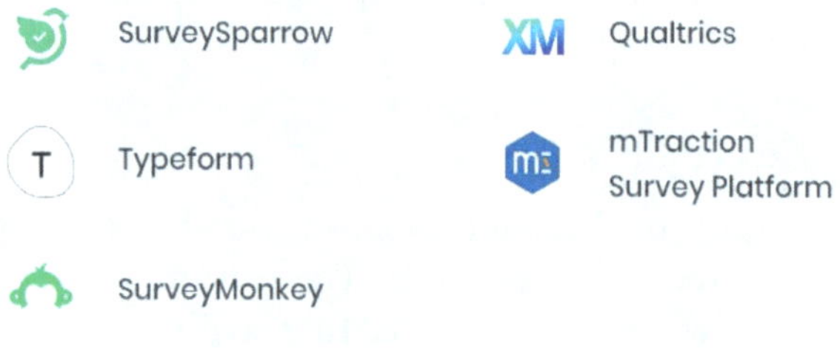

Figure 2.11. Online survey platforms.

Source: https://www.mtractionenterprise.com/blog/wp-content/uploads/2022/11/Top-Five-Online-Survey-Platforms.jpg.

2.4.2. Data Analytics Software

Data analytics software is at the heart of present data collection means and technology and significantly enhances the efficiency of how organizations acquire, analyze, and make meaning of data. The software facilitates simple joining of varied sources of data, making possible instant analysis and decision-making knowledge. With organizations increasingly relying on data-driven decisions, effective analytics tools have skyrocketed in demand. These programs not only automate the gathering process but also enable the user to visualize complex data sets in order to help identify trends, outliers, and patterns. Through the transformation of raw data into forms that can be understood, data analytics software enables stakeholders with the means necessary to make decisions that are data-driven and can enable strategic goals as well as operational optimization. Besides, the creation of data analytics software has been spurred on by advances in technology that assist in enhancing efficiency in managing data. Upcoming developments point to trends of incorporating artificial intelligence and machine learning into the platforms, which enhance their analytical capability. Using these technologies, organizations can automate the repeated collection of data and develop predictive models that provide insights into future patterns using historical data. Not only does this save time but also minimizes human error, with greater precision. With organizations handling complex datasets, the use of new tools becomes increasingly vital to sustain competitive advantages in multiple businesses, emphasizing the need for continued investments in the creation of tools and in the users' training (Bassi et al., 2013). Also, the function of data analytics software is not just mere collection and analysis but also an essential tool for evaluation and improvement of iterative processes. Current evidence confirms the interaction between new methods and technological progress in measuring program and intervention effectiveness (Weston et al., 2017). Such ongoing evolution in data practices emphasizes the necessity for companies to adopt data analytics tools that align with contemporary best practices. As technology advances, not only do data-gathering processes improve but also provide assistance with integrated program design and evaluation strategies. Through an emphasis on utilizing analysis outcomes to inform decision-making, organizations are positioned to improve outcomes and react adaptively to an ever-evolving environment, indicating the imperative role of data analytics in their organizational framework.

2.4.3. Mobile Data Collection Applications

Technological innovation has revolutionized data collection, particularly with the application of mobile data collection software. Software exploits the extensive usage of smartphones so that individuals can effectively and efficiently gather information. For instance, citizen science initiatives heavily depend on mobile applications that allow for real-time input and feedback of data. These instruments are advantageous not only to user interaction but also to the quality and amount of data collected because they provide instant ways of analysis and interactive possibilities for acquiring knowledge. Projects like the Networked Naturalist demonstrate how mobile applications can transform participant experience by making data collection simpler and fun. Through the assistance of organizations such as the National Park Service, these endeavors demonstrate the strength of mobile applications to engage public involvement and awareness towards science endeavors, ultimately producing improved data and increased interaction with environmental issues (Graham, 2010). Mobile data collection apps play a particularly critical role in the field of international development and evaluation. With budgets becoming more limited and stakes for successful results higher, there is always greater demand for robust mechanisms that are capable of measuring the endeavors of development. Mobile applications can bridge the gap between old-school and modern demands by giving stakeholders the ability to monitor and view projects in real-time. Also, the trend toward the use of more participatory evaluation approaches underscores the importance afforded to user-generated data, where citizens actively engage to provide feedback and observations. Data democratization enhances a sense of participation and responsibility among participants, pivotal in ensuring that development initiatives are effective and sustainable (Raftree et al., 2016). The convenience and flexibility in gathering data through mobile applications have the ability to address many of the operational problems encountered in various industries. Under traditional methods, there are some problems that are universally encountered such as limited time, insufficient resources, and limitations on data quality. Under mobile technology, organizations have the ability to automate the gathering of data and minimize error from manual entry. In addition, such apps can also make measurements timelier, allowing practitioners to refine their strategies based on real-time responses and emerging trends. As development agencies and practitioners face increasingly complicated settings, mobile data collection software is a cutting-edge solution that addresses methodological requirements and the need for high-quality, timely data. Hence, the integration of these applications can significantly enhance the effectiveness of monitoring and evaluation processes (Raftree et al., 2016).

2.5. ETHICAL CONSIDERATIONS IN DATA COLLECTION

Ethical data collection concerns are central to ensuring that research procedures maintain the rights and dignity of participants. Intrinsic to this ethical system is the informed consent principle, which necessitates that individuals understand fully their participation in a study and the possible consequences of their data being utilized. This necessitates that researchers present clear information about the purpose, methodology, and possible effects of the research. A failure in informed consent can lead to mistrust between communities and researchers, undermining future research. Furthermore, ethical data collection must be mindful of cultural backgrounds so that the tools employed do not exploit and misrepresent the participants. Adherence to such ethical standards not only establishes credibility but further improves the research itself. As crucial as ethical data collection is the consideration of data privacy and security. Researchers must implement strong controls to safeguard participants' personal data from misuse or loss by unauthorized parties.

Figure 2.12. Ethical considerations in data collection.

Source: https://www.promptcloud.com/wp-content/uploads/2024/12/Ethical-Considerations-in-Data-Collection.jpg.

The threat of data breaches has been heightened with the growing use of computer-based tools for data collection, hence the imperative to adopt proactive cybersecurity measures. Compliance with legal guidelines, such as the General Data Protection Regulation (GDPR), is a must to maintain personal rights and avoid legal repercussions. Furthermore, researchers need to take into account long-term effects of data storing and utilization, and such transparent procedures for what data would be utilized, distributed, and stored. Keeping privacy paramount maintains participant confidence and ethical value in the data-collecting process. The ethical consequences extend beyond the immediate concerns of privacy and consent to encompass the overall impact of research outcomes on society. Researchers must critically assess the possible ways in which results of their research might affect data populations from which data are collected. For instance, in health-focused research, like in some trials, data collection impacts local healthcare policy or interventions (Becher et al., 2008). Hence, researchers have a moral obligation to analyze if their research will unintentionally marginalize or misrepresent certain groups. In return, this ethical consideration can generate more reflective and responsible conclusions that are socially useful. Moreover, integrating advanced data tools, such as those applied in narrative projects (Byrne et al., 2011), holds the potential to illuminate individual experiences without violating ethical principles, whereby human aspects of data remain at the forefront of research discourse (Figure 2.12).

2.5.1. Informed Consent

Informed consent is a ground rule of research ethics that ensures participants are adequately aware of the nature, intent, and potential risks of taking part in a study. The element is particularly critical where data collection methodology entails personal or sensitive information. Not just should researchers provide detailed information regarding the study, but they ought to ensure that participants can understand it, thereby being able to exercise informed choices regarding participation. The importance of informed consent goes beyond compliance with ethical standards; it builds trust between participants and researchers, which is critical for the research process to be of sound integrity. Without informed consent, the validity of the evidence gathered could be compromised since participants will be coerced or not fully aware of their rights, and this would affect the general quality of the research findings. Having a proper process of informed consent is particularly critical in environments with socio-economic disparities, like in studies conducted in developing nations like Pakistan. For example, during the evaluation of demand-side financing approaches toward increasing the utilization of modern contraceptives by low socio-economic status women, researchers ought to make sure the participants are properly informed about the consequences and the benefits expected from the study. Women from such groups are often faced with information barriers and access to health care services, thus the need for informed consent to enable them to make decisions in harmony with their health needs and life circumstances. Through effectively communicating the objectives of the study's as well as that of voluntary consent, researchers can ensure ethical standards while enhancing the credibility of their research results and thus their contribution towards improved health outcomes among disadvantaged populations (Abbas et al., 2014). Also, advancements in data collection technologies as well as methods have created new avenues for providing informed consent. The availability of digital platforms makes it possible to convey study information in a more efficient and transparent way, with participants better informed. It is particularly valuable where, as part of evaluation and program development, novel data collection tools can have items that optimize participant engagement and understanding. These tools can be utilized by researchers to provide interactive consent forms or video explanations outlining the aim and process of the study. These advances are in line with advice in existing literature, highlighting the importance of informed consent as more than a legal requirement but also as a ethical responsibility when it comes to research. By incorporating these new methods into their process of consent, researchers can raise the quality and ethical basis of their research, ultimately leading to better methods of collecting data (Weston et al., 2017).

2.5.2. Privacy and Confidentiality

In the rapidly evolving landscape of means to gather information, privacy and confidentiality are issues that must be given serious examination. Businesses are increasingly compelled to harvest massive amounts of data in an attempt to gain knowledge and enhance services. This activity, however, has profound moral consequences for the protection of individuals' personal information. The test here is not only in the technical aspects of data management but also in the policy frameworks within which such activity is conducted. Whereas many institutions, including libraries, have established policies regarding data management, few of them have sufficient strategies in place that effectively safeguard user privacy. This policy gap in inconsistency renders it imperative to promote inter-institutional conversation for the purpose of understanding and addressing learning data complexities, such as acknowledging concerns for student privacy as well as inappropriate use of information collected (Atwood et al., 2019). The roll out of data collection apparatus which needs to achieve a delicate balance between leveraging data to drive innovation and preserving ethical standards of privacy. For instance, the concept of "black-box medicine" illustrates this tension very nicely, especially in medical applications where dependence on sophisticated algorithms can generate dramatic leaps in diagnosis and treatment options. To benefit from these advances, however, researchers are often forced to require large numbers of patient data, a state that occurs at enormous threat to confidentiality and privacy. Transparency and accountability are the topmost concern as the users must be assured that their data are handled with utmost care. For addressing these concerns, stakeholders must implement robust systems with privacy-preserving components to ensure data sharing but compromise no personal data. These include applying substantive limitations to the collection and processing of data as well as requiring independent audit to guarantee the validity and impartiality of the algorithms used (Ford et al., 2016). Further, the nexus of confidentiality and data collection procedures is putting more scrutiny in academic and professional circles. Since organizations have access to sensitive data, the applicability to research integrity and participant trust rises. Confidentiality is greater than just encrypting information; it is creating a culture of respect for users' privacy, which enhances public trust in data-driven initiatives. With regard to an academic library, the lack of set best practices to manage learning data further complicates the issue. Libraries must engage positively with administrators and the broader library profession to develop sweeping policies that address privacy issues as well as enable efficacious uses of data. The application of specific procedures can help resolve student privacy issues, ultimately leading to better data governance practices that respect individual rights and further the mission of knowledge dissemination and education improvement by the library (Atwood et al., 2019).

2.5.3. Data Security Measures

Increasing volume and sensitivity of information collected by organizations need robust data security measures to protect against unauthorized access and potential exposures. A simple method is the application of encryption technologies, which convert sensitive information into encrypted forms that are not readable without related decryption keys. This is particularly important in the situation where data collection procedures are involving personal information, where compliance with regulations such as the General Data Protection Regulation (GDPR) is mandatory. Organizations that employ encryption not only secure users' information but also enhance their credibility and trust factor. In critical infrastructure security support frameworks, effective encryption can prevent grave threats from data exposure, aligning with risk management best practices in sophisticated technological systems as highlighted in recent literature (Procházka et al., 2018). Access controls also play a key role in data protection practices. Strenuous user authentication processes, such as multifactor authentication (MFA), significantly reduce the likelihood of unauthorized access. This process requires users to verify their identity using different ways so that only approved staff can see sensitive data. Coupled with role-based access control (RBAC), access to data is limited according to a user's role within an organization, these procedures altogether minimize threats from within and enhance overall data integrity. The relationship between access controls and data safety reveals the necessity of frequent checks and revisions of security policies, particularly in environments where decision-making is data-informed. Such innovative risk management practices are featured in technical facility and safety policy debates, with the focus on rigorous checks (Procházka et al., 2020). Regular audits and active monitoring are essential components of effective data security strategies. Through the intensive scanning of security protocols and monitoring data access patterns, organizations can gain insights into vulnerabilities and respond quickly to danger. Such a proactive strategy is not only necessary for regulatory compliance but also for maintaining customers' trust in organizations' data handling. Minimizing risks in data processing and collection, thereby ensuring high safety standards, reflects organizational accountability and respect for society. The technical system risk management standards specify that end-to-end monitoring ought to be capable of addressing varied vulnerabilities, thereby addressing the dynamic character of data attacks in real time. With this, organizations are able to successfully safeguard prime data assets so that they can enjoy robust defense against more sophisticated cyber attacks (Procházka J et al., 2018, 2020) (Figure 2.13).

Figure 2.13. Data security measures.

Source: https://kanerika.com/wp-content/uploads/2024/02/data-security-measures--1280x720.jpg.

2.6. CHALLENGES IN DATA COLLECTION

Data gathering is laden with many challenges, particularly in dynamic environments such as humanitarian contexts. The requirement for timely data on reproductive maternal, newborn, and child health (RMNCH) emphasizes the need for effective data collection methods. Research attests that inadequacies in available data can hinder planning and the delivery of essential health programs. In a study, respondents carried out data gathering in crises always identified severe obstacles. Among these, the unstable nature of humanitarian crises, where security concerns usually exclude access to affected populations, was the general concern. Second, the emergency acute phase exacerbates the difficulties of gathering credible information, and it is therefore necessary for researchers to design adaptable strategies that recover quickly to changing circumstances while ensuring the accuracy and suitability of gathered data (Dickinson et al., 2015). Differences in languages create another critical issue in data collection, particularly in scenarios where multi-lingual settings prevail in the context of humanitarian crisis. Support for multiple languages in tools used to collect data is integral in order to enhance communication as well as accurate interpretation of findings. Participants cited that without tools that are adapted to cover local dialects and cultural nuances, the success of data collection is significantly undermined. Secondly, tools that are currently available being fitted into the specific environments of different communities require more local capacity building efforts.

Figure 2.14. Challenges in data collection.

Source: https://www.hitechbpo.com/wp-content/uploads/2023/10/challenges-of-data-collection-768x675.jpg.

These involve the training of local data collectors who are able to manage cultural sensitivities and have the ability to change the method of data collection in real time, thereby improving the overall quality of data obtained at every step of humanitarian response (Dickinson et al., 2015). In other technology advanced environments, such as in astronomy research, other issues also arise, such as data processing and analysis from large data sets. Long Wavelength Array Software Library (LSL) is an example of specialized software needed to manage complex data sets, such as those obtained from radio astronomy equipment. Owing to complex data formats and the large number of inputs, conventional data processing software might be less than sufficient, resulting in possible delays or errors in analysis. LSL provides a solution by simplifying processes like filter synthesis and incoherent de-dispersion so that effective management of data is possible. This adaptability points towards the necessity of tailor-made tools in resolving issues of data collection in various fields, showing that regardless of the field, infrastructure and tools must adapt to meet the unique requirements of data collection and analysis (Figure 2.14).

2.6.1. Sampling Issues

Sampling is a fundamental process of data collection that is very critical in establishing research findings to be valid and reliable. Statistical laws say that good sampling makes research findings generalizable to the population, while bad sampling can result in biased data and wrong interpretations. Both of the two main sampling strategies—probability and non-probability sampling—also possess various advantages and limitations. Probability sampling, in which methods such as random sampling are used, benefits in providing each of the target population members with an opportunity of selection, thus minimizing bias. Non-probability sampling is easy and in most instances cost-saving but can lead to overgeneralization or under-representation, which is very crucial in developing research environments such as Sierra Leone where the low literacy rate limits ethical considerations and application of sampling strategies (Jackson et al., 2018). Also, the importance of selecting an appropriate sampling strategy extends beyond mere representation; it is paramount in facilitating ethical research practice. In low literacy settings, researchers may face certain challenges, and there is a requirement for using methods of engaging participants in an ethically informed and efficient way. Ethics take precedence, as researchers must tread the thin line of voicing marginalized viewpoints with maximum precision while giving participants consent and understanding (Jackson et al., 2018). To this end, researchers tend to not utilize the best available technological advancements that can ease data collection and better sampling techniques. For instance, computer-based tools can identify overlapping subgroups in populations, thus improving sampling precision and allowing researchers to make more relevant conclusions from their findings. The data collection terrain is constantly changing due to innovative research tools and methods that are capable of overcoming numerous barriers present in sampling procedures. Conferences, as highlighted in the UNICEF reports, are where scientists share information on good tools to employ in exploring specific challenges and evaluating interventions across settings. The use of mixed-method approaches to triangulate the findings in order to achieve a better appreciation of the target population is an example. Through the combination of qualitative and quantitative information, researchers are able to evade certain limitations of conventional sampling methods, especially in culturally heterogeneous environments. For that reason, dealing with sampling problems through strict methodologies and creative research devices is most important for creating robust datasets that benefit the scientific community and guide practical applications across a wide range of fields (Figure 2.15).

Figure 2.15. Sampling issues of data collection.

Source: https://wallstreetmojo-files.s3.ap-south-1.amazonaws.com/2023/05/Sampling-Error.png.

2.6.2. Data Accuracy and Reliability

In the methodology and means of data collection, data accuracy and reliability are priorities. Data accuracy refers to the degree to which data reflects the phenomenon under measurement. Incorrect data will lead to wrong analysis and deceptive conclusions with devastating impacts on decision-making processes in all sectors, including business, healthcare, and the social sciences. There is a need to establish tight measurement protocols that minimize uncertainties and cross-check received data to align with desired metrics. In manufacturing environments, as referred to, where accuracy is a priority, the challenge lies in balancing the need for accurate measurement against productivity and efficiency functionality needs, thereby inadvertently ignoring necessary measurement processes (Fletcher et al., 2015). This subtle balance underscores the significance of developing valid data collection instruments that prioritize accuracy as a means of gaining credible results. Reliability, on the other hand, is used to mean constancy of data over time and conditions. An instrument of data collection that is reliable will yield similar results under equal circumstances and hence establish trustworthiness. Reliability is particularly crucial in the application of tests or surveys, where response variability can be indicative of rooted issues with data quality. Researchers may use established processes such as the Define-Measure-Analyze-Improve-Control (DMAIC) method to enhance reliability by identifying and streamlining any inconsistencies through the process of data collection. Such a systematic approach serves to emphasize training programs aimed at ensuring data collectors, regardless of position, adhere to standardized procedures, thus enhancing a culture of reliability and accuracy in data practices (McClellan, 2012). Ultimately, the effectiveness of data collection techniques is ultimately reliant on the accuracy and reliability of the data generated. Furthermore, the intersection of dependability and accuracy is particularly critical in high-risk environments, such as factory production, where the consequences of data error can have dire economic effects. The combining of Lean and Six Sigma philosophies offers a framework for addressing the complex issues of maintaining data integrity. Through the spirit of ongoing improvement, organizations can systematically evaluate measurement tools and techniques to reduce uncertainty and optimize production potential. It recognizes that while seeking higher accuracy may at times stifle production quantity at times, neglecting to prioritize measurement accuracy can facilitate waste and inefficiency. Hence, building a culture that values proper mechanisms of data collection is not only healthy but necessary in ensuring operational excellence in any data profession (Fletcher et al., 2015).

2.6.3. Resource Limitations

Insufficiency of resources considerably impacts the accessibility and efficiency of data collection tools across various areas of study. Budgetary constraints might restrict the selection of optimum tools and means of data collection, leaving researchers with less-than-optimum alternatives. For instance, insufficient funds to avail advanced software or hardware can undermine the potential to gather intact datasets, consequently affecting the validity of the resulting research findings. This is a typical situation in schools of learning, where budgetary constraints have the propensity to affect the volume and scope of research. The implications of such constraints have a potential to propagate throughout the research process, since insufficiency of proper data collection will distort conclusions and diminish the overall integrity of research. Therefore, understanding resource limitations is crucial while planning research methods, resulting in a careful consideration of the instruments and potential sacrifices that are associated with employing them. In addition to funding constraints, human resources are also pivotal in establishing data-gathering activities. Having access to qualified resources noticeably affects the effectiveness of the chosen methods and tools. Inability to analyze data or work with sophisticated technologies can lead to inefficient interpretation of data, and the collected information becomes less valuable. This can be addressed by investment in training and development of staff to enhance the level of data collection team members' skills. Besides, researchers may be constrained to utilize methodologies within their existing skill set, hence foregoing innovation and the adoption of new methodologies. Recognizing these human resource limitations, researchers can better organize their strategy in a way that there is adequate support and expertise made available to handle the complications of data collection well. Additionally, the technology environment has its own limitations of resources which hamper data collection activities. The rapid rate of change at which software and tools undergo developments results in complications in regard to making research approaches current and applicable. In most cases, researchers may be employing ancient technology due to their limited resources or lack of experience in utilizing advanced technologies. This is at the cost of embracing techniques that employ cutting-edge methods of data extraction, like in the situation of web data extraction. Upon researching techniques of web data extraction, it is apparent that there are various approaches, among them those that employ unsupervised machine learning to enhance data gathering efficiency (Banos et al., 2013). Nonetheless, the efficacy of these methods relies on the availability of adequate technological resources to researchers and willingness to adopt new technologies. Thus, overcoming technological resource limitations is a central element in optimizing data collection tools and acquiring quality research outcomes.

SUMMARY

In summarizing the main conclusions of the essay Data Collection Methods and Tools, it is apparent that choosing methods plays a tremendous role in data quality and interpretability gathered across various regions. Strategic choice between data collection quantitative and qualitative methods facilitates researchers to adjust their approaches in accordance with their specific objectives. For instance, while quantitative methods can confer statistical representativeness and generalizability, qualitative methods confer rich contextual inscriptions that allow for greater insights into social phenomena. Such dualism awards high importance to the choice of appropriate methodologies by researchers where, not only do they fit research questions, but they also help towards more refined conclusions. Also, the use of sophisticated tools like hybrid systems like VEX-93 illustrates the potential for enhancing data analysis on the basis of sophisticated reasoning models. The capabilities of such systems enable users to employ diversified analytical processes ranging from rule-based logic to fuzzy reasoning, thereby enabling insight generation from complex datasets. These tools emphasize the importance of marrying perception with reason in data analysis to facilitate richer understanding beyond just numeric return. For this reason, the application of flexible means of data collection alongside advanced analytical tools is required to effect inclusive conclusions in diversified research environments, particularly those that involve complex decision-making systems. Finally, the application of contextual approaches to data collection still reshapes paradigms of traditional research. As the analysis of health planning and utilization indicates, it is clear that some kinds of knowledge can only be made apparent by contextualizing research methods to the local worlds and practices. The use of a variety of data collection methods makes it possible for researchers to remain attentive to local contexts, employing contextual insights to inform the production of non-contextual tools as and when needed. This adaptive and integrated approach ultimately enhances the robustness of research findings, leading to more effective interventions and policy-making towards resolving pressing societal issues, thus justifying the central role of diverse data acquisition methods in leading effective conclusions (Martínez et al., 1997; Hentschel et al., 2025).

Under the context of data collection tools and methods, there is a systematic approach behind the quality of research results. The methods used in data gathering, data analysis, and data interpretation go a long way in ensuring the quality of evidence gathered. For example, the UK Department for International Developments commissioned review of fisheries and aquaculture science is a classic example of an integrated data collection strategy. The assessment included a scoping review that comprehensively searched available literature on the contribution of the field to economic growth and food security. Through systematic exclusion of 202 documents and classification of findings into two primary strands—developmental outcomes and mediating factors—this project illustrates the effectiveness of a robust framework in data collection that provides meaningful responses to primary research questions. This methodological precision enhances the quality of findings and contributes to clearer understanding of the subject (Béné et al., 2005). Moreover, inclusion of environmental components in data collection cannot be overlooked. The Data Management Plan for addressing land degradation mitigation in Africa also emphasizes monitoring biodiversity as well as ecological sustainability. By creating major sites as field laboratories, the DMP lays the groundwork for comprehensive evaluations of biodiversity condition and natural resource management alternatives. Not only is data acquisition prioritized through this method but evidence acquisition is also intertwined with operational applications, including the creation of sustainable livelihoods and policy recommendations. As a result, the initiative plays a critical role in the preservation of biodiversity and the reduction of environmental issues. Employing strategic research methods of data collection, the DMP perfectly demonstrates how research can contribute normative principles and recommendations, thereby address wider environmental problems as well as promote sustainable development in nations (Leclerc et al., 2005). Lastly, the integration of effective data collection techniques and their applications is the essence of efficacious research outputs. Both the examination of aquaculture and fisheries and the DMP reveal how laser-focused data gathering can lead to comprehensive knowledge in both economic and environmental realms. By meticulously mapping interventions already established and identifying areas of evidence gaps, researchers can inform stakeholders and policymakers alike, ushering in practices that resonate across domains as diverse as they are far-reaching. This interdependence thus of data gathering and aftermath fuel more research, allowing the advancement that comes from clearer grasp of complex global issues. Consequently, sustained enhancement in data collection activities remains significant in provoking evidence-based decision making and development promotion in various sectors (Béné et al., 2005; Leclerc et al., 2005).

REFERENCES

1. Abbas, G., Ali, M., Azmat, S. K., Bilgrami, M., et al. (2014). A study protocol: Using demand-side financing to meet the birth spacing needs of the underserved in Punjab Province in Pakistan. https://core.ac.uk/download/55764863.pdf (accessed on 21st October 2025).
2. Adams, A., Baptista, R., Beck, A., Blatter, B., Castanho, R., Clark, D., et al. (2018). Sustainable planning of cross-border cooperation: A strategy for alliances in border cities. https://core.ac.uk/download/225163008.pdf (accessed on 21st October 2025).
3. Al Mutawa, A., Andriotis, P., Backes, M., Barghouthy, N., Chung, J., Do, Q., Enck, W., et al. (2015). Conceptual evidence collection and analysis methodology for Android devices. http://arxiv.org/abs/1506.05527 (accessed on 21st October 2025).
4. Asriati, N., Ranny, T., & Rustiyarso, R. (2017). Peran Pusat Pelayanan Terpadu Pemberdayaan Perempuan dan Anak "Maharani" dalam proses pembinaan korban trafficking. https://media.neliti.com/media/publications/216276-peran-pusat-pelayanan-terpadu-pemberdaya.pdf (accessed on 21st October 2025).
5. Atwood, G., Benson, D., & Sherriff, G. (2019). Practices, policies, and problems in the management of learning data: A survey of libraries' use of digital learning objects and the data they create. https://core.ac.uk/download/215154984.pdf (accessed on 21st October 2025).
6. Back, L., Becker, H., Bollier, D., Bowker, G. C., Danowski, J. A., Duhem, P. M. M., Weltevrede, E., et al. (2013). Scraping the social? Issues in live social research. https://core.ac.uk/download/2794969.pdf (accessed on 21st October 2025).
7. Banfill, P. F. G., Khasreen, M. M., & Menzies, G. F. (2009). Life-cycle assessment of buildings: A review. https://core.ac.uk/download/29083877.pdf (accessed on 21st October 2025).
8. Banos, V., Davis, R., Gkotsis, G., Pincent, R., et al. (2013). BlogForever D2.6: Data extraction methodology. https://core.ac.uk/download/30436955.pdf (accessed on 21st October 2025).
9. Bassi, R., Caballé Llobet, S., Daradoumis, T., Xhafa, F., et al. (2013). A review on massive e-learning (MOOC) design, delivery and assessment. https://core.ac.uk/download/187768781.pdf (accessed on 21st October 2025).
10. Beadle, C., Carswell, M., Maginness, M., Shehab, E., et al. (2013). Principles for aerospace manufacturing engineering in integrated

new product introduction. https://core.ac.uk/download/74409972.pdf (accessed on 21st October 2025).

11. Becher, H., Bridger, J. M., De Allegri, M., Dong, H., et al. (2008). Step-wedge cluster-randomized community-based trials: An application to the study of the impact of community health insurance. https://core.ac.uk/download/337060.pdf (accessed on 21st October 2025).
12. Béné, C., Little, D., Arthur, R., & Leschen, W. (2005). Synthesis of FMSP experience and lessons learned for fisheries co-management, final technical report. https://core.ac.uk/download/75777504.pdf (accessed on 21st October 2025).
13. Bilen, V. M. (2016). Developing a framework for creating mHealth surveys. https://core.ac.uk/download/213054950.pdf (accessed on 21st October 2025).
14. Bouhineau, D., Cazes, C., Cerulli, M., Chaachoua, H., et al. (2007). TELMA cross experiment guidelines. https://core.ac.uk/download/47118478.pdf (accessed on 21st October 2025).
15. Bowman, A., Scott, E. M., Crawford, E., & Gibson, I. (2025). Interactive teaching tools for spatial sampling. https://core.ac.uk/download/pdf/6340261.pdf (accessed on 21st October 2025).
16. Bryers, H. M., Pitchforth, E., & van Teijlingen, E. (2014). Advocating mixed-methods approaches in health research. https://core.ac.uk/download/29050976.pdf (accessed on 21st October 2025).
17. Byrne, D., Jones, G. J. F., & Kelliher, A. (2011). Life editing: Third-party perspectives on lifelog content. https://core.ac.uk/download/11310340.pdf (accessed on 21st October 2025).
18. Chorley, A. H., Anderson, A. H., Bernstein, A., Chorley, A., et al. (2009). e-Social science and evidence-based policy assessment: Challenges and solutions. https://core.ac.uk/download/109223.pdf (accessed on 21st October 2025).
19. Clarke, T., Dowell, J., Ray, P. S., Stovall, K., et al. (2012). The Long Wavelength Array Software Library. http://arxiv.org/abs/1209.1576 (accessed on 21st October 2025).
20. Creswell, J. W., Curry, L. A., & Fetters, M. D. (2013). Achieving integration in mixed methods designs—Principles and practices. https://core.ac.uk/download/127441332.pdf (accessed on 21st October 2025).
21. Dabbert, S., Gonzalvez, V., Lampkin, N., Michelsen, J., et al. (2008). Development of criteria and procedures for the evaluation of the

European Action Plan for Organic Agriculture. https://core.ac.uk/download/10930262.pdf (accessed on 21st October 2025).

22. Denniss, A., Doolan, B., Fink, J., Jones, R., et al. (2019). TB or not TB? That is the question regarding TB treatment in a remote provincial hospital in Solomon Islands. https://core.ac.uk/download/286085122.pdf (accessed on 21st October 2025).
23. Dicker, A., Gressen, E., et al. (2020). Utilizing digital health to collect electronic patient-reported outcomes in prostate cancer: Single-arm pilot trial. https://core.ac.uk/download/288847196.pdf (accessed on 21st October 2025).
24. Dickinson, F. M., Pyone, T., & van den Broek, N. (2015). Experiences from the field: Maternal, reproductive and child health data collection in humanitarian and emergency situations. https://core.ac.uk/download/132609480.pdf (accessed on 21st October 2025).
25. Dowsland, S., Hanson, V. L., Hiden, H., Simpson, A., et al. (2017). A platform for the analysis of qualitative and quantitative data about the built environment and its users. https://core.ac.uk/download/96883181.pdf (accessed on 21st October 2025).
26. Edwards, J. A., Ismail, N., & Kinchin, G. (2017). Pilot study, does it really matter? Learning lessons from conducting a pilot study for a qualitative PhD thesis. https://core.ac.uk/download/153570255.pdf (accessed on 21st October 2025).
27. Elliott, K. (2008). Varieties of exploratory experimentation in nanotoxicology. https://core.ac.uk/download/11921904.pdf (accessed on 21st October 2025).
28. Fletcher, S., Longstaff, A. P., & Shagluf, A. (2015). A preliminary study of applying Lean Six Sigma methods to machine tool measurement. https://core.ac.uk/download/30732249.pdf (accessed on 21st October 2025).
29. Ford, R. A., & Price, W. N., II. (2016). Privacy and accountability in black-box medicine. https://core.ac.uk/download/84123604.pdf (accessed on 21st October 2025).
30. García Martínez, A., García, Y., Hernández, D., Hita Rensoli, Y., et al. (1997). The VEX-93 environment as a hybrid tool for developing knowledge systems with different problem-solving techniques. https://core.ac.uk/download/81577408.pdf (accessed on 21st October 2025).
31. Gienapp, A., Reisman, J., & Stachowiak, S. (2012). A handbook of data collection tools: Companion to "A guide to measuring advocacy and policy." https://core.ac.uk/download/84911833.pdf (accessed on 21st October 2025).

32. Goffin, K., Michea, A., van der Hoven, C., Varnes, C., et al. (2013). Giving the customer a voice: A study of market research methods and their perceived effectiveness in NPD. https://core.ac.uk/download/14343201.pdf (accessed on 21st October 2025).

33. Graham, E. (2010). The networked naturalist: Mobile phone data collection for citizen science and education. https://core.ac.uk/download/pdf/289574.pdf (accessed on 21st October 2025).

34. Hentschel, J. (2025). Distinguishing between types of data and methods of collecting them. https://core.ac.uk/download/pdf/6521437.pdf (accessed on 21st October 2025).

35. Hukal, P., & Stelmaszak, M. (2017). When data science meets social sciences: The benefits of the data revolution are clear but careful reflection is needed. https://core.ac.uk/download/80784660.pdf (accessed on 21st October 2025).

36. Istiandini, W., Linda, N. S., & Nandar, I. (2014). Fungsi Tari Bukung dalam upacara kematian pada masyarakat Suku Linoh Desa Nobal Kabupaten Sintang. https://core.ac.uk/download/326760103.pdf (accessed on 21st October 2025).

37. Jackson, E. A. (2018). Critical review of sampling procedures in the context of Sierra Leone's low literacy (and under-resourced) research communities. https://core.ac.uk/download/286357495.pdf (accessed on 21st October 2025).

38. Johnman, C., Kennedy, I. T. R., & Takah, N. F. (2016). Impact of approaches in improving male partner involvement in the prevention of mother-to-child transmission (PMTCT) of HIV on the uptake of PMTCT services in sub-Saharan Africa: A protocol of a systematic review and meta-analysis. https://core.ac.uk/download/42372377.pdf (accessed on 21st October 2025).

39. Kann, C. W. (2014). Digital scholarship: Applying digital tools to undergraduate student research papers, a proposal for a freshman seminar. Part I: Definition of student research methodology. https://cupola.gettysburg.edu/cgi/viewcontent.cgi?article=1022&context=csfac (accessed on 21st October 2025).

40. Kohnle, A. (2014). Interactive simulations for the learning and teaching of quantum mechanics concepts. https://core.ac.uk/download/30318650.pdf (accessed on 21st October 2025).

41. Lamont-Doherty Earth Observatory, Columbia University. (2007). Using data in undergraduate science classrooms. https://core.ac.uk/download/pdf/5206792.pdf (accessed on 21st October 2025).

42. LeBlanc, F., Letellier, R., & Schmid, W. (2008). Recording, documentation, and information management for the conservation of heritage places: Guiding principles. https://core.ac.uk/download/71354261.pdf (accessed on 21st October 2025).

43. Leclerc, G. (2005). Experimenting With the Design of Policies on Sustainable Resources Management: Phase 1 Final Report, January 2004–February 2005. http://agritrop.cirad.fr/545377/1/document_545377.pdf (accessed on 21st October 2025).

44. Liu, J. (2014). Developing a life cycle assessment model for measuring sustainable performance of buildings in China. https://opus.lib.uts.edu.au/bitstream/10453/30396/1/01front.pdf (accessed on 21st October 2025).

45. McClellan, C. (2012). Teacher evaluator training & certification: Lessons learned from the Measures of Effective Teaching Project. https://core.ac.uk/download/71359050.pdf (accessed on 21st October 2025).

46. Procházka, J., & Procházková, D. (2020). Analysis, management and trade-off with risks of technical facilities. https://core.ac.uk/download/323214553.pdf (accessed on 21st October 2025).

47. Procházka, J., Procházková, D., & Rusko, M. (2018). Principles for management of risks of critical infrastructure. https://core.ac.uk/download/287889332.pdf (accessed on 21st October 2025).

48. Raftree, L., & Bamberger, M. (2016). Emerging opportunities: Monitoring and evaluation in a tech-enabled world. https://core.ac.uk/download/83111958.pdf (accessed on 21st October 2025).

49. Randall, S., & Koppenhaver, T. (2025). Qualitative data in demography. https://core.ac.uk/download/pdf/7127276.pdf (accessed on 21st October 2025).

50. Smith, C., & Goncalves, M. (2018). Social media as a data gathering tool for international business qualitative research: Opportunities and challenges. https://open.bu.edu/bitstream/2144/40303/2/20180601%20-%20PRE-PRINT%20-%20Social%20media%20as%20a%20data%20gathering%20tool%20for%20international%20business%20qualitative%20research%20opportunities%20and%20challenges.pdf (accessed on 21st October 2025).

51. Stephenson, C. (1999). Recent developments in cultural heritage image databases: Directions for user-centered design. https://core.ac.uk/download/4817615.pdf (accessed on 21st October 2025).

52. Sukmawati, S., & Zainuddin, Z. (2013). Peningkatan aktivitas siswa dalam pembelajaran matematika dengan metode inkuiri di kelas V Sekolah Dasar Negeri 08 Padang Pio. https://media.neliti.com/media/

publications/217357-peningkatan-aktivitas-siswa-dalam-pembel.pdf (accessed on 21st October 2025).

53. Vasquez, E., Sommer, M., Sahin, M., Worthington, N., & Dooley, T. (2014). WASH in Schools empowers girls' education: Proceedings of the Menstrual Hygiene Management in Schools Virtual Conference 2013. https://core.ac.uk/download/75783637.pdf (accessed on 21st October 2025).

54. Weston, A., Kazimirski, A., Kenley, A., McLeod, R., & Gripper, R. (2017). Global innovations in measurement and evaluation. https://core.ac.uk/download/87085547.pdf (accessed on 21st October 2025).

CHAPTER 3

DATA MANAGEMENT AND PROCESSING

Contents

INTRODUCTION

With the data-intensive global scenario today, effective data processing and handling are critical to the success of various industries, ranging from business to scientific research. Organizations need the capability to harvest, store, and process humongous quantities of information in order to make effective decisions. With exponentially increasing data, the intricate mechanisms set up to manage this data need to evolve as well. This evolution necessitates the integration of new processing techniques that do not just enhance efficiency but also ensure data security and integrity. Full understanding of data processing and management is thus crucial for practitioners in this intricate environment since these are supporting functions that underlie strategic planning and operational effectiveness. Moreover, emergence of technologies such as wireless sensor networks has revolutionized the way data is obtained and processed. These networks facilitate real-time data aggregation of multivariate sources into a huge volume of data that can be utilized for analysis.

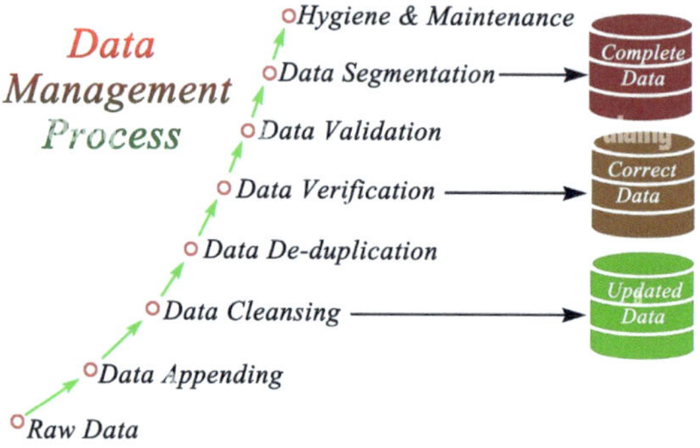

Figure 3.1. Data management and processing.

Source: https://c8.alamy.com/comp/2AGF2GB/components-of-data-management-process-2AGF2GB.jpg.

The convergence with artificial intelligence has also opened up new battlefields in optimized data processing. For instance, neural networks increasingly address complex tasks like path discovery, extending past the limitations of power consumption and processing capability in traditional systems. Scholars have thoroughly studied several facets of this new field, such as the use of power, MAC protocols, and routing protocols, all of which introduce a heightened awareness of the means by which data processing and management can be optimized in a host of situations (Concejero et al., 2006, 2007). The applications of effective data processing and management extend far beyond technical optimization; they also have an influence on organizational strategy and societal advancement. As businesses make use of data analytics, they can gain competitive advantage by understanding market trends and customer behavior. Moreover, advances in data processing technologies can lead to such paradigm shifts in human quality of life as smart cities and improved healthcare systems. Such continued R&D not only propel industries forward but also bring forth questions about the proper utilization of data and limitations in privacy. As corporations traverse these issues, the debate about managing and processing data will certainly be at the forefront of technology development and policy creation, marking its paramount significance in today's society (Figure 3.1).

Definition of Data Management

Data management is a general discipline addressing how data is organized, stored, and utilized to effectively enable decision-making in myriad organizational settings. At its core, data management aims at developing systematic approaches that are able to define and examine various ways through which effective data processing and management systems can be established. The core of this involves the combination of various techniques and systems engineering tools that enable early definition of desired services as well as their integration with existing operation concepts. Through the provision of information in outline form, graphs, and charts, data management enables stakeholders to visualize complex landscapes of data, facilitating better understanding, which makes it easy to have an informed strategy in addressing organizational requirements. This focus on systematic analysis is a testament to the inherent significance of data management in assisting in the guarantee of data moving as smoothly as possible over its lifetime, hence improving organizational effectiveness and efficiency (Moe et al., 2025). Data management has been influenced by increasing needs for sophisticated database systems, particularly in areas like Computer-Aided Design (CAD) and Computer-Aided Manufacturing (CAM). These systems are integral integration factors that need advanced data modeling and retrieval capabilities to manage the complexities of geometric data. Technical advances like spatial databases have hence emerged to cope with the unique challenges of geometric entities such as freeform curves and surfaces. Traditional database systems typically find it difficult to handle geometric objects like multidimensional points, but advances in database management have offered improved mechanisms for effectively managing complicated representations like B-spline and Bezier curves. This highlights the critical role of data management in pushing the advancement of technology to accommodate increasing demands of contemporary engineering applications (Fahldiek et al., 1993). In addition to addressing the technical requirements of storing and retrieving data, effective data management also has an equal emphasis on governance and compliance. Data quality, security, and privacy are maintained by policies and procedures adopted by organizations, since regulatory environments become increasingly more complex. By establishing a framework for handling data assets throughout their lifecycle, organizations can minimize the threats of data breaches and adhere to law and ethics. Secondly, the emergence of big data analytics calls for a more advanced approach to data management, where organizations harness massive amounts of data to derive meaningful insights. While corporations make this transition, the importance of good data management practices becomes increasingly self-evident because it is still the context in which strategic programs invoking innovation and competitive advantage in the market are built. A complete understanding of data management is thus critical for organizations who want to thrive in an increasingly data-driven economy.

Importance of Data Processing

The significance of data processing is not only transactional efficiency but is intrinsic to organizations' strategic decision-making processes. Data processing, in a big data and advanced analytics world, enables the identification of actionable knowledge out of huge datasets. Not only is this form of data processing effective in improving organizational effectiveness, but it also fosters innovation. As the Special Issue of Sustainability points out, convergence of technology and social sciences becomes a crucial factor in studying the manner in which organizations utilize data for competitive reasons. Sophisticated information and communication technology (ICT) helps organizations manage complex variables and address market demands, vindicating the demand for competent handling of data to facilitate sustainable policy-making and overall competitiveness. Through the establishment of a strong data processing infrastructure, organizations are in a position to continue dominating their areas, converting raw data into meaningful knowledge that informs strategy and fuels growth (Lytras et al., 2018). Additionally, data processing effects are profound in terms of resource management and organizational structure. Standard arguments on managing data tend to center around effectiveness vs. efficiency, but what becomes a more evolved comprehension is that the organizational politics are central to defining how the computing resources are allocated. The contemporary debate alludes to the requirement for both centralized and decentralized models of managing data, analyzing the particular requirements of specific organizations. As low-cost and mobile computing options become more prevalent, managers are faced with new challenges in optimizing data processing approaches without relinquishing control of information systems. As a result, embracing an adaptive strategy allows organizations to be able to foresee changes in their data processing mechanisms to keep pace with shifting organizational goals and resource abilities (King et al., 1984). Lastly, the strategic implications of adequate data processing cannot be downplayed in light of present-day business landscapes. Organizations dealing with data processing acquire better risk management, higher compliance, and ultimately better competitive advantage. By controlled data processing, businesses can recognize patterns and trends that aid market change and customer preference forecasting. This predictive method not only minimizes the risk but also enhances the company's responsiveness in responding to shifts in the business environment. Considering the immense learnings gained through studies of knowledge management and big data, it is possible to observe directly that the adoption of robust data processing mechanisms is essential for any organization to thrive in this age of big data. This holistic perspective of data management ultimately inspires an attitude of innovation, pushing organizations towards greater sustainability and long-term success (Lytras et al., 2018) (Figure 3.2).

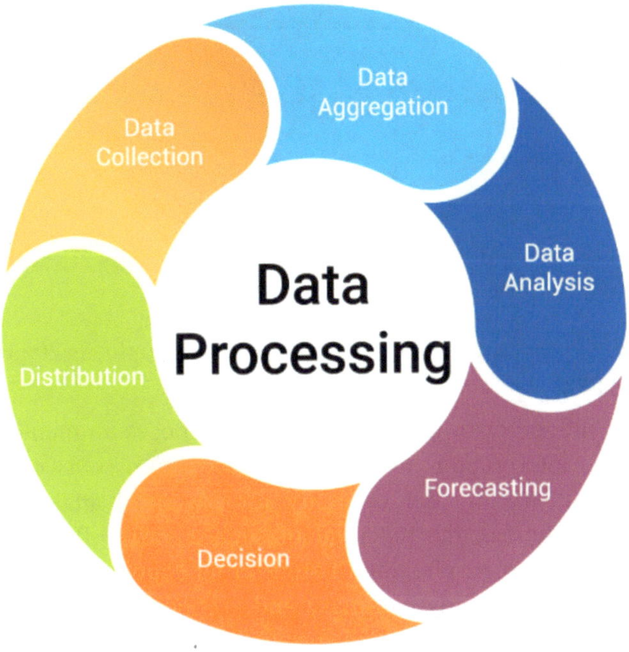

Figure 3.2. Importance of data processing.

Source: https://external-content.duckduckgo.com/iu/?u=https%3A%2F%2Fwww.dataentryindia.in%2Fblog%2Fwp-content%2Fuploads%2F2022%2F04%2Fdata-processing.jpg&f=1&nofb=1&ipt=8fed04cc7541ae5a7133c6253a2bd809f617a991bbae6f3e39338cd40a71c275.

3.1. TYPES OF DATA

Understanding the various types of data is the foundation for efficient data management and processing. In general, data is broadly divided into two categories: structured and unstructured data. Structured data means data that is neatly organized and easily search engine-able within a database, i.e., customer data within relational databases that will adhere to an established schema. This type of data facilitates efficient querying and reporting, so organizations can make informed decisions through clear-cut measures. Unstructured data, whose formats consist of text files, images, and videos, does not have a pre-existing model or schema.

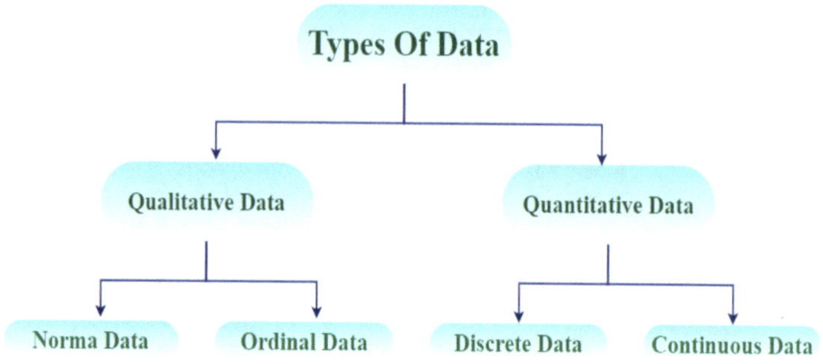

Figure 3.3. Types of data.

Source: https://media.geeksforgeeks.org/wp-content/uploads/20231011122244/Types-of-Data.png.

Handling and processing unstructured data poses immense challenges, particularly in interpreting meaningful information out of humongous data, thereby emphasizing the necessity for advanced analytical tools and methodologies to unlock the potential of such a vast collection of data. In addition, data can also be classified based on its source and type, leading to another important dichotomy—primary and secondary data. Primary data is collected firsthand regarding a specific research objective or purpose, providing the most accurate depiction of the topic under research. Interviews, surveys, and experiments are examples because they allow researchers to create insights as a direct response to their research. Secondary data, however, are derived from external sources, such as previous research, records of statistics, or databases. While secondary data may be relevant and time-efficient, its application is typically subject to detailed investigation with respect to context and quality of the data. With proof in the form of coal fire monitoring programs in northwest China, the integration of various types of data—from remote sensing images to geological survey data—recommends the holistic approach of working with complex datasets in a cost-effective manner (Genderen et al., 1999). In addition to these types, quantitative and qualitative data sets are types of data that exist, in essence impacting analytical methods. Quantitative data is measurable or countable and is numerical in nature, with the ability to do statistical analysis and objective comparison. Quantitative data is particularly helpful in empirical research and is needed for hypothesis testing and verification. On the other hand, qualitative data, normally characterized by descriptive attributes, provides depth and context to research findings. For example, the application of qualitative data can be seen in categorization of Wi-Fi enabled devices where analysis of probe request frames provides insight into usage without intrusive methods (Cesana et al., 2016). Both quantitative and qualitative data are required in the analysis continuum, providing researchers with a wide array of tools to tackle complexity in their field of study (Figure 3.3).

3.1.1. Structured Data

In data management and processing, structured data constitutes a crucial foundation for effective organization and analysis. Structured data refers to data that is highly organized and readily searchable with algorithms, usually stored in relational databases. The predictability of structured data makes it easy to enter, access, and manipulate the data, hence making it easy to conduct operations in most organizational environments. Most importantly, structured data is vital to optimizing the capabilities of Big Data management. As companies and organizations find themselves making more decisions based on data analytics, embracing structured data enables them to leverage its inherent organization to arrive at conclusions rapidly and efficiently. This systematic approach not only improves data integrity but also enables analytics applications to boost competitive advantage and operational efficiency, as underscored in available literature by the agency of Big Data as being transformative (Nda et al., 2019). The presence of structured data in higher education also points to its influence on institutional efficiency and innovation. Universities tend to generate massive amounts of structured data from student information, research results, and financial transactions. Proper management of data is essential to support informed decision-making and strategic planning. In that sense, utilization of structured data enables better reporting, compliance, and performance measurement. Higher education Big Data management issues such as modeling and storing present the need for solutions scalable enough to cater to both structured and unstructured data. Institutions that are successful at organizing their structured data are better able to respond to emerging education technologies and analytics deployments, thereby improving the learning experience and operational efficiency, a point pointed out in the context of Big Data's deployment to higher education (Nda et al., 2019). In the intersection of structured data and modern organizational forms, new platforms are emerging that leverage complex-event processing for optimal utilization of data. Such integration should work to support traditional data management strategies with more intelligent organizational behavior. The SELFSERV system is a prime example, in that it integrates community-based strategies with advanced data analysis to be able to transcend limitations of ancient operating systems. As companies develop in accordance with new, more vulnerable, and uncertain social environments, structured data emerges as an asset that facilitates the flexibility and rapid action of such new systems. Under these circumstances, organized data management as a strategic initiative is critical since it allows organizations to leverage real-time intelligence, provoke efficiency, and maintain resilience amid changing demands and expectations in the digital age.

3.1.2. Unstructured Data

The explosion of unstructured data in the current digital world creates enormous challenges for the processing and management of data. Unlike structured data, which can be easily ordered into known structures such as tables, unstructured data does not have a pre-defined data model and therefore becomes more difficult to process and analyze. It is complicated by the variety of shapes that the unstructured data can take, from text files, images, social media messages, and multimedia files. As our reliance on the internet grows, so does the volume of this unstructured data, which creates information retrieval and query performance issues. Users are likely overwhelmed with the volume of data, facing challenges of identifying relevant information in the presence of irrelevant noise, semantic inefficiency in comprehension, and difference in data quality, which requires innovative solution and robust structures in processing and managing the type of data in an efficient manner (Yafooz et al., 2015). Strategies for limiting unstructured information have stimulated work on sophisticated methods aimed at optimizing data retrieval and analysis efficiency. The methods employed include text indexing, full-text searching, and applying complex database management models to address the nuances of unstructured material. While relational databases offer a platform for organizing structured data, they can also be applied to manage large amounts of unstructured data. However, users generally experience difficulties in obtaining meaningful insights from such heterogeneous data. To address these challenges, scientists are exploring models like the Textual Virtual Schema Model (TVSM) that attempts to enhance organization and retrieval of unstructured text data in relational databases through enabling automatic semantic linking and clustering (Yafooz et al., 2015). This paradigm reflects a paradigmatic shift toward convergence of theoretical paradigms and emerging technological means of data management. Moreover, the potential use of unstructured data on the Semantic Web is greatly promising for streamlining the extraction of knowledge and semantic processing. The Semantic Web is an extension of the World Wide Web that concerns itself with interlinking and contextualizing data by means of a semantic framework of structure. By using Semantic Web approaches, organizations can carry over unstructured sources of data into forms that are more usable, allowing rich semantic relationships to flourish. This is a beneficial task in areas such as codicological scholarship, where the capability to tap digital collections of hand-written manuscripts has been a topic of great interest. But to effectively integrate unstructured data into the Semantic Web is to require novel extraction methods and robust information management techniques, particularly those adhering to models like the CIDOC Conceptual Reference Model (CRM). This confluence of unstructured data and semantic technology facilitates enhanced knowledge discovery across various fields and, ultimately, drives research and development efforts forward.

3.1.3. Semi-Structured Data

In the context of data processing and management, semi-structured data poses specific challenges and opportunities that differentiate it from the more disciplined structured and unstructured paradigms. Compared to structured data, which is arranged systematically within pre-defined fields (e.g., databases), semi-structured data offers a more flexible format, often characterized by the occurrence of tags or markers that allow for organization while not necessitating strictly defined structure. They include JSON or XML files, which may vary in length and kind of data but still possess a discernible schema. As companies are faced with increasing complexities of holding information, analyzing semi-structured data is a requirement since it can offer more detailed analysis when treated properly. This type of information is very relevant in the context of modern-day applications that require information handling agility and flexibility to enable business organizations to respond rapidly to shifting operational needs. Emerging technologies and frameworks created to manage semi-structured information have gained significance, particularly as data volume increases exponentially. Indeed, as chronicled by existing literature, with growing need for accessing and leveraging different sources of data at optimal efficiency — ranging from business analytics to social media analysis — arises a requirement for systems that can store, manage, and query as easily at least structured records but all data types, including semi-structured and unstructured data. With technology like appliance, which combines hardware and software to answer such sophisticated data requirements, organizations are able to make better use of semi-structured data in a better way. The ability to analyze data more comprehensively, combining various types of data, is the foundation to fuel data-driven business decisions in the future (Bhattacharjee et al., 2006).

Moreover, the arrival of the Internet and advanced data-extraction techniques rendered semi-structured data management increasingly important. Web Data Extraction technology is an example of this, with numerous data sources like social networks and other websites on the web generating vast amounts of semi-structured data. This data takes on immense significance in identifying business trends and social behavior because it is possible to process and integrate it so that it can be converted into useful information. Effective semi-structured data systems thus allow organizations to conduct advanced analyses, tailoring data extraction techniques to meet specific needs in both enterprise and social environments. As more scholars and practitioners investigate this field further, the potential for cross-fertilization of techniques previously disconnected within specific domains still expands, representing a fertile ground for future applications in data processing and management (Abel et al., 2014).

3.2. DATA COLLECTION METHODS

The reliability of data gathering methods is at the core of the broader paradigm of data processing and management. Digital imaging is one of the primary methods in contemporary data gathering, which has gained popularity over visual methods due to its efficiency and accuracy. Visual methods, although commonly employed in the past to measure vegetation and land use, tend to be prone to observer bias and tend to be very time-consuming. Digital imagery, particularly with threshold segmentation methods, enables more methodical tracking of ground vegetation. The methodology allows for information collection and interpretation over very large areas without the inherent subjectivity of naked-eye surveys. This development significantly contributes value to well-informed decision-making in such fields as agriculture, where knowledge of vegetation dynamics is crucial to effective resource management (de Oliveira et al., 2013). Other than digital imaging, the landscape of data collection techniques involves other IT solutions that enable effective capture and processing of data. For instance, the integration of supervisory control systems and wireless sensor networks is paramount for real-time data capture in most applications. Such technologies form the core of ensuring that data collected is timely and relevant, thereby enhancing the quality of information obtained from such data. In addition, integrating techniques such as data cleaning can greatly enhance the validity and consistency of gathered data, hence impacting follow-up analyses. Thorough scrutiny of these IT methods indicates their significance not only in niche projects but also in general scenarios where the integrity of the data holds maximum importance, as evidenced in the survey conducted by EMILI (Brodt et al., 2010). Lastly, the choice of data collection methods must be compatible with the specific needs of a project and consider scale, cost, and desired level of accuracy. Specialized settings and question types may necessitate special strategies; in ecological studies, for example, the typical patchiness of plant communities can be tracked effectively by using calibrated digital image analysis on high-resolution imagery. This adaptability reflects the dynamic nature of data collection methodologies, which must be continually assessed and updated to meet the challenges posed by evolving technological advancements. Emphasizing robust data collection will invariably enhance the management and processing phases, thereby leading to richer analyses and more reliable outcomes across various disciplines. Engagement with such evolving methods is inevitable for researchers who want to take full advantage of the maximum potential of information in their areas of research (Oliveira et al., 2013) (Figure 3.4).

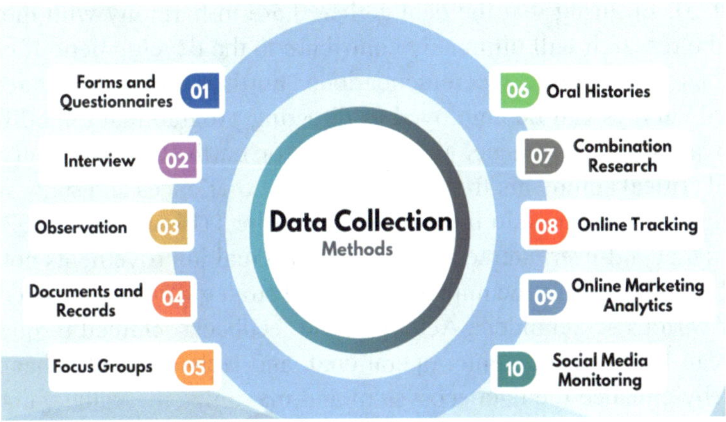

Figure 3.4. Data collection methods.

Source: https://www.globalpatron.com/images/10-best-data-collection-forms-1024x576.png.

3.2.1. Surveys and Questionnaires

Questionnaires and surveys play a significant role in information processing and management, as they are the primary tools of information gathering across different disciplines. They are designed to collect formatted data from the respondents, providing meaningful information on behavior, opinion, and demographics. The format of a questionnaire or survey must be carefully conceptualized to ensure the questions are unambiguous, relevant, and comprehensible. Carefully formulated questions do not only result in higher response rates but also validate the data collected. As such, identification of the population in question and adherence to question formats—open-ended or closed ones—are imperative steps in survey preparation. The success of a survey in totality rests on whether or not it is able to collect the sought information without bias, thus paving the way for successful analysis and interpretation of results. Apart from efficient data collection, questionnaires and surveys also need to face serious examination in the data processing phase. This entails undergoing stringent statistical analysis to extract meaningful patterns and connections from raw data. A successful data processing plan includes cleaning of the data to remove inconsistencies or errors and applying applicable analytical methods to derive insights that inform decision-making. For instance, disaggregated data can be quite helpful in offering more insight into specific user trends and preferences, as shown in household energy survey guidelines where methodologies are noted for their role in driving detailed analysis (Leitmann et al., 2025). Ensuring that the data gathered are in harmony with the overall aims of the research will ultimately contribute to the development of powerful conclusions and feasible recommendations. Furthermore, explicit results of conducted surveys can be employed in directing programmatic modifications and innovations. For instance, a user survey for Earth Resources Survey data identified critical comments from users on their experiences and suggestions on enhancing data utilization in future implementations (Ebbert et al., 2025). This illustrates that end-user interaction can lead to actual improvements not only in survey design but also in the implementation of strategic programs that meet the needs of various stakeholders. Acting on the feedback obtained through these surveys can lead to geographic, operational, and technological enhancements and thereby enhance the data ecosystem and maximize the value gained from information gathered. Therefore, the cyclical process of data collection, analysis, and application of survey data is at the heart of efficient data management and processing systems.

3.2.2. Automated Data Capture

The evolution of data management has now come to largely rely on automatic data capture systems, which are fundamental aspects in the processing and management of sets of information. Automated data capture systems can facilitate the seamless collection and integration of information from diverse sources, enabling organizations to improve their efficiency in operations. For instance, automated data capture technology not only eliminates errors but also slashes the time spent in manual data capture to a great extent. These efficiencies are important to organizations faced with big data inflows because they help secure data integrity and the realization of real-time processing capabilities. As a result, the use of automated data capture solutions realizes enhanced data accuracy and availability, which are vital to high-quality decision-making processes.

As the technology environment keeps changing, the level and sophistication of data generated continually strain conventional data management practices. Organizations like NASA are representative of this trend, where programs like the Earth Observing System (EOS) are anticipated to produce data volumes of over 2 Terabytes per day. This explosion of data demands advanced solutions in data capture, distribution, and processing, demonstrating the central role of automation in current data management systems. The need for scalable solutions is also depicted by the development of systems like the E-Systems Modular Automated Storage System (EMASS) that is designed to cater to such extensive data requirements. With the use of advanced technologies, such systems provide a robust infrastructure that enables effective data management and storage, thus allowing for sound use of data in various applications in industries. Besides, the integration of automated data capture systems not only optimizes internal operations but also optimizes stakeholder engagement by providing access to information in real-time. By enabling real-time data flow, organizations can respond better to changes and trends around their operating environment, thus enabling agility and adaptability. This talent is essential in sectors such as healthcare, finance, and research, where quick access to accurate information can mean improved outcomes and competitive advantage. As stated in a number of studies, the use of automated systems is not a technical innovation; it is a paradigm shift in thinking with regard to how information is regarded and utilized in an organization. Hence, these changes are imperative in pushing data management practices to evolve in order to meet the growing demands of data-centric environments, ultimately transforming the way organizations operate in a data-centric world (Figure 3.5).

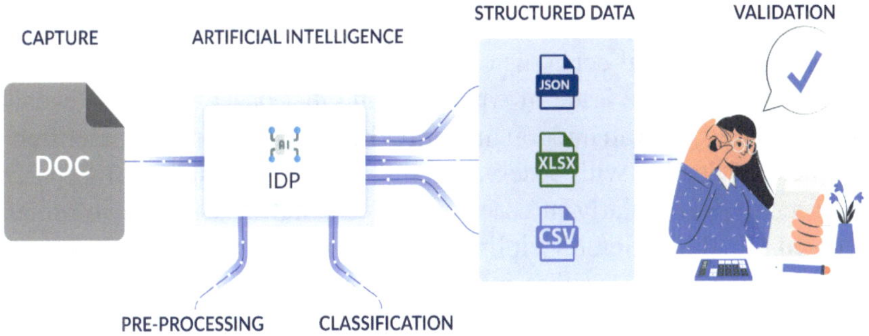

Figure 3.5. Automated data capture.

Source: https://assets-global.website-files.com/5f689f82910c6b4f1ffb855b/61320fb851d1ac2758709af1_Intelligent%20Document%20Processing%20(IDP)%402x-min.png.

3.2.3. Observational Techniques

Observational methods are very important in data management and processing, as they allow researchers to collect a large amount of data from different disciplines. It becomes possible to use these methods only if one can develop sophisticated tools to be used for collecting and sharing the data. Technological advancements allow researchers to merge different sources of information smoothly, enabling them to make more detailed analyses of complex data. The evolution into interactive environments emphasizes the importance of not just collecting information but that it should also be understandable and distributable across various sciences. Integration becomes increasingly essential in multidisciplinary environments where epistemological and methodological variations hinder information dissemination. As observed, the issue is one of cooperation among scientists under differing epistemological practices and principles (Brasher et al., 2006). Besides, observational methods are more than mere data collection; they are fundamental in answering specific societal issues such as environment monitoring and climate prediction. The application of these methods in space observation has helped shed key insights into climatic fluctuations and weather patterns, thus aiding in disaster preparedness and response. Through the use of observation data obtained through satellites and other remote sensing, scientists are able to help curb adverse impacts caused by natural disasters and therefore minimize human and economic losses. Such a capability can be used towards greater aims, such as improvement in forecasting systems as well as developing solutions for managing air pollution levels. Observation techniques, thus, not only aid improved capability for handling data but also provide real-world solutions towards several world issues (N/A). The placement of observation-gathered data within holistic management styles also brings ethical concerns of privacy and data protection. As data gets shared more widely, the issue of when and on what basis to publish sensitive information emerges as an integral matter for both researchers and practitioners. The nuances of dealing with observational data require weighing transparency against secrecy, particularly in cases where personal or sensitive information is involved. Enabling interdisciplinary collaboration without compromising confidentiality necessitates that institutions implement transparent practices and policies regarding data sharing. These challenges must be handled because developing a culture of trust and cooperation among various scientific groups has the potential to enable better exchanges of knowledge and expertise. This, in turn, enhances the final purpose of advancing scientific inquiry and that observational approaches are adequately leveraged for society's benefit (Brasher et al., 2006).

3.3. DATA STORAGE SOLUTIONS

In the context of data management and processing, the choice of data storage products is pivotal for sustaining the organization, availability, and longevity of data. With businesses generating enormous amounts of data each day, traditional storage practices are no longer able to meet increasing demands adequately. Cloud storage is a widely adopted option, offering flexible and cost-effective solutions that accommodate evolving storage demands. With cloud-based infrastructure, organizations can streamline data access and leverage more collaborative attributes. The shift in the storage of data is a part of the entire movement toward digitalization of industries, emphasizing the need for flexible solutions responsive to the evolving uses and holding of data.

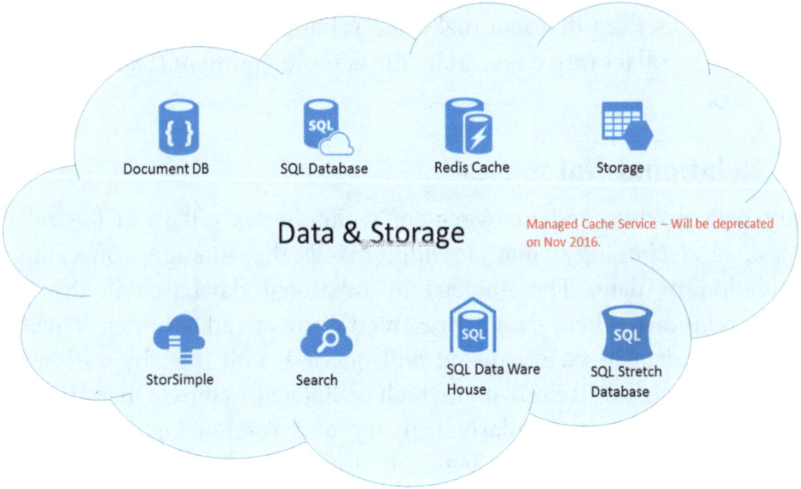

Figure 3.6. Data storage solutions.

Source: https://www.dotnetcurry.com/images/azure/azure-platform-overview/data-storage.png.

The National Space Science Data Center (NSSDC) is one instance of the significance of new-age storage solutions addressing massive amounts of data. Established in 1966, the NSSDC archive, which holds over 120,000 data tapes and 4,000 data sets, illustrates the issues related to effectively managing valuable datasets. With a computer archive that is estimated to exceed 28,000 gigabytes, the NSSDC is examining various methods, including mass storage technologies and data compression techniques, to minimize the management of data. These initiatives not only seek to make information more accessible to gain but also complement the requirement to protect this information against potential doomsday loss. These developments mirror a growing sensitivity to the importance of not merely storing information but keeping it in order within a system anticipating future growth and usage patterns (Green et al., 2025). Furthermore, the adoption of cloud computing solutions in education has revolutionized research data management, notes recent research. With increasingly complex and large research data sets, optimized storage and retrieval are necessary. Cloud systems provide mechanisms for automatic capture of data and easy metadata tagging, enabling researchers to maintain data organized and give convenient access for collaborative use. Although apparent advantages of cloud storage are present, data standardization and security issues still prevail. Issues such as these must be resolved to form robust data management systems that will help meet the evolving demands of academic research. Future studies are warranted to evaluate these measures effectively and to determine standardized practices that can make the reliability and effectiveness of data storage within collaborative research efforts more optimum (Karin et al., 2023) (Figure 3.6).

3.3.1. Relational Databases

Among data process and management technologies pillars is the relational database, a technology that revolutionized the storage, querying, and manipulation of data. The nucleus of relational databases is the idea of tables, or relations, where data is presented in rows and columns. This tabular setup enables data to be organized and queried with ease by various query languages, the most well-known of which is SQL (Structured Query Language). Application of keys, particularly primary and foreign keys, facilitates the establishment of relationships between different tables, thereby enhancing the data consistency and integrity. As software becomes more complex, relational databases provide a good framework for programmers to ensure that information can be retrieved and manipulated efficiently, thereby handling demands of ever-evolving business and technology environments (Sakama et al., 1993). Efficiency in data processing depends not only on the database schema but also on the strategies employed to access data as well as conduct integrity testing. Strategies such as bottom-up and top-down query answer techniques enable relational databases to optimize the execution of queries, hence improving performance and reducing latency. These strategies allow the database management system to automatically choose the best course of information retrieval, utilizing indexing and caching strategies. Furthermore, the integrity constraints are enforced to ensure that data follows set rules so that no anomalies occur in processing transactions. Following these principles, relational databases not only provide healthy data storage and retrieval but also contribute to strengthening the systemic consistency of the overall data management system (Sakama et al., 1993). As data now increasingly appear in forms other than the conventional, the application of relational databases is further extended. The advent of these advanced applications requires low-cost storage infrastructure and effective query provision optimized for certain data types such as images and multimedia. Relational databases are hence presently required to provide flexible query languages that are able to accommodate the needs of heterogeneous data representations and remain effective. The demand for effective retrieval and manipulation techniques determines the imperative of revamping and enhancing relational database systems to advance with emerging challenges in data storage and accessibility. With the application of scalable architecture and enhanced querying, relational databases remain a fundamental driver in maintaining complex data operations efficiently and enabling organizations to optimize their data assets.

3.3.2. NoSQL Databases

The advent of NoSQL databases has revolutionized the processing and management of data in numerous domains, particularly in those datasets that are dominated by huge volumes of unstructured and semi-structured data. Although relational databases are robust when handling structured data, when they have to handle increasing varieties and velocities of incoming data streams, they face tremendous challenges. This absence of relational models is especially felt in scenarios such as the Internet of Things (IoT), where enormous sensor networks generate real-time information in diverse fields from urban surveillance to environmental monitoring. Attempting to address these requirements, NoSQL databases offer a schema design that is adaptable, supporting dynamic data types and structures that can be altered as the users' requirements evolve. This adaptability is squarely to blame for efficiently collecting and processing the high amount of data produced by massive collections of networked devices in real-time environments (Küçükkeçeci et al., 2017). An excellent use of NoSQL databases is in the field of genomics where Next Generation Sequencing (NGS) has a massive amount of data handling problems. Relational databases cannot support the vast datasets generated by genomic analysis due to their structured schemas, making structuring and retrieval of information cumbersome. The example of the use of NoSQL systems such as GIGwA depicts how non-relational databases are tailored for biological researchers' purposes. Not only do they speed up storage and querying of complex genomic data, but they also enhance accessibility through simple-to-use interfaces, thereby democratizing data utilization by biologists. Consequently, with the volume of data generated in this field increasing, the flexibility and speed of NoSQL databases position them well to play a role in enhancing genomic studies and data processing operations (Dereeper et al., 2015). Moreover, the natively adaptable nature of NoSQL databases extends beyond the mere handling of large amounts of data, giving birth to new forms of data modeling and retrieval that traditional systems cannot match. For instance, graph-based NoSQL databases such as Neo4j and OrientDB demonstrate how interrelated data can be more effectively handled and even analyzed through intricate relationships between data points that users can design. This is a key feature in applications involving large sensor networks of multimedia data, where understanding the correlation between various data streams so that useful information could be gleaned is vital. Organizations can optimize their data management architecture to keep it responsive and flexible to various analytics demands by using the NoSQL technologies. With the landscape of handling data continually shifting in favor of larger and more complex datasets, the role played by NoSQL databases is increasingly crucial in modern data processing frameworks (Küçükkeçeci et al., 2017) (Figure 3.7).

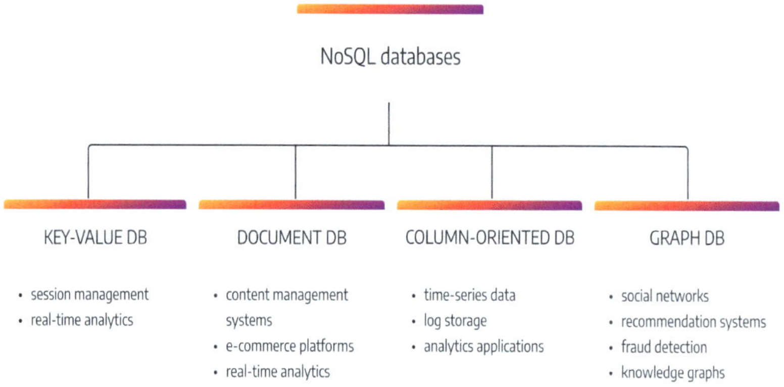

Figure 3.7. NoSQL databases.

Source: https://assets.website-files.com/6141c89a3874c3702674a1c0/6488d7dcbce3e56559dd2fc2_nosql-databases.png.

3.3.3. Cloud Storage Options

In the digital age, cloud storage services have evolved into key components of data management and processing strategies. The boom in data generated by individuals and organizations necessitates efficient storage capabilities that are at once expandable and secure. Cloud storage offers an elastic platform that adjusts to evolving data needs, enabling users to store massive amounts of data without the constraint of traditional hardware. This is particularly vital in work settings, where the pace of technological advancements and evolving market demands can completely transform storage requirements. Cloud-based services enable organizations to streamline processes, reduce infrastructure expenses, and render data more readily available, thereby powering innovation and efficiency. The character of cloud storage varies considerably based on service providers and what they provide. An in-depth understanding of these features is necessary for organizations seeking the best solutions for their needs. Various cloud storage models, i.e., public, private, and hybrid, possess inherent advantages depending on usage habits and organizational constraints. Public clouds tend to offer low-cost solutions to small and medium-sized enterprises, while private clouds may be a suitable option for big organizations that are more security-sensitive and require more control over data. Besides, hybrid cloud storage allows organizations to maximize flexibility and security, allowing for seamless consolidation of public and private cloud resources. This categorization of cloud storage options is to blame for the need for taxonomy to allow users to properly browse through their options (Höfer et al., 2011). Cloud storage management of data is also enhanced with the incorporation of sophisticated methods of data processing, including data mining and analytics. As organizations go on implementing cloud computing environments, the ability to gain useful information from data stored is paramount. The combination of data mining technologies with cloud-based storage not only guarantees managed structured data but also enables users to derive useful information from big data with efficiency. This reduces infrastructure and storage costs associated with traditional data processing environments, consequently boosting overall productivity. As noted in current research, cloud services harmonization and data processing capabilities strengthen an organization's ability to generate insights, automate processes, and support ongoing, secure service delivery for users.

3.4. DATA PROCESSING TECHNIQUES

With information management increasingly necessary in various industries, effective data processing techniques are core in enabling precise analysis and application of information. One of the fundamental data processing techniques is having a structured arrangement and scrutiny of raw information to break it down into formative information. Techniques such as data cleaning, transformation, and aggregation allow companies to standardize datasets, remove inaccuracies, and find useful information. Besides, these procedures are important in guaranteeing the quality of the produced data, which is crucial for quality decision-making. Reliable and consistent processing is required in environments such as the Apollo program, where processing telemetry data was critical in post-flight analysis, with the emphasis on the need for reliable processing techniques in handling large volumes of complex information (Foster et al., 2025). In addition to traditional data processing methods, emerging technologies have introduced radical practices such as machine learning and artificial intelligence into the mix. The new approaches enable automatic analysis of data, thus making the work much more efficient and accurate. Through application of algorithms capable of recognizing patterns in large datasets, companies are able to extend their capability to derive valuable insights that cannot be easily discovered through traditional methods.

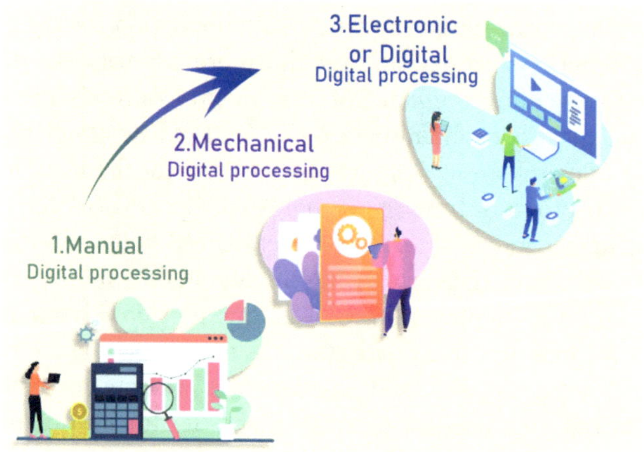

Figure 3.8. Data processing techniques.

Source: https://d2ds8yldqp7gxv.cloudfront.net/Blog+Explanatory+Images/data+processing+2.webp.

For instance, in space vehicle radar control, end-to-end data systems use intelligent algorithms to transform unprocessed radar measurements into geophysical values, hence allowing for proper interpretation and use of data (N/A). With the implementation of such sophisticated methods, not only is the process workflow minimized, but also new analysis opportunities are created, hence enhancing overall data management efforts. Besides, the employment of data compaction techniques has become a key field of attention in regard to data transmission and storage. Through minimizing the quantity of data that needs to be stored, organizations can considerably minimize storage costs as well as increase processing speed. Techniques such as lossless compression ensure the integrity of the original data is preserved while ensuring space requirements are minimal. This is particularly useful for mass industries dealing with colossal datasets—such as aerospace and healthcare—were speed matters most. The call to further explore compaction techniques, especially for aircraft radars, illustrates the constant need for creative data processing methods so that data remains tractable and accessible for effective analysis in various applications (N/A) (Figure 3.8).

3.4.1. Data Cleaning

Data cleaning is a fundamental step in the broader context of data management and processing that ensures accuracy and consistency in datasets. It involves the identification and resolution of errors, inconsistencies, and inaccuracies in data inputs that can compromise analytical outcomes. Widespread prevalence of poor-quality data can lead to suboptimal business decisions, flawed research conclusions, and operational inefficacy. As companies increasingly rely on data insights to drive strategy, the importance of systematic cleaning cannot be overstated. Essentially, cleaning data is not merely a hygiene function to cleanse datasets but also enhances the quality of information as a whole, thereby facilitating better decision-making processes in industries.

The data cleaning process involves a series of processes, such as deduplication, normalization, and validation, all aimed at ensuring data consistency and use. Deduplication means removing duplicate records whose incidences might distort analyses, while normalization ensures data structures are normalized for consistency. Validation checks ensure data meets pre-specified standards and permissible ranges, thereby ruling out the possibility of incorporating erroneous entries into analyses. Besides, use of automated tools has significantly eased these steps, with quicker turnaround and improved data management. Increasing complexity of data at large, especially big data, demands robust cleaning measures to address not only precision but also data's underlying integrity (Amezquita et al., 2009).

Furthermore, the broader implications of successful data cleansing extend beyond internal processes; they are resonated in the wider arenas of regulatory compliance and public trust. In healthcare and finance, and other sectors, attention to accuracy in data management is just as vital for operational success as it is to avoid violating legal boundaries and ethics. For instance, crime intelligence derived from data needs rigorous cleaning measures to make the information being passed effective in facilitating actionable strategy (Chainey et al., 2006). With the use of effective data cleaning measures, organizations are able to significantly improve their data governance approaches, make interdepartmental coordination more effective, and ensure that their decision-making processes are facilitated by solid and accurate information. Thus, data cleaning is not only a technical requirement but also a strategic support for effective data processing and management.

3.4.2. Data Transformation

Data transformation is a key component of data management and processing, serving as the bridge between raw data and actionable information. Data transformation refers to the conversion of data from one form or structure to another and preparing it for analysis and interpretation. Data transformation is necessary to ensure the integrity and usability of data, particularly for diverse data sources that are susceptible to inconsistency, error, or redundancy. By the scientific execution of methods of transformation, organizations are capable of achieving improved data quality, which is imperative in decision-making. Effective data transformation is necessary when the data being processed is highly complex information. In addition to cleansing the data, effective transformation also renders the data more pertinent and logical, thereby ensuring more precise outcomes in various applications of data analytics (Athanasiadis et al., 2008).

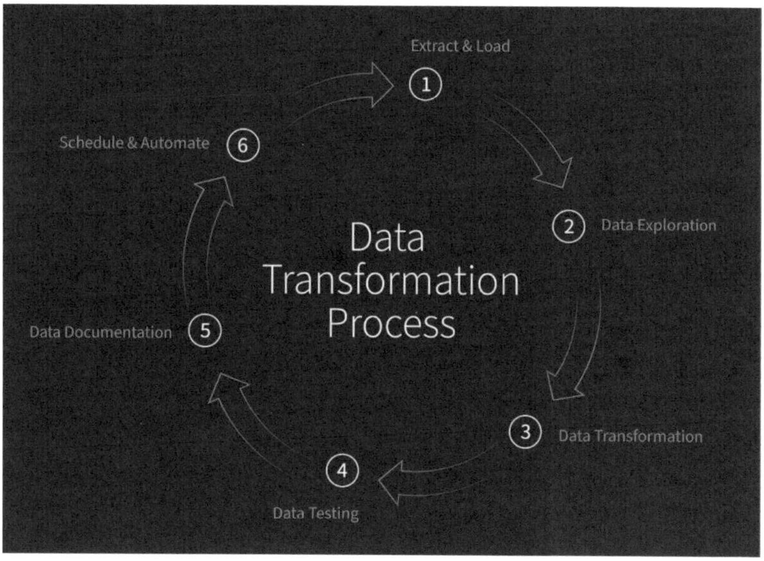

Figure 3.9. Data transformation.

Source: https://www.getdbt.com/_next/image?url=https:%2F%2Fcdn.sanity.io%2Fimages%2Fwl0ndo6t%2Fmain%2F156286804651 4fa4e343475bbc3af7820d07f592-2560x1440.png%3Ffit%3Dmax%26auto%3Dformat&w=3840&q=75.

Furthermore, data quality acquired through data transformation is also interdependent with pre-processing and post-processing data analysis processes. Pre-processing constitutes vital processes such as data cleaning, normalization, and integration, which all in all prepare data for transformation. Without these initial steps, data in the transformation process could compromise the accuracy of insights gained. After data has been converted, post-processing guarantees that the outcomes are delivered briefly and essentially. This is especially crucial in complicated domains, like environmental systems, where all stakeholders require precision in the conclusions. When data conversion is done with careful consideration, not only does it facilitate knowledge discovery but also enables organizations to make sound judgments based on clearly stated conclusions (Athanasiadis et al., 2008). In the case of accounting information systems, data transformation also holds great significance, particularly in the wake of emerging technologies and techniques. With industries moving towards the provision of big data and analytics, related transformation processes have to be evolved as well. Existing research focuses on the importance of accounting information systems embracing new topics such as enterprise resource planning (ERP) and IT audits, in line with the change in data handling dynamics. Converting accounting information not only enables operational efficiency but also enables better detection abilities for anomalies and patterns that could indicate fraud. Because of this, it is necessary that accounting educational programs place emphasis on these changes in curriculum while educating the future professionals for the data-centered world of the profession (Figure 3.9).

3.4.3. Data Analysis

In contemporary data processing and management, data analysis is an integral component that enables companies to derive useful information from large volumes of data. This complex dance of data collection, cleaning, processing, and interpretation is essential in decision-making across sectors. Organizations more and more apply advanced analytical techniques, such as machine learning and statistical modeling, to derive trends and patterns that might be imperceptible in raw data sets. Effective data analysis not only enhances operational efficiency but also facilitates organizations to predict future outcomes, as testified by the rapid growth of condition-based maintenance (CBM) supporting technologies. With effective data analysis, organizations can implement productive CBM programs, overcoming the obstacles presented by innovative prognostics and health management technologies, thereby fulfilling their strategic objectives (Márquez et al., 2016). The validity of information has a direct impact on the usefulness of analysis output, and hence data preprocessing is an essential process in the data processing workflow. With greater reliance on data-based analysis, the importance of providing high-quality input information cannot be overstressed. Data in real-world scenarios is often contaminated with noise, inaccuracies, and redundant information that are detrimental to the analysis process and may lead to defective conclusions. To overcome these challenges, approaches such as Knowledge Discovery from Databases (KDD) provide systematic methodologies for pre-processing and refining data before analysis. This strict preparation includes cleansing, normalizing, and arranging data in a manner that enhances the quality of generated knowledge. Additionally, given that environmental systems are defined with particular complexities, stakeholders tend to require accuracy and clarity in analytical results, thus emphasizing the importance of good data processing techniques (Athanasiadis et al., 2008). Moreover, the use of advanced data analysis techniques has widened the scope of information that organizations can derive from their data. Using sophisticated analytical software, businesses are not only able to identify past trends but also forecast future changes and traps. Predictive analytics, of value in manufacturing and green space management industries, enables managers to act ahead of time rather than react. For instance, application of universal standards like ISO 13374 facilitates efficient management of the CBM program by structuring failure symptom analysis and enabling the organization to adjust its maintenance strategy accordingly. Analytical steps based on a process also ensure that recommendations are actionable and tailored to organizational needs, hence facilitating overall performance and robustness in today's data-oriented era (Márquez et al., 2016).

3.5. DATA SECURITY AND PRIVACY

In contemporary digital environments, data privacy and security are paradigm pillars when it comes to data processing and management. With companies relying increasingly on digital platforms for their activities, there has been an urgent need to protect sensitive data. Identity management is an issue of particular concern considering its connection to privacy and protection of personal information. The Centre for Democracy and Technology set the main privacy principles required to establish a balance between technology development and individuals' expectations of their own data. The principles are not guidelines but a clarion call to organizations to embrace strategies of growing compliance with legislations. With increased digital services, the challenge of balancing identity-aware technology and privacy needs must come first in a bid to offer secure and responsive data handling in everyday practices (Savirimuthu et al., 2008). Also, the need for comprehensive privacy frameworks is augmented by the implementation of technologies like Radio Frequency Identification (RFID) technology, which has a different privacy challenge. The initiative of European Commissions on privacy impact assessment (PIA) guidelines for RFID implementations is a case in point to consider the future privacy vulnerabilities and take measures to thwart them beforehand. This proactive approach gives the stakeholders the tools to deal with the complexities of modern-day data affairs. The PIA approach not only educates RFID operators but also sets a template for broader application during the Internet of Things period. As technology advances, the method outlined in the PIA will continue to be important in upholding privacy standards designed to protect user data as well as encourage innovation in data management and processing (Spiekermann-Hoff et al., 2020). In short, the complexity of data processing and handling calls for robust data privacy and security frameworks that match developments in emerging technologies. Organizations should be vigilant and proactive in their adherence to established privacy principles and guidelines like the PIA, which addresses the privacy issues presented by emerging technologies. The success of these frameworks is not only predicated on compliance but also on establishing a culture of openness and accountability when it comes to personal data. Effective governance around security features for data can mitigate risk associated with data breaches while embedding public confidence. Lastly, commitment to best practice in information handling will lay the groundwork for sustainable growth and a dignified digital culture that respects the privacy of citizens and their information, placing all individuals in a safer digital space (Savirimuthu et al., 2008; Spiekermann-Hoff et al., 2020) (Figure 3.10).

Figure 3.10. Data security and privacy.

Source: https://cynicaltechnology.com/wp-content/uploads/2022/04/securityvsprivacy-1024x768.jpg.

3.5.1. Importance of Data Security

In the modern age of the digital world, data security cannot be overstressed, particularly in the fields of data management and processing. Organizations these days are quite reliant on gigantic amounts of sensitive information, thus making them lucrative targets for cybercriminals. The effects of data breaches can be catastrophic, ranging from financial losses to loss of reputation and legal penalties. Moreover, the evolving regulatory landscape, since it places greater emphasis on the protection of people's data, compels organizations to institute robust security measures. As corporations integrate data more into their operational systems, prioritizing security routines becomes intrinsic not just for regulatory compliance purposes but also to secure stakeholder trust. Hence, data security must be treated as a part of data management, which is crucial to safeguard organizational integrity as well as customer privacy. In addition, data usage patterns and processing determine the connection between business efficiency and effective data security. Organizations tend to implement numerous data management practices with the aim of maximizing the usefulness of their information assets. However, these practices provide entry points for loopholes, if data security practices are not implemented. Protection against violation requires a proactive solution in the detection of potential violation threats, integrating security into the system development process. For instance, the integration of business process patterns in system design can facilitate operational reusability as well as enhanced security features, further prompting an approach to data management in a holistic manner. In this regard, adopting an approach to a methodological framework that emphasizes early-stage data security in system development is crucial. Aldin et al. (2009) emphasizes the importance of harnessing these patterns to deliver a solid data handling environment capable of repelling future security threats. Secondly, the concept of control of personal information has emerged as a key principle of data security that is governing both law and organizational behavior. Responding to the demands of stakeholders for greater control in managing their data, the discussion around data control is highlighting openness in the manner in which organizations handle data. This particularly applies to various regulatory requirements, which seek to empower the data subject and enhance privacy. However, the impact of control extends beyond conformity; it extends to the ethical aspect and personal agency in the context of their data. By critiquing the idea of control offered in academic sources, organizations can view how to navigate data security challenges to preserve user autonomy and foster a culture of responsibility. Lazaro et al. (2015) calls out this interdisciplinary approach, which can eventually lead to more innovative and safe data handling.

3.5.2. Regulatory Compliance

Regulatory compliance relating to handling and processing of data has also gotten more complex with the intersection of many laws and standards. The organizations need to handle a scenario that involves local, national, and international regulations governing the collection, storage, and processing of data. The regulations not only protect consumer rights and privacy but also define the legal jurisdictions in which companies shall operate. Compliance requires a clear understanding of the effects of non-compliance, which will lead to legal penalties, loss of reputation, and substantial monetary loss. The ever-changing nature of these guidelines along with increasing technological advancements requires organizations to keep changing and refining their data governance policies in alignment with regulatory demands so as to safeguard themselves from potential compliance risk and establish stakeholder confidence. As the experience of food safety and agricultural health standards proves, compliance with regulations often amounts to a double-edged sword for developing countries engaged in data-driven businesses. Regulations may initially appear to be impediments, but they can become drivers of enhanced capacities of the locals.

Figure 3.11. Regulatory compliance of data management.

Source: https://msl-ltd.co.uk/wp-content/uploads/2019/05/MSL-Compliance.jpg.

The countries can struggle to meet international standards due to poor technological and administrative capacity; but the ones that achieve overcoming such requirements use compliance as their competitive advantage. The countries that succeed in complying typically enter high-value markets, which is an indicator that they must be aware of and adhere to regulatory frameworks that can propel them to higher levels in the global economy. This offers significance to the need to support and provide adequate resources to developing nations in order to empower them to use regulatory compliance as a vehicle of growth, particularly within data-driven industries like agriculture and food production (Henson et al., 2025). Additionally, the creation of certification schemes, particularly within organic farming, offers significance to the role of regulatory compliance in upholding consumer confidence and market differentiation. Presence of various assurance and certification schemes is reflective of a regulatory climate attempting at food quality and safety assurance. As outlined by the report, legislations such as Council Regulation (EC) 882/2004 lay the groundwork for compliance work, highlighting that upholding stringent standards is imperative for producers as well as consumers. In addition, the role played by third-party certificates shows how compliance can enhance voluntary standards' legitimacy, indicating to consumers that the products purchased are in line with their needs and beliefs. As the need for transparency grows, not only must organizations comply but also announce their compliance in a manner that involves stakeholders, further supporting regulation within the broad context of data treatment and processing (Belliere et al., 2010) (Figure 3.11).

3.5.3. Best Practices for Data Protection

Comprehensive data protection is in itself based on the twin pillars of control and transparency. With data breaches now on the increase, organizations need to make it their agenda to offer individuals meaningful control over their data. This means not only informing data subjects about what data is being collected but also clearly stating how it will be used and safeguarded. As highlighted in recent work, there is an urgent need to refine the concept of control so that it guarantees users full control of their data. Such transparency establishes a trusting and responsible culture since users are more likely to engage with systems that value their rights and acknowledge their agency within data practice (Lazaro et al., 2015). Thus, firms can minimize risks of data misuse by ensuring that their procedures satisfy legal requirements and ethical criteria which reinforce data safeguards. Apart from that, the application of strong technical protection is important in order to secure data. Firms need to install widespread protection steps that consist of encryption, access control, and regular observation of data processing practices. The measures have the benefit of shielding confidential data from unauthorized users and potential leaks. Encryption, for instance, ensures that even if data is intercepted, it remains unreadable to attackers. Additionally, establishing strict access controls—where only authorized personnel can access specific data sets—prevents internal breaches and maintains data integrity (Lazaro et al., 2015). Regular audits allow organizations to assess their compliance with established data protection policies, identifying gaps that need addressing. By implementing these technical safeguards within their business practices, organizations not only ensure compliance with legislative demands but also embed best practices reflecting the growing expectation of robust data protection. Finally, embedding a culture of ongoing learning and data protection sensitivity must be addressed. Organizations must constantly educate and train staff about best practices and why user data must be protected. These training sessions should include familiarization with legal requirements such as GDPR and CCPA so that all members of the organization are clear regarding their roles in managing data. In addition, building a culture where employees feel at liberty to report potential security incidents without fear of retribution helps enforce organizational security. With new developments in the law concerning data privacy, it is crucial that employees stay informed about existing regulations and security practices. By focusing on continuous learning and accountability, organizations can highly reduce the threat of data breaches as well as enhance their capability for resisting malicious cyber-based activity, thereby ensuring their commitment to data protection.

3.6. EMERGING TRENDS IN DATA MANAGEMENT

With the threats of a more digital world confronting organizations, emerging data management trends are transforming data collection, processing, and utilization. Among them is the expansion of data democratization, which provides users in various levels of an organization with access to and utilization of data without requiring a plethora of technical know-how. This shift is instrumental in building a data culture, as it facilitates teamwork on data analysis and allows for more informed decision-making. Moreover, the integration of other advanced analytics solutions, including artificial intelligence (AI) and machine learning, allows organizations to derive meaningful information from vast data collections. With these technologies, businesses can make predictions about market trends and conduct, and they can gain a competitive advantage over others. Thus, even as information is made more accessible, the role of data management itself has to shift so that quality and security levels are preserved in a landscape that is ever more democratized. Cloud-based data management platforms are also at the cutting edge of new trends, enabling firms to store and retrieve data in flexible and cost-effective manners. These systems enable scalability and allow firms to scale their information storage and processing capacity in sync with shifting business requirements. As has been deliberated in recent times, that level of flexibility is even more critical in an ever-changing market situation, where the capacity to quickly change in order to suit evolving needs can significantly affect organizational success. Interestingly, cloud migration of data tends to make between-team collaboration more effective because offsite access offers instant data exchange and communication. However, the organizations need to deal with matters such as data privacy, regulation, and cyber security threats when implementing cloud-based solutions. Ultimately, benefits of cloud computing call for a strategic approach of data governance that serves to strike a balance between security and the promotion of innovation. Further, the evolving landscape of data protection and privacy legislation is another emerging trend, which has a significant influence on data management practice. As there has been greater awareness of data security weaknesses and protections of consumers, legislation like the European General Data Protection

Regulation (GDPR) and special initiatives elsewhere in the world highlight the necessity for organizations to give high priority to compliance. Good data management has to come with robust data governance frameworks that address organizational needs as well as those of law. Organizations are increasingly interested in tools and training that enhance transparency and accountability in the manner in which they deal with customer information. Coupled with these regulatory needs is increasing concern for corporate social responsibility, particularly the use of the data in an ethical manner. In the context of this, firms are realizing that it is not only a matter of law but also a strategic benefit to handle data in a responsible manner and develop trust and loyalty towards consumers while minimizing risks of non-compliance (Adegbie et al., 2012; Pierson-Balik et al., 2005) (Figure 3.12).

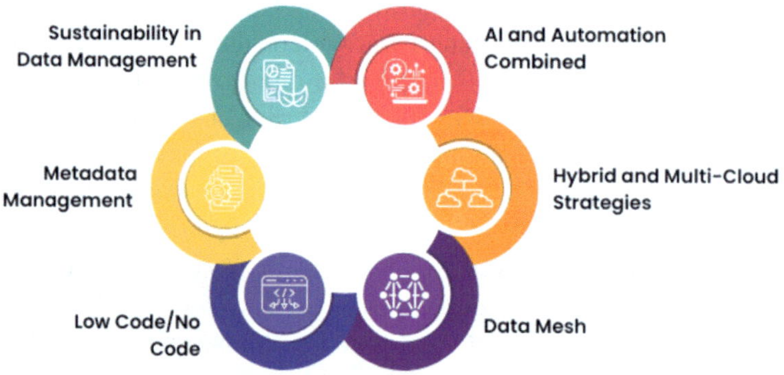

Figure 3.12. Emerging trends in data management.

Source: https://www.solzit.com/wp-content/uploads/2025/01/Key-Trends-in-Data-Management-Services-in-2025-and-Beyond-1.png.

3.6.1. Big Data Analytics

The emergence of Big Data Analytics (BDA) has completely transformed the management of data within organizations, especially as they struggle with an increasingly complex decision-making landscape. Through the application of massive datasets, companies are able to pull out valuable insights that inform strategic decisions. Based on studies, organizations utilizing BDA are more effective than others since they are capable of pulling out actionable intelligence from their data assets. This synergy not only drives efficiencies in operations but also drives a culture of being data-driven where decisions are backed by numerical data and not by intuition. Thus, incorporating BDA into business frameworks is crucial to sustain competitive edges in today's data-intensive environments. Thus, organizations should create robust analytical strategies that enable them to optimize the application of BDA to ensure that their processes align with their strategic objectives and goals (Cao et al., 2017).

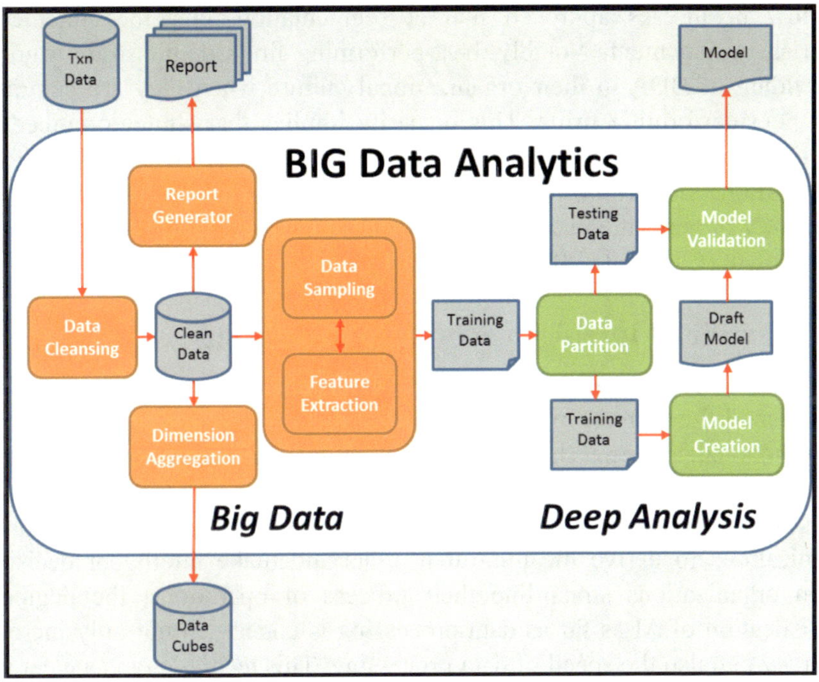

Figure 3.13. Big data analytics (BDA).

Source: https://i.pinimg.com/originals/43/a3/f3/43a3f30ac8157fa7212929e481445123.png.

As organizations work to harness the potential of BDA, the management practices change, particularly in decision-making. In southern Africa, for instance, qualitative research has shown that BDA integration with management accounting has revolutionized traditional practices. By facilitating a move from manually processed information to mechanized processes, BDA attains precision and punctuality in decision-making. The shift allows the management accountants to focus on higher-level analysis and strategic planning, ultimately enriching organizational strategy through improved customer service and operational effectiveness. In addition, the necessity for professionals to gain data analytics skills grows increasingly urgent, indicating a broader trend where analytical capabilities are at the heart of effective management practice (Buys et al., 2025). BDA carries implications that extend well beyond optimization of operations to have significant impacts on company strategy and long-term viability. Companies that can incorporate data analytics not only expand decision-making capability but also optimize resource utilization and financial performance. Effective adoption of BDA tools can have a profound impact on how businesses approach market segmentation, customer interaction, and risk management. Notably, best-performing firms demonstrate a greater embedding of BDA in their organizational culture when they are contrasted with poor-performing firms. This disparity implies that strategic embedding of BDA capabilities plays a crucial role in superior performance. Finally, the organizations must embrace BDA as part of their data management strategy to effectively traverse the nuances of the modern business world and drive long-term growth (Cao et al., 2017) (Figure 3.13).

3.6.2. Artificial Intelligence in Data Processing

The integration of Artificial Intelligence (AI) into data processing has really transformed the manner in which organizations process and analyze data. AI algorithms accelerate data handling by automating repetitive tasks, thus allowing human employees to focus on more strategic initiatives. The technologies apply robust machine learning techniques to find patterns in large datasets to allow organizations to derive meaningful insights and make intelligent decisions. When organizations streamline their process of operations, the degree of sophistication of AI as far as data processing is concerned not only increases accuracy but also the speed of data processing. This technology is a clear shift towards a data culture, and companies are utilizing insights from AI to influence performance and remain competitive in the ever-changing marketplace. But AI implementation in data processing is also with profound ethical dimensions, and that comes with fairness, accountability, and transparency (FAT) principles. With more organizations applying AI models to process individual data, algorithmic bias and data protection are the most serious challenges. That is why organizations must harmonize their AI practices with the General Data Protection Regulation (GDPR) to shun the dangers of data abuse. Notably, a survey of data protection professionals revealed that the majority of organizations were less compliant with GDPR regulations than they could be, and this calls into question their ability to act responsibly where AI technologies are concerned. By developing greater understanding of FAT principles, however, organizations can improve their data governance and ensure that AI adds capabilities rather than subtracts them from human ability, making their operating environment more trustworthy. Furthermore, the integration of AI, big data, and knowledge management presents new avenues for developing innovation and sustainability within organizations. As explained in the Special

Issue of Sustainability dedicated to this subject matter, advanced information and communication technologies (ICT) can have a significant contribution to facilitate more effective data management activities through optimized team collaboration and decision-making. Visuals acquired through AI-powered data analysis not only help organizations attain a stronger grasp of market dynamics but also help them become more rapid and responsive to emerging challenges. Moreover, as organizations grapple with the nature of complex modern-day business environments, the addition of AI simplifies sustainability practices by reducing waste and streamlining resources. This general trend emphasizes the need to embrace AI as one of the pillars of contemporary data processing strategies, thus enhancing organizational competitiveness and individual well-being (Addis et al., 2020; Lytras et al., 2018) (Figure 3.14).

Figure 3.14. Artificial intelligence in data processing.

Source: https://media.licdn.com/dms/image/v2/D4E12AQHFMBb-Q6NJAQ/article-cover_image-shrink_720_1280/B4EZXqtEZZHMAI-/0/1743399438932?e=2147483647&v=beta&t=mCuXRjJ81Pf912YJwIjl_Suoa59IbLjDYr-Y_q30c1M.

3.6.3. Data Governance Frameworks

The application of effective data governance models is essential for organizations that want to optimize their data management and processing operations. These models establish an orderly regime of policies, procedures, and guidelines that control the collection, storage, use, and dissemination of data. Through the provision of detailed guidelines, organizations ensure that their data is always accurate, secure, and compliant with regulatory requirements. A successful data governance model, apart from enhancing data quality, also fosters trust among stakeholders by demonstrating that it is invested in the right data management practices. The principles embedded within these models are such that they align with overall organizational goals, allowing for a culture of data-driven decision-making that ultimately leads to improved results across functions. With organizations grappling with the complexities of data-driven environments, the alignment of data governance models with organizational objectives grows increasingly vital. Imposition of cross-functional data stewardship responsibilities forms one of the core principles of successful data governance models. This function enables stakeholders across the organization to participate in data governance initiatives, creating an environment where data handling is everybody's business and not the function of a single department. Through cross-functional teams—IT to compliance, and business units to legal—organizations can leverage varied talent and perspectives to develop end-to-end governance strategies. Collaboration through this allows organizations to better steward data so that they can identify possible risks, analyze data quality, and implement necessary changes in real time. Besides, communication channels facilitate transparency and accountability among stakeholders whose work is coordinated in alignment with the overall data governance objectives that enhance operational effectiveness and regulatory compliance. Monitoring and evaluation are equally essential components of data governance frameworks for sustaining effectiveness and adaptability. Through performing systematic analysis of data governance practices, organizations will be able to maximize improvements and address evolving regulatory landscapes and technology. The provision of continuous assessment mechanisms allows organizations to segment the impact of their data governance processes to determine whether or not they remain permanently relevant and effective in achieving their data management goals. For instance, organizations can utilize trends based on various analyses to refine their strategy, as N/A (2014) and Timmers et al. (2012) suggest evaluation processes that emphasize collaborative frameworks for specific industries. Lastly, ongoing monitoring and evaluation foster a culture of innovation and responsiveness that is vital in ensuring the actualization of the entire value of data management and processing in today's rapidly changing digital world.

SUMMARY

In review of the conversation on data handling and management, there is a need to emphasize the significance of solid data governance structures that improve organizational performance. With rigorous methodologies, organizations are equipped to manage the complexities that define data handling, where qualitative and quantitative evidence informs decision-making processes. Past studies have explained how organizational problems mostly stem from a lack of objective evidentiary foundation, therefore limiting the potential for real performance improvements. As Ferreira et al. (2018) has explained, the absence of rigorous data authentication weakens the efficiency of quality management systems, casting the requirement for systems that not only collect data but also examine and verify it for beneficial feedback. This groundwork is essential for the creation of a culture in which data-driven decision-making is central and overall operational effectiveness is enhanced. Underpinning the necessity for robust data administration, the addition of top-of-the-range platforms such as AEGIS demonstrates how systematic approach can advance paltry datasets to significant conclusions. As demonstrated in Auzoux et al. (2020), this platform allows effective sharing, interoperability, and continuous improvement in data quality throughout its lifecycle. With adherence to FAIR principles, organizations can ensure that their data assets are not only discoverable but also usable and valuable in various applications. This potential is of paramount importance to stimulate collaborative research and development in agronomic areas, thereby paving the way towards innovative sustainable activities. The discipline that has been brought to data processing operations in terms of methodology is merely an endorsement of a path forward in which systematic data management can enable productivity and environmental sustainability breakthroughs. For that reason, the outcome of the discussion on data processing and data management is merely an appeal to organizations to strategically invest in inclusive and flexible data infrastructures. Such infrastructures should be created with the flexibility to evolve and expand with the evolving data landscape while still remaining in compliance with global standards and best practices. As laid out, methodological flaws associated with qualitative data gathering procedures can cloak valuable insights requiring a shift to more quantitative analyses affirming qualitative findings (Ferreira et al., 2018). In addition, systems like AEGIS demonstrate how proper data management not only preserves the value chain of the data but also optimizes the efficiency of modeling and analysis, ultimately guiding decision-makers across various sectors to thoughtful and sustainable decisions (Auzoux et al., 2020). By prioritizing optimal data management, organizations can get ahead of their time in an age of data, unveiling the tremendous potential of their data assets in the process.

In information management and processing, computer-based information systems have gained increasing importance. A project at the Timber Technology Centre (TiTC) illustrates the application of technology to extract operating data about the head rig handsaw in a sawmill. This is an instance of modular system design whereby analysis modules and sensor hardware could be replaced without disrupting continuous operation. The utilization of client-server architecture improves data analysis and storage efficiency, and users can easily fetch and interpret the information. This system, thus, emphasizes the need for flexible and expandable solutions for modern data processing because it facilitates not only existing operational needs but also future expansions into full-mill-wide programs (Bayne et al., 1999). Backing such advancements, an IT technique survey on some particular use cases highlights the importance of data integrity and efficiency in handling. The research delves into seminal areas like data cleansing, supervisory control, and complex event processing that are most essential to the guarantee of robust data environments. Examining these IT areas, the survey highlights the importance of employing advanced methodologies in addressing various issues in data administration. Although the content falls within reach, readers who are not aware of such concepts may find the technical content to be challenging. That being said, being aware of these techniques is important in order to properly harness data, as they are the basis for decision-making strategies in many industries (Brodt et al., 2010). Additionally, the effectiveness with which data management systems have been implemented in industrial settings, like the TiTC sawmill, reflects the capability of technology to revolutionize operational effectiveness. Through the collection and analysis of data in real-time, such systems not only eliminate unnecessary intermediaries and eliminate delays but also offer insight into what drives results in production. The capacity to automate the collation and analysis of data guarantees that institutions are able to act fast in response to changing conditions. As organizations increasingly rely on data-driven methods, the use of cutting-edge data management technologies is imperative in achieving maximum productivity and innovation. Continual exploration of these technologies will further enhance data processing and management strategies in various industries (Bayne et al., 1999).

REFERENCES

1. Abel, T., Amalfitano, D., Balduzzi, M., Baumgartner, L., Baumgartner, R., Baumgartner, J., Baumgartner, S., et al. (2014). Web data extraction, applications and techniques: A survey. http://arxiv.org/abs/1207.0246 (accessed on 21st October 2025).
2. Addis, M. C., & Kutar, M. S. (2020). AI management: An exploratory survey of the influence of GDPR and FAT principles. https://core.ac.uk/download/322492373.pdf (accessed on 21st October 2025).
3. Adegbi, F. F., & Fakile, A. S. (2012). Globalization and the precarious state of public finance in Nigeria. https://core.ac.uk/download/18294904.pdf (accessed on 21st October 2025).
4. Aldin, L., de Cesare, S., & Lycett, M. (2009). Semantic discovery and reuse of business process patterns. https://core.ac.uk/download/8018.pdf (accessed on 21st October 2025).
5. Amezquita, T., Amezquita, T., Bras, B., Gehin, A., Graedel, T., Ijomah, W., Ishii, K., et al. (2009). Addressing decision making for remanufacturing operations and design-for-remanufacture. https://core.ac.uk/download/9032076.pdf (accessed on 21st October 2025).
6. Athanasiadis, I., Comas, J., Gibert, K., Holmes, G., et al. (2008). On the role of pre- and post-processing in environmental data mining. https://core.ac.uk/download/29197625.pdf (accessed on 21st October 2025).
7. Auzoux, S., Christina, M., Ripoche, A., Soulie, J. C., et al. (2020). Coupling of cropping system models with the AEGIS platform [S4-O.04]. https://core.ac.uk/download/288641768.pdf (accessed on 21st October 2025).
8. Barbancho Concejero, A., Barbancho Concejero, J., León de Mora, C., Molina Cantero, R., et al. (2007). Using artificial intelligence in routing schemes for wireless networks. https://core.ac.uk/download/161255840.pdf (accessed on 21st October 2025).
9. Barbancho Concejero, A., Barbancho Concejero, J., León de Mora, C., Molina Cantero, R., et al. (2006). Giving neurons to sensors: An approach to QoS management through artificial intelligence in wireless networks. https://core.ac.uk/download/161255291.pdf (accessed on 21st October 2025).
10. Bayne, P. (1999). Development of a Head Rig Process Information System: A Thesis Presented in Partial Fulfilment of the Requirements for the Degree of Masters of Technology in Computer Systems Engineering at Massey University. https://core.ac.uk/download/148645188.pdf (accessed on 21st October 2025).

11. Belliere, S. R., Compagnioni, A., Huber, B., Jespersen, L., et al. (2010). The European regulatory framework and its implementation in influencing organic inspection and certification systems in the EU. https://core.ac.uk/download/10930577.pdf (accessed on 21st October 2025).
12. Bensbaa, S., De Florio, V., De Florio, V., Le-Phuoc, D., Rakay, R., & Yalgashev, O. (2016). Towards a smarter organization for a self-servicing society. http://arxiv.org/abs/1701.04616 (accessed on 21st October 2025).
13. Bhattacharjee, B., Ercegovac, V., Glider, J., Golding, R., et al. (2006). Impliance: A next generation information management appliance. http://arxiv.org/abs/cs/0612129 (accessed on 21st October 2025).
14. Brasher, A., Joinson, A., McAndrew, P., Minocha, S., et al. (2006). Smart labs and social practice: Social tools for pervasive laboratory workspaces: A position paper. https://core.ac.uk/download/84342132.pdf (accessed on 21st October 2025).
15. Brodt, S., Bry, F., Eckert, M., Hausmann, R., et al. (2010). A survey on IT-techniques for a dynamic emergency management in large infrastructures. https://core.ac.uk/download/12175104.pdf (accessed on 21st October 2025).
16. Buys, P. W., Gulko, N., Masuke, R., Middelberg, S., et al. (2025). Big data analytics and its influence on management accounting: Evidence from Southern Africa. https://core.ac.uk/download/658996401.pdf (accessed on 21st October 2025).
17. Cao, G., & Duan, Y. (2017). How do top- and bottom-performing companies differ in using business analytics? https://core.ac.uk/download/130975349.pdf (accessed on 21st October 2025).
18. Cardwell, P. L., & Peterson, A. L. (2025). EMASS (tm): An expandable solution for NASA space data storage needs. https://core.ac.uk/download/pdf/42809896.pdf (accessed on 21st October 2025).
19. Cesana, M., Redondi, A. E. C., & Sanvito, D. (2016). Passive classification of Wi-Fi enabled devices. https://core.ac.uk/download/74314036.pdf (accessed on 21st October 2025).
20. Chainey, S., & Smith, C. (2006). A review of GIS-based information sharing systems. https://core.ac.uk/download/1679791.pdf (accessed on 21st October 2025).
21. Crespo Márquez, A., Guillén López, A. J., Gómez Fernández, J. F., Sanz, J. A., et al. (2016). A framework for effective management of condition-based maintenance programs in the context of industrial development of E-Maintenance strategies. https://core.ac.uk/download/190375887.pdf (accessed on 21st October 2025).

22. Dereeper, A., Larmande, P., Philippe, F., Ruiz, M., et al. (2015). Genotype investigator for Genome Wide Analysis (GIGwA) [P1111]. http://agritrop.cirad.fr/575381/1/document_575381.pdf (accessed on 21st October 2025).
23. Ebbert, T. F., Landis, L., Smail, H. E., Stephan, J., et al. (2025). Survey of users of earth resources remote sensing data. https://core.ac.uk/download/pdf/42882653.pdf (accessed on 21st October 2025).
24. Fahldiek, A., Heep, S., Kriegel, H. P., Mysliwietz, B., et al. (1993). Query processing of geometric objects with free form boundaries in spatial databases. https://core.ac.uk/download/12165116.pdf (accessed on 21st October 2025).
25. Ferraz de Oliveira, M. I., Rato, L., & Sales-Batista, E. (2013). Pasture digital image analysis as a support tool for grazing management in montados: A preliminary approach. https://core.ac.uk/download/62460154.pdf (accessed on 21st October 2025).
26. Ferreira, O., Novas, J. C., Pires, A. R., Saraiva, M., et al. (2018). ISO 9001 Quality Management Systems: Critical analysis of literature review. https://core.ac.uk/download/187759793.pdf (accessed on 21st October 2025).
27. Foster, G. B., Jr. (2025). Apollo experience report: Data management for postflight engineering evaluation. https://core.ac.uk/download/pdf/42896260.pdf (accessed on 21st October 2025).
28. Green, J. L. (2025). Space data management at the NSSDC (National Space Sciences Data Center): Applications for data compression. https://core.ac.uk/download/pdf/42828322.pdf (accessed on 21st October 2025).
29. Gudivada, V. N. (2025). Image databases: Problems and perspectives. https://core.ac.uk/download/pdf/42817102.pdf (accessed on 21st October 2025).
30. Henson, S., & Jaffee, S. (2025). Standards and agro-food exports from developing countries: Rebalancing the debate. https://core.ac.uk/download/pdf/6243028.pdf (accessed on 21st October 2025).
31. Höfer, C. N., & Karagiannis, G. (2011). Cloud computing services: Taxonomy and comparison. https://core.ac.uk/download/pdf/11481589.pdf (accessed on 21st October 2025).
32. Karin, J. (2023). Optimizing data capture and storage for research data management: A study on cloud computing solutions in academic institutions. https://core.ac.uk/download/658869642.pdf (accessed on 21st October 2025).

33. King, J. L. (1984). Centralized vs. Decentralized Computing: Organizational Considerations and Management Options. https://core.ac.uk/download/287623541.pdf (accessed on 21st October 2025).

34. Küçükkeçeci, C., & Yazıcı, A. (2017). Big data model simulation on a graph database for surveillance in wireless multimedia sensor networks. http://arxiv.org/abs/1708.03878 (accessed on 21st October 2025).

35. Kummer, R. (2011). Semantic technologies for manuscript descriptions–Concepts and visions. https://core.ac.uk/download/12010960.pdf (accessed on 21st October 2025).

36. Larkin, T. P. (2020). The transformation of accounting information systems curriculum in the last decade. https://core.ac.uk/download/322959194.pdf (accessed on 21st October 2025).

37. Lazaro, C., & Le Métayer, D. (2015). The control over personal data: True remedy or fairy tale? https://core.ac.uk/download/49532918.pdf (accessed on 21st October 2025).

38. Leitmann, J. (2025). How to collect data on household energy consumption. https://core.ac.uk/download/pdf/6521915.pdf (accessed on 21st October 2025).

39. Lytras, M., & Ordóñez de Pablos, P. (2018). Knowledge management, innovation and big data: Implications for sustainability, policy making and competitiveness. https://core.ac.uk/download/pdf/162571906.pdf (accessed on 21st October 2025).

40. Moe, K. L. (2025). Systems analysis methodology for large scale systems. https://core.ac.uk/download/pdf/42835138.pdf (accessed on 21st October 2025).

41. N/A. (2014). United Nations Development Assistance Framework for Kenya. https://core.ac.uk/download/75781829.pdf (accessed on 21st October 2025).

42. N/A. (2025). Data management, chapter 5, part C. https://core.ac.uk/download/pdf/42886157.pdf (accessed on 21st October 2025).

43. N/A. (2025). Weather and climate. https://core.ac.uk/download/pdf/42884419.pdf (accessed on 21st October 2025).

44. Nda, R. M., & Tasmin, R. B. (2019). Big data management in education sector: An overview. https://core.ac.uk/download/220112831.pdf (accessed on 21st October 2025).

45. O'Sullivan, C. S., Labriola, M., Rempel, M., Moore, S., & Bradley, S. (2009). A national portrait of domestic violence courts. https://core.ac.uk/download/71361444.pdf (accessed on 21st October 2025).

46. Pierson-Balik, D., Kass, H., Cleary, J., Dixon, K. A., Harrington, L., Berger, P., & Gyarfas, S. (2005). Understanding occupational and skill demand in New Jersey's finance industry. https://core.ac.uk/download/71341811.pdf (accessed on 21st October 2025).

47. Sahu, A. K. (2016). Cloud computing with reference to data mining. https://core.ac.uk/download/539905289.pdf (accessed on 21st October 2025).

48. Sakama, C., Chimenti, D., Codd, E. F., Komorowski, J., Lloyd, J., Nicolas, J. M., & Lei, L. (1993). Towards intelligent databases. https://core.ac.uk/download/12169280.pdf (accessed on 21st October 2025).

49. Savirimuthu, J., & Wong, R. (2008). Identity principles in the digital age: A closer view. https://core.ac.uk/download/30649021.pdf (accessed on 21st October 2025).

50. Spiekermann-Hoff, S. (2020). The RFID PIA–Developed by industry, agreed by regulators. https://core.ac.uk/download/286370697.pdf (accessed on 21st October 2025).

51. Timmers, B. (2012). Impacts of climate change and variability on fish value chains in Uganda. https://core.ac.uk/download/11023461.pdf (accessed on 21st October 2025).

52. van Genderen, J. L., Vekerdy, Z., & Wang, F. (1999). Database management and implementation for coal fire detection and monitoring in the Ruijgou Coalfield, Northwest China. https://core.ac.uk/download/11857789.pdf (accessed on 21st October 2025).

53. Yafooz, W. M. S. (2015). Application of clustering in managing unstructured textual data in relational database/Wael Mohamed Shaher Yafooz. https://core.ac.uk/download/154828403.pdf (accessed on 21st October 2025).

CHAPTER 4

STATISTICAL AND ANALYTICAL METHODS

Contents

INTRODUCTION

The combination of analytical and statistical methods in transport engineering is a basic building block in the solution of modern transport issues. As the demand for the efficient and safe transportation of the world grows, the importance of engineers using these strict methods of analyzing, modeling, and optimizing infrastructure and traffic systems is undeniable. In an environment where decisions directly impact public safety and mobility, statistical tools guarantee that engineers are able to interpret complex data sets and recognize patterns that may not be apparent otherwise. Based on such analyses, projections and improvements can be formulated, which are crucial for the sustainability and development of transportation networks. Further, the use of analytical tools also goes beyond statistical numbers; it involves a multidisciplinary endeavor where engineering principles are integrated with statistical computation. While the literature stresses, as has been highlighted above, the importance of critical safety case evaluation towards assessing reliability and maintainability in transport systems (Bedford et al., 2005), either quantitative analyses of accident data or qualitative assessments of user experience facilitate through these tools vital information. As engineers navigate through the intricacies of various modes of transport and their interactions, reliance on established analytical models allows making an enlightened choice that balances efficiency and safety. Lastly, dynamically shifting transport infrastructure calls for continuous monitoring to enhance statistical and analytical practices. As technology changes, increasing resources and complexity in data collected demand complex analytical tools that can adapt to the shifting environment (Hutchinson et al., 2013). Transportation engineers must be flexible in their methods, open to embracing new techniques to enhance their analysis and conclusions. By constantly updating their acquaintance with statistical and analysis methods, the engineers are not only enhancing their own procedures but contributing to designing smarter, safer, and more efficient transportation systems for the future (Figure 4.1).

Figure 4.1. Statistical and analytical methods.

Source: https://wallstreetmojo-files.s3.ap-south-1.amazonaws.com/2022/05/Methods-of-Statistical-Analysis.jpg.

Definition of Transportation Engineering

As a discipline, transportation engineering is primarily concerned with designing, planning, operating, and managing transport systems. It synthesizes a number of scientific and engineering concepts to develop infrastructure that meets society's needs while ensuring safety, efficiency, and environmental stewardship. Transportation engineering can encompass a wide spectrum of transport modes like roadways, railroads, air transport, and marine systems. Each mode contains its own difficulties and characteristics that call for analysis methods unique to assess traffic movement, structural behavior, and human factors.

Figure 4.2. Transportation engineering and management.

Source: https://www.uab.edu/degrees/images/hero/transportation-engineering-gc.jpg.

In this regard, statistical tools have become necessary to perform the complexities of modeling along with forecasting transport phenomena. Additionally, transport engineering holds a valued position in maintaining infrastructure, particularly in the context of present environmental concerns. For instance, standard maintenance activities such as winter road maintenance (WRM) operations are prone to using road salt, which is not just dangerous for vehicles and infrastructure but also for the environment. Using a Life Cycle Sustainability Assessment (LCSA) framework illustrates how economic, environmental, and social factors are taken into account when evaluating such maintenance activities. Through the understanding of these features, engineers can choose the appropriate decision on the trade-off among various materials and techniques, ultimately embracing a more sustainable approach in the field. The numerical outcomes derived from these analyses highlight the necessity of the use of statistical methodology in transportation engineering as a means to facilitate an equitable, sustainable future. Finally, the interplay between engineering practice and statistical techniques reaffirms the innovative potential of data-driven decision-making in transportation systems. Engineers increasingly are employing advanced analytics to optimize traffic efficiency and safety protocols while managing operating costs. The impacts of these techniques become evident in innovative solutions to longstanding transportation issues, such as congestion and accident prevention. Emphasizing the call for interdisciplinarity, collaboration among various stakeholders—decision-makers, engineers, and the people—is paramount in the success of transportation system development. While the evolution of transportation engineering unfolds, constant incorporating of statistical and analytical methods will be central in addressing newly emerging issues, thereby shaping future urban and rural efficient and sustainable mobility frameworks (Cui et al., 2016; N/A, 2017) (Figure 4.2).

Importance of Statistical and Analytical Methods

The significance of statistical and analytical methods in transport engineering is primarily their capability to enhance decision-making. The use of reliable data collection and analytical frameworks allows engineers to systematically assess multiple transport networks. For instance, the systematic analytical framework employed in analyzing the Triangle Expressway effectively scrutinized its impact on regional traffic based on travel time reliability measurements. This approach highlights the importance of spatial and temporal changes in traffic patterns, which are to be known in order to tackle efficient transportation systems. Evidence-based knowledge that is derived through such approaches facilitates refinement of operational strategies and evidence-based planning for future projects. Finally, integrating statistical methods into transportation engineering operations builds a more reliable and accurate infrastructure system capable of accommodating shifting urban mobility needs. Besides, statistical methods are useful in establishing risk factors and estimating crash risks in transport systems. A good example is found in bicycle-motor vehicle crash studies, where spatial and temporal analysis utilizes statistical data to determine safety risk. The study analyzes 4120 geocoded bicycle crashes and finds that urban arterials are high-risk areas due to the mixed patterns of motor vehicles and bikers. Researchers can identify risk factors and patterns of crashes through weighted kernel density estimation and negative binomial regression. Not only does this increase understanding of crash dynamics, but it also informs a successful approach to bicycle facility design. As a result, it is critical to understand these risks through robust statistical models to ensure safety and overall performance in transportation systems. Besides improving decision-making and safety, analytical and statistical methods allow evaluating the effectiveness and performance of transportation projects over time. With the help of advanced analytical procedures, transportation stakeholders can analyze shifts in travel time distributions and reliability measures as exposed by the analysis of the Triangle Expressway. In this case, the stabilizing nature of travel time distributions over a couple of years indicates that traffic reliability can be used to give insight into the long-run performance of toll roads. Such determinations are essential to planners and engineers as it enables them to forecast and react to changes in travel demand patterns correctly. These future-oriented examinations, complemented with statistical modeling, assist in guiding planning decisions, shedding light on the essential nature of these methodologies to shaping sustainable transportation systems that are adaptable to contemporary urban needs.

4.1. HISTORICAL CONTEXT OF TRANSPORTATION ENGINEERING

The history of the field of transportation engineering traces its roots back to the community need for efficient mobility as early as ancient times when primitive roads and paths were employed to carry goods and correspond. The earliest engineers addressed transportation by focusing on the construction of roads, bridges, and canals, laying the way for modern engineering activities. Over the centuries, innovative technological innovations such as the creation of the steam engine and later, the internal combustion engine, transformed transport means. This provides a clear indication of how societal demands were merged with engineering innovations to create infrastructure systems that responded to the evolving needs of growing populations. Consequently, the history displays the beginnings of transportation engineering as an area largely characterized by its response to human mobility demands, laying the groundwork for modern analytical approaches to dealing with transportation systems. With transportation systems becoming increasingly complex, there arose a requirement for systematic analysis and statistical methods to deal with rising complexities in travel behavior, congestion, and safety. Early 20th-century advances in methods of collecting data enabled engineers to start measuring transportation phenomena. Traffic studies formalized the establishment of essential information regarding user activity and road performance, which influenced design and policy-making. In addition, techniques like regression analysis started to take center stage when it came to determining variables influencing transportation systems. It was here that engineers valued the role of integrating statistical techniques into transportation planning, as is evident from studies comparing traffic volumes and flow characteristics. The discipline, therefore, inherited a robust set of analytical instruments that continue to guide engineering practices to date and highlight the necessity of making decisions on the basis of data in effective transportation management. The addition of advanced statistical and analytical methods to transportation engineering has been necessitated by the demands for enhanced operating efficiency and reliability. For instance, large-scale studies of traffic management for major events such as sports events have shown the crucial role that analysis of real-time data has in predicting congestion patterns. The use of such technologies as the Multi-EigenSpot algorithm is indicative of the capability of today to both detect and respond to event-conditioned traffic hotspots. Further, methods like Dynamic Bayesian Networks (DBN) enable predictive modeling of congestion, enabling planners to anticipate traffic behavior with astounding precision. Such developments are indicative of the shift away from older engineering paradigms to present-day methodologies incorporating analytical methodology in order to ultimately impact direction for the future in transportation systems. This emphasis on statistical technique reveals a crucial advancement in a field with its very existence grounded in the provision of legitimate and effective mobility solutions, as seen in studies examining traffic behavior on game days (Bedford et al., 2005; Ahsani et al., 2019).

4.1.1. Evolution of Transportation Systems

The evolution of transport systems has long been shaped by the quest for efficiency and technological progress. The transition from animal-powered transport to the invention of steam engines was a critical milestone, as transport of commodities and humans became faster and more reliable. The invention of automobiles and airplanes further revolutionized transport, enabling globalization and opening up areas to more seamless interconnection. Over the past few years, the convergence of statistical and analysis methods has taken center stage in overseeing such advanced systems. Through data analytics, traffic engineers are able to examine traffic flow, optimize routes, and predict demand, hence maximizing user experience as well as operational performance. Ride-hailing apps have become an essential component of urban transportation in our modern era. Demand analysis for such services in the short term reveals intricate interdependencies among various factors affecting them such as weather, prices, and traffic. Machine learning-based studies have found that models such as artificial neural networks and boosted decision trees have been more accurate predictors in such cases. The ability to manage big data with high accuracy allows transport systems to respond dynamically to changing patterns of demand, which ensures that resources are being utilized efficiently. Thus, such advanced analytical methodologies not only enhance the quality of service but also assist in ensuring the overall sustainability of urban transport networks (Cools et al., 2017). Further, the application of network science to the analysis of public transport networks, particularly bus networks, yields insightful details concerning their structural properties and patterns of development. Graph theory analysis of these networks reveals their vulnerabilities as well as their resistance to random failures. As evident in current research, the bus systems exhibit small-world features, which means that they possess the capability to facilitate fast information exchange between distant nodes. Understanding how geographical and socio-economic constraints influence the evolution of bus networks can inform the planning of more efficient and robust networks. This statistical method allows engineers to analyze the performance of these networks under varying conditions, leading to improved concepts for the growth of connectivity in cities as well as for the construction of a believable transportation network (Chatterjee et al., 2016) (Figure 4.3).

Figure 4.3. Evolution of transportation systems.

Source: https://www.shutterstock.com/image-vector/evolution-transportation-past-present-future-260nw-2369907329.jpg.

4.1.2. Early Use of Statistical Methods

The early use of statistical methods was instrumental to the construction of analytical methodologies in transport engineering. Resulting from the need to quantify and interpret complex transport phenomena, these methods allowed engineers to assess variables such as traffic flow, crash rates, and road condition in a systematic manner. The employment of statistical tools in transport studies facilitated practitioners to make reliable conclusions on the basis of empirical data, allowing for informed decision-making. For instance, traffic model building was largely based on statistical principles, which helped engineers to predict traffic flow and identify possible safety issues. As they continued, they began to influence transportation system design and operation, and pave the way for sophisticated analytical methods and applications in later years. Moreover, statistical methods were applied for the first time as a valuable move from anecdotal procedures to transportation engineering. Rather than relying mainly on professional practice or experience, practitioners attempted to build their decisions on an examination of data, offering a more scientific approach to problem-solving. This trend was depicted in the scientific research of speed limits and the way speed limits are related to safety measures. The discovery of the 85th percentile rule, whereby it was argued that the speed limit should be set by the speed at which 85% of motorists drive, was a popular method that represented the initial application of statistics to policy-making. However, there have been subsequent criticisms regarding empirical justification for the rule, and thus arguments for regularly updating and refining statistical methods according to evolving transport problems (Anthoine et al., 2020). In the modern transportation engineering context, the historical foundations set by the early statistical methods are still being experienced in today's practice. With new challenges arising, such as the requirement of zero traffic accidents, data-driven decision-making is still essential. Existing projects, such as those undertaken by the California State Transportation Agency, illustrate a clear evolution of statistical methods from their inception to present-day application. The application of future-based alternative approaches to traditional practices and improved road designs constitutes a transformation of analytical methods employed by transport engineers. The emphasis on early statistical research suggests the imperative for ongoing research and development, showing that novel solutions can create the basis of practical solutions to modern transportation challenges (N/A, 2017).

4.1.3. Impact of Technology on Data Analysis

The impact of technology on transportation engineering data analysis has introduced hitherto unseen levels of understanding and manipulation of traffic flow. Advances such as real-time data capture and advanced analysis methods have equipped engineers with advanced tools for ascertaining road conditions more efficiently. For instance, the use of data from sources such as INRIX has enabled analysts to compare historical traffic flow with large events affecting congestion levels. By leveraging sophisticated algorithms such as Multi-EigenSpot, as identified in research, traffic hotspot behavior can be analyzed more deeply to facilitate more effective traffic management plans. Such analysis allows for accurate interventions during rush hours, ultimately towards optimizing traffic flow and reducing delays, pointing to the imperative link between technological advancements and data-driven decision-making in transportation planning. Besides, technology application in making travel time more predictable is core in transportation engineering. Tools of analysis that were designed to assess the effects of toll roads, like studies exploring the Triangle Expressway, demonstrate how travel time data processing becomes easier by the application of technology. By analyzing travel time distributions, spatial and temporal trends impacting general reliability are determined by engineers. The positive trend in managed lanes' reliability indicates that with sophisticated data analytics being employed to install and maintain toll roads, data exists that can be utilized to make decisions on future infrastructure decisions. In this way, technology guarantees traffic behavior and reliability are understood in great detail, which results in informed policymaking to enhance the transportation system's overall efficiency. Last but not least, the potential of advanced analytics to transform the management of transportation systems and future infrastructure development is huge. Data-driven methods allow for real-time monitoring of travel time volatility and their resulting reliability indicators, providing an understanding of the temporal dynamics of transportation demand. This analytical capacity is crucial when evaluating the performance of newly implemented projects or infrastructure improvement, as highlighted in previous work. As understood by the findings of the study, the operational data reveal patterns that can guide future investment in transportation facilities such that projects are consistent with changing transportation demand trends. Such consistency between technology and data analysis, therefore, impacts current traffic patterns as well as equips transportation engineers with the insights to plan robust and sustainable transport systems for the future (Figure 4.4).

Figure 4.4. Impact of technology on data analysis.

Source: http://cdn.getmidnight.com/50062ddd13358c02dea31c6e82a524c5/2023/09/AdobeStock_600314909.jpeg.

4.2. KEY STATISTICAL METHODS IN TRANSPORTATION ENGINEERING

Statistical methods play a central role in planning and decision-making in transportation engineering. One of the rudimentary methods is regression analysis, which engineers use to model relationships between variables, say, traffic volume and road condition. From past data, these models are used to predict future flows of traffic, which is crucial in infrastructure planning and maintenance. In addition, Bayesian methods allow transportation engineers to incorporate prior information or expert opinion into their statistical models, thus making predictions more precise. Moreover, the adaptability offered by such models also allows transportation engineers to efficiently respond to condition changes and emerging trends, which leads to safer and more efficient transport systems. The second significant area of statistical application is the area of traffic safety statistics. Techniques such as control charts and time series analysis are used to monitor and examine crash data across time, and identify patterns and potential risk factors. This assists engineers in determining high-risk sites and intervening to decrease accidents, aimed directly at those areas. Moreover, the use of survival analysis in ascertaining the efficacy of interventions makes it possible to more accurately estimate how certain measures of road safety influence accident incidence over the long run. In this case, statistical methods not only aid the development of more secure road networks but also provide the basis for continued refinement and adjustment to the evolving issues of safety. In addition, simulation methods, particularly Monte Carlo simulations, are becoming powerful tools in transportation engineering. These methods enable engineers to consider uncertainty in parameters such as traffic demand and weather conditions, thereby enabling more efficient decision-making processes. By generating a range of conceivable situations and their resulting consequences, Monte Carlo simulations provide insights into potential gains and risks associated with various transport projects. Such a probabilistic approach enhances the credibility of planning processes by quantitative evaluation of the chances of success under different circumstances. Hence, the application of such advanced statistical methods to transportation engineering bears testimony to their significance in the formulation of innovative and evidence-driven solutions to complex transport issues.

4.2.1. Descriptive Statistics

Descriptive statistics is one of the key elements in transport engineering, particularly in the interpretation and analysis of data. By summarizing and organizing data into meaningful information, the statistics allow engineers to identify patterns, trends, and outliers in transport systems. Statistics such as mean, median, mode, variance, and standard deviation are usually employed to quantify characteristics of traffic flow, vehicle speeds, and accidents that occur, thus enabling engineers to make decisions backed by empirical evidence. The statistical method not only assists in effectively handling existing transport infrastructures but also informs the design and planning of future ventures such that they are synchronized with safety and efficiency targets. In transportation engineering, descriptive statistics are also vital in assessing safety interventions and identifying high-risk areas. For example, analysis of accidents can reveal important information on spatial and temporal patterns of crashes that can guide interventions and resource allocation.

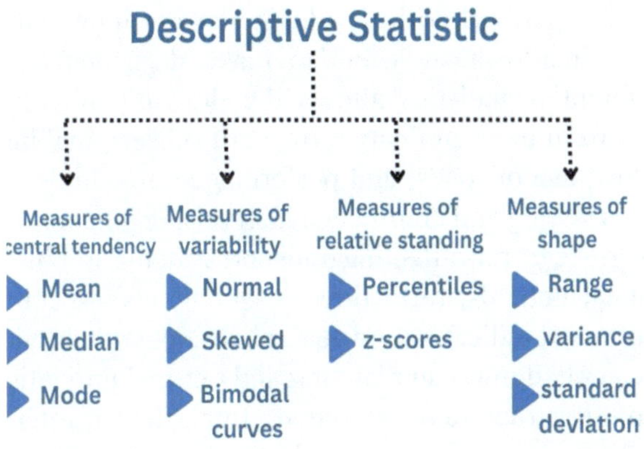

Figure 4.5. Descriptive statistics.

Source: https://files.codingninjas.in/article_images/descriptive-statistics-1-1693334431.webp.

Descriptive statistics research can easily represent trends in bike crashes and motor vehicle crashes, as seen from research in urban areas. With statistical techniques such as kernel density estimation, it is possible for analysts to chart hotspots of crashes and provide stakeholders with a clearer picture of where the safety enhancements are most needed. Data-driven methodology has the potential to go a great distance in transforming transportation safety programs and policymaking, ultimately reducing the accident rate on the roads. Also, the application of descriptive statistics in transportation engineering is extended to inventory management strategies in areas such as the pharmaceutical supply chain where optimal levels of stocks need to be held. Statistical analysis of inventory configurations can reveal inefficiencies and identify factors contributing to inventory levels, thereby empowering supply chain managers with valuable intelligence. By the understanding of complex interplay among production operations, regulatory requirements, and external pressures, companies can optimize their businesses and reduce associated costs of unnecessary inventory. This descriptive statistic application strategy not only maximizes operational effectiveness but also contributes to enhancing the total transportation supply chain resilience facing issues due to unstable demand and regulatory circumstances (Andreu et al., 2017). In addition, this analytical approach helps reconcile service levels with an organization's financial goals to achieve both customer satisfaction and economic sustainability (Figure 4.5).

4.2.2. Inferential Statistics

Inferential statistics plays a critical role in the discipline of transportation engineering since it allows engineers to make decisions based on sample information. Inferential statistics allows the drawing of conclusions about large populations from the examination of small subsets, making it possible to predict traffic flow, use of roads, and performance of vehicles. With methods such as hypothesis testing and confidence intervals, engineers can ascertain the significance of various factors affecting transport systems. For instance, applying inferential statistical methods, traffic flow measurements at specific intersections can be utilized to make predictions and generalize about trends within a complete city. This can be applied in urban planning and optimal utilization of resources in transportation infrastructure development. In particular, inferential statistics can be utilized to ascertain how variables in transportation systems interact and lead to insights vital to strategic planning. Statistical methods such as regression analysis enable transport engineers to study the interrelationships of a number of factors, such as rental rates, site location, and revenues of shopping centers, which are pertinent to urban transportation planning. For example, one study comparing financial performance indicators across shopping centers revealed that strategic site location has significant influence on the return on investments, whereas other aspects such as parking revenues can have less influence. This is the significance of combining inferential statistical methods as they identify significant relationships that are not noticeable through descriptive statistics alone, hence allowing stakeholders to make informed choices based on data (Magdalena et al., 2017). Moreover, the utilization of inferential statistics also comes in handy for analyzing transport systems' performance and how it evolves over time. Using these statistical tools, engineers can analyze data collected using various transport modes, including public transport, cars, and bikes. In so doing, travel demand can be comprehended in terms of commuter behavior and decision-making, with implications on policy changes or infrastructure that result in radical shifts in travel demand. For instance, hard evidence of travel patterns can indicate correlation with higher ridership and the availability of bicycle lanes, hence the investment in them is justified. In such a manner, inferential statistics not only provides information to current evaluations but also provides projections that can guide the planning for future transport schemes, so urban areas can adapt and adapt to changing mobility needs (University, 1990).

4.2.3. Regression Analysis

Under the broader context of transport engineering, regression analysis is one of the most crucial tools employed in the study and prediction of complex relationships between various operational factors. With the use of statistical analysis, transport engineers can model how multiple factors impact performance measures such as travel time, safety, and system reliability. Such an approach is particularly relevant in ascertaining the influence of outside disturbances on transport infrastructure. For instance, inland waterway disruption can significantly affect commodity flow and have financial impacts. Statistical models that will take lock use and lack of availability into account can provide insight regarding these connections, enabling decision-makers to allocate resources more effectively for upgrades and maintenance. Using regression analysis, engineers are able to quantify these relationships, thus promoting strategic planning and system resilience against operational mishaps (McGee et al., 2018). Moreover, the integration of advanced technologies into transport systems has made regression analysis more applicable to real-world uses. AVL and APC systems provide a wealth of data, which can be utilized to substantiate stringent analytical models. For instance, regression models can be built not just to predict travel time but also to capture uncertainties inherent in such predictions. It has been observed that standard linear regression models perform fine, but accelerated failure-time survival models are useful as they capture the entire distribution of travel time and produce lower variances in estimation. This responsiveness to modeling intricacies assists transit operators in offering better estimations, hence satisfying users' needs for improved consistency of service and enhancing overall satisfaction with public transport systems (Gayah et al., 2016). Furthermore, the applications of regression analysis extend beyond mere prediction to the optimization of the data-integration processes for public transport systems. Based on various regression models, transport analysts can examine real-time data to create actionable intelligence on passenger occupancy along the stops. Such information is useful in optimizing the users' experience to the maximum since it allows the passengers to board optimally in line with expected seating capacity or crowding. The difference between count and linear regression techniques also underscores the flexibility of regression analysis to offer specific transportation problem solutions. Ultimately, such modeling procedures demonstrate how regression analysis not only enhances the operational effectiveness but also facilitates a data-driven solution in public transit problem-solving, thereby improving strategic planning practice in transportation engineering (Gayah et al., 2016) (Figure 4.6).

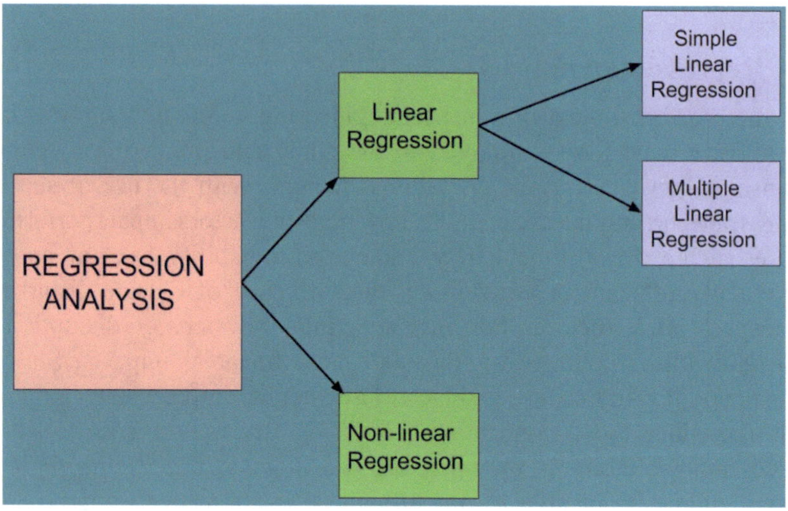

Figure 4.6. Regression analysis.

Source: https://www.erp-information.com/wp-content/uploads/2021/01/regression-analysis.png.

4.3. DATA COLLECTION TECHNIQUES

Accurate data collection techniques play an imperative role in transportation engineering to inform decision-making and enhance operation performance. The exercise typically begins by describing the study purpose, such as from traffic conditions analysis through the evaluation of infrastructure performance. Observational studies, sensor data gathering, and surveys are employed to identify quantitative and qualitative data pertinent. For instance, real-time traffic counters allow engineers to track car volumes and speeds in real time, allowing them to analyze congestion patterns and plan intervention. Moreover, well-designed questionnaires can provide information on commuter behaviors and preferences that are essential in developing user-focused transportation systems. By using various tools of data collection, engineers are able to build detailed datasets that support data-driven decision-making for transportation projects. The integration of statistical analysis and data collected enables transportation engineers to arrive at meaningful conclusions and evidence-based suggestions.

Data collected are examined using a number of statistical methods to determine trends, correlation, and factors of prediction. For example, project information collected from the Florida Department of Transportation revealed major factors influencing construction project time performance, such as contract types and project sizes (Park et al., 2010). Use of statistical tools like regression analysis facilitates quantitative measurement of the relationships between such variables so that there is better understanding of how different factors contribute to project outcomes. Apart from being essential for improving project planning and execution, these insights are also useful for predictive model development that can guide future transport projects. Engineers are then able to justify their decisions and optimize resource usage with the application of robust analytical methods. Besides, technological progress has also transformed data gathering processes within transport engineering. Modern technology such as Geographic Information Systems (GIS) and mobile data collection software enables simpler assembling of spatial and temporal data. GIS, for instance, enables engineers to visualize and analyze the spatial pattern of various transport indicators and enable identification of areas that need improvement or investment. Also, as discussed in a study on project time and cost performance modeling, the continuous updating of data collected enhances predictive models' reliability and accuracy (Park et al., 2010). Adaptive approach ensures that the models and analyses reflect the changing nature of transportation networks, ultimately giving better planning outcomes. With improvements in data gathering techniques, their integration with analytical procedures continues to be essential towards addressing the complex issues faced in transportation engineering (Figure 4.7).

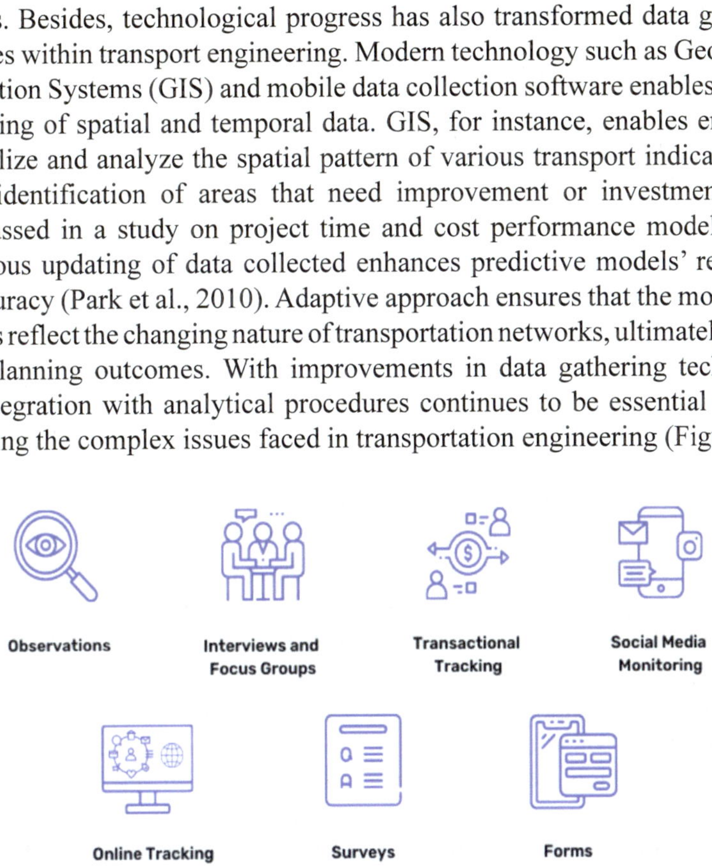

Figure 4.7. Data collection techniques.

Source: https://safetyculture.com/_next/image/?url=https:%2F%2Fwp-website.safetyculture.com%2Fwp-content%2Fuploads%2Fsites%2F3%2F2023%2F12%2F7-Data-Collection-Techniques.png&w=1920&q=75.

4.3.1. Surveys and Questionnaires

Questionnaires and surveys are core tools in transportation engineering that assist in obtaining useful information to shape future plans and policy actions. Categorized questions gather remarks from diverse groups of stakeholders such as transport users, policymakers, and affected communities via these tools. Data that is collected using surveys can be able to reveal patterns in user behavior, preferences, and attitudes that aid engineers in creating responsive systems to human needs. The process of developing such surveys must therefore be elaborately defined to ensure that the collected data is valid and reliable. This underscores the necessity of applying statistical analysis to examine survey outcomes to facilitate engineers to make analytical decisions from qualitative and quantitative investigations. Furthermore, how effectively the surveys are carried out can greatly influence the acceptance and satisfaction with the transport system by the users. As per investigations, ride quality, which is significantly affected by a large number of environmental factors, has a dominant role to play in user satisfaction in air transport, as argued in Jacobson et al. (2025). Development of models to measure ride quality and passenger comfort illustrates how survey information is convertible into quantitative measures to assess various transportation scenarios. As engineers review survey responses, they can convert user-perceived experience into quantifiable design specifications. Such understanding not only optimizes the transportation experience but also offers deeper insight into user behavior determinants ultimately leading to better system design and community participation. With the constant changes in demands being put on the transportation system, generalized methods need to be used that can adapt to local community requirements as they optimize resource use. The establishment of precise measures to enhance transport systems, as outlined in Deptula et al. (2025), highlights the importance of conducting extensive surveys considering the diversity and constraints of the population. By employing surveys and questionnaires, transport engineers can evaluate transport projects based on tailor-made criteria, which consider intermodal and intramodal optimality. In addition, this approach ensures that a number of perspectives are considered, enhancing collaborative planning processes that utilize public opinion within the transport decision-making process. Last but not least, effective application of surveys ensures that the gap in between technical engineering solutions and real users' needs in everyday life is bridged, resulting in an efficient and sustainable transport environment.

4.3.2. Traffic Counting Methods

Traffic counting methods are the foundations of knowing and interpreting transportation systems since they provide critical data necessary for planning, safety analysis, and infrastructure development. Methods used in traffic counting include different processes such as manual counts, automatic counters, and advanced simulation models. Manual count is labor intensive but allows qualitative data collection where researchers have a chance to study some behavior and road user characteristics. On the other hand, computerized traffic counting devices like inductive loop sensors and camera-based technologies offer higher efficiency and real-time data capture. However, initial costs as well as ongoing maintenance on these devices can be quite significant. As such, choosing a proper traffic counting process is usually a compromise between accuracy, cost, and utilization of the data gathered. Application of statistical methods comes in handy while interpreting traffic counting data to transform raw figures into meaningful information.

Figure 4.8. Traffic counting methods.

Source: https://tagmaster.com/wp-content/uploads/2023/05/example_counting-bus-1024x470.jpg.

One top approach is the use of regression models in analyzing patterns of traffic flow and predicting future traffic conditions. For instance, understanding how traffic volumes fluctuate with seasonality or time of day can guide city planners in designing more responsive road systems. Moreover, spatial analysis techniques can assist researchers in identifying high traffic accident zones, hence highlighting road safety and the effectiveness of existing traffic management measures. Such techniques come into significant focus when analyzing the interactions between bicycles and motor vehicles, according to research using both spatial and temporal approaches for identifying crash risk from different road environments (Chang et al., 2019). By combining statistical methods with traffic counts, engineers may be able to gain more detailed insights into transportation behavior. Traffic counting methods also serve as a basis for more comprehensive evaluative research that attempts to understand mode share changes and overall travel conduct. For instance, the application of Monte Carlo simulation can be utilized to place confidence intervals around traffic statistics, thereby enabling one to identify statistically significant changes (Clark et al., 2005). This is particularly vital with the advent of changing urban dynamics and evolving transportation patterns because it enables planners to actually modify and make informed decisions based on solid facts. By systematically applying these advanced statistical techniques along with traffic counts, engineers can not only research the current context, but also plan strategically for future infrastructure evolution. The engagement of such modes of counting with analytical tools eventually enhances the efficiency of transport systems and promotes safer transport environments within urban areas (Figure 4.8).

4.3.3. Use of Geographic Information Systems (GIS)

Geographic Information System (GIS) application in transportation engineering is a transportation data processing and application revolution. GIS facilitates combining spatial data, leading to the visualization of complex relationships between various components of transportation. Geospatial analysis allows traffic patterns to be identified by engineers, identify the needs for transportation infrastructure, and create models for forecasting future transportation demand. This analytic capability is joined by the interactive geospatial processes that GIS can model, making it easier to understand how spatial elements influence transportation systems. Next-generation GIS technology research emphasizes not just enhanced visualization but also potent analytical methods drawn from areas such as artificial intelligence and cognitive science. This syncretism of various disciplines enhances the analysis foundation on which transportation engineering decisions are based, and is precisely aligned with goals towards smarter urban mobility and sustainability. Public Participation Geographic Information Systems (PPGIS) are becoming an integral component in incorporating community input into transportation planning. While transportation engineers work on making eco-friendly transport solutions, integrating PPGIS within them allows stakeholders to have a direct role in mapping and determining their transport preferences and needs. The success of PPGIS methods in participation is showcased in case studies in urban areas like Poznań and Łódź, where contributions from people influence the result of plans directly. Not only does this participatory method enhance the collected information, but it also boosts the legitimacy and acceptability of planned transport projects. By situating PPGIS in the various stages of the planning process, engineers are able to enable data collection and public participation simultaneously, thereby enabling informed decisions that truly reflect the needs of people. Such methodology is required to unite individuals towards promoting cooperation and better and sustainable transportation results. However, using GIS and PPGIS effectively in transportation engineering is plagued by numerous challenges, primarily reviving around incorporating data and involving participants. The need for spatial consistency and qualitative examination is paramount to ensuring the data drawn from various sources lead to consistent outputs. Recruitment challenges for PPGIS participants affirm the need for advance planning to ensure provision of a representative sample for input within the community. Furthermore, increased development of GIS capabilities requires provision of fundamental questions regarding the analytical computational processes underlying their use in transport settings. Researchers and practitioners must join forces across disciplines to tackle these challenges so that GIS may achieve its complete potential as an analysis method. By focusing on the development of qualitative spatial reasoning and iterative enhancement of effective public participation, the field of transportation engineering may better meet the evolving requirements of urban mobility (Figure 4.9).

Figure 4.9. Geographic information systems (GIS).

Source: https://1.bp.blogspot.com/-TxfluLf0uy0/XOu75pEjtLI/AAAAA AAAA A8/k6TFC9n3Cc4xtF79zGvGkIJ6fYbpNTs_gCLcBGAs/s1600/introfig1.jpg.

4.4. ANALYTICAL MODELS IN TRANSPORTATION ENGINEERING

The use of analytical models in transport engineering is an important role to enhance efficiency, safety, and operation reliability. Using sophisticated mathematical models, engineers can simulate various situations, predict traffic flow, and design facility plans optimally. The models enable analysis of transport systems under various conditions, including varying levels of traffic and environmental conditions. Apart from this, they assist in identifying potential points of failure and bottlenecks, allowing preventive interventions. The application of analytical models, as shed light on by recent research and development, indicates their applicability to the reliability and maintainability of transport infrastructures, thereby contributing to a safety-based approach to policy-making and design (Bedford et al., 2005). Another essential aspect of analytical models is that they may be written with statistical methods, which further increases the precision of predictions being produced under the field of transportation engineering. By integrating data-driven approaches, these models can analyze historical traffic data, identify patterns of accidents, and predict future needs of the transport system. This convergence of statistics and analytics not only advances the analysis of existing systems but also drives planning innovations and resource allocation. In the field of transportation policy, this analytic acuity informs investment and development planning decisions. Refining these models further is significant in managing future challenges, particularly as urbanization and technological transformation continually redefine transportation contexts (Hutchinson et al., 2013). Ultimately, the employment of analytical models in transportation engineering systematically enhances the capability of transportation systems to adapt to evolving demands. They are an essential tool in reconciling technical, economic, and environmental trade-offs in transportation system planning and operation. In addition, these models facilitate the development of a system for gauging system performance, risk management, and strategic investment in infrastructure. With stakeholders targeting increased mobility and decreased adverse environmental impacts, the findings based on analytical models will become increasingly significant. Their roles extend beyond mere short-term optimization of operations to long-term sustainability of transport networks, with indicators of in-depth understanding of mobility issues in modern society (Bedford et al., 2005).

4.4.1. Traffic Flow Models

Traffic flow models are really valuable in analyzing and predicting vehicle movement on roads. They develop from a set of statistical and analytical methods that help engineers design better transport systems. One of the most significant advances in traffic modeling has been the application of geostatistical methods, which take into account the spatial dependence among traffic variables rather than assuming that the data points are random samples. For instance, research has shown that certain relationships, such as the entering flow (Q sub e) and circulating flow (Q sub c) in roundabouts, are more accurately described if the inherent data structures are considered. This perspective alters the methodology from simple interpolation methods to a more realistic view of traffic flow, so model predictions conform to observations. These advancements not only enhance the predictability of traffic flow but also inform safer and more efficient road planning (Mazzella et al., 2011). Furthermore, traffic flow dynamics become complex during special events when timed congestion is capable of disturbing normal traffic patterns. The inclusion of big data analytics in such environments has been worth its weight, according to evidence drawn from studies using INRIX data to assess traffic congestion surrounding significant events such as sports games. Based on observation of travel behavior over a number of years, such research offers insights into trends and congestion hotspots specific to specific venues. For example, the Nebraska Cornhuskers football games were employed as a case study in how crowd density affects traffic congestion along major arterials. The hotspots—those areas of ongoing congestion—provide transportation engineers with essential data for infrastructure planning and event management. Consequently, uncovering these spatiotemporal patterns makes preventive strategies possible for minimizing congestion during large events (Ahsani et al., 2019). Lastly, the technological advancement in traffic flow modeling has allowed for greater depth of predictive ability, which is a necessity in managing normal as well as abnormal traffic conditions. Techniques like Dynamic Bayesian Networks (DBN) have become a valuable resource in forecasting congestion patterns based on several factors that affect traffic behavior, such as game timing and competitor teams. The resulting models enable congestion prediction, which facilitates transport authorities and city planners to implement intervention strategies that are targeted. As recipients of these models, engineers have improved ability to maximize roadway productivity and minimize disturbance. By consistently refining these statistical and analytical methods, the field of transportation engineering not only improves traffic management operations but also helps to facilitate a more secure and effective urban mobility environment overall (Ahsani et al., 2019).

4.4.2. Demand Forecasting Models

Demand forecast models play a central role in transportation engineering as they decide infrastructure planning and resource planning. Statistical and analytical methods are utilized by such models to forecast future demand for various transport services, ranging from passenger travel to air cargo. Transportation engineers can plan adequately for changing patterns of demand due to good knowledge of the techniques of forecasting. Currently, there are methods ranging from qualitative examination to quantitative models, such as time series analysis and regression techniques, each with advantages and disadvantages. With the application of these techniques, practitioners can make more precise policy adjustments and reduce overcapacity or under provision costs of services, resulting in an overall more efficient transport system. Moreover, empirical studies also showed the efficiency of different methods of forecasting demand, with specific application to niche markets like air freight. A thorough review of current literature shows a rich variety of methods employed today, all with specific market conditions and operational requirements in mind. As Kuhlthan et al. (2025) points out, these forecasting models get reviewed on a regular basis to check their advantages and disadvantages, with those results having a direct impact on future study directions and practice developments. Although others are more accurate, the others could raise the usability for decision-makers, and the demand for tailored techniques in demand forecasting ultimately. The reality that such models keep improving is an active research field that can significantly enhance transport systems by being capable of predicting demand changes in real time.

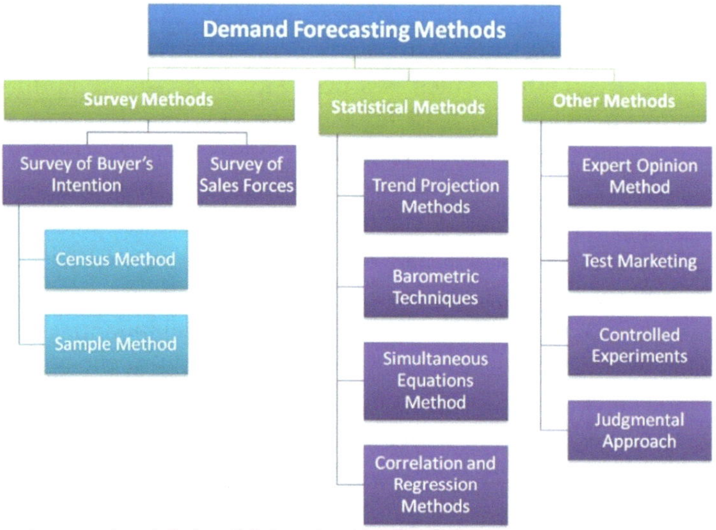

Figure 4.10. Demand forecasting models.

Source: https://image.slidesharecdn.com/class6definitionofmethodsdemandforecasting-140725025347-phpapp01/95/class-6-definition-of-methods-demand-forecasting-4-1024.jpg?cb=1406256855.

The inclusion of simulation tools also enhances demand forecasting, particularly in supply chain management use. They provide transport engineers with a way to compare different circumstances and derive lessons from complex relationships between data. To illustrate, simulations can mimic the impact of unstable demand on the performance of supply chains and help with making strategic decisions. As explained in Åšlusarczyk et al. (2025), the use of simulation in transportation uses not only provides an environment for the validation of theoretical models but also helps bridge the gap between analytical forecasts and practical usage. By the integration of simulation with traditional forecasting methods, transport experts are able to develop stronger and more adaptive methodologies that react to the dynamic behavior of travel demand, ultimately improving service delivery and efficiency in the transport industry (Figure 4.10).

4.4.3. Safety Analysis Models

Safety analysis models are a very critical instrument in transportation engineering, particularly in identifying and averting road use risks. They employ various statistical and analytical approaches to establish the safety performance of transportation systems and inform policy development and engineering design. Through systematic examination of past crash data and other variables of relevance, safety analysis models can detect trends and patterns that are not always apparent. Such in-depth examination can assist planners and engineers in a greater understanding of root causes of accidents and consequently better interventions and improvement in the infrastructure. Moreover, safety analysis models can be tailored to specific contexts so that interventions are designed uniquely for the particular needs of different roads and cities according to their unique characteristics, thus enhancing transport safety as a whole. New strategies complementing traditional safety analysis models are introduced that leverage developments in temporal and spatial analysis. For instance, studies that account for bicycle-vehicle crashes typically use a combination of spatial methods, including weighted kernel density estimation, and time-based methods, including negative binomial regression. These methods allow researchers to estimate crash risks both at the macro and micro levels and outline an elaborate description of safety concerns. Specifically, the spatial approach is capable of identifying high-risk areas, such as city arterials, where congestion and design restrictions form severe safety issues. By means of an apt study on 4120 geocoded crashes of bicyclists in Antwerp, researchers were able to successfully integrate data from different government agencies to construct a large database, which underlines the significance of interagency coordination in models of safety analysis (Chang et al., 2019). Application of safety analysis models is more than the mere data analysis; they also facilitate proactive measures in transportation planning and design. Through the identification of key risk indicators and assessment of the effectiveness of existing countermeasures, such models influence the development of safer infrastructure and traffic management strategies. Most importantly, the reliability and maintainability of these models are critical components in their effectiveness, as suggested by the lessons of the case study (Bedford et al., 2005). Insights from safety analyses can be used to allocate resources, identify places to target interventions, and establish performance metrics. Thus, by integrating different statistical and analytical strategies, safety analysis models do not only offer solutions to transportation safety issues but also facilitate the development of durable and secure transportation infrastructures.

4.5. APPLICATIONS OF STATISTICAL METHODS

Statistical methods have an important role to play in the transportation engineering field to enhance safety and make proper decisions. Greater focus on pedestrian and cycling safety, for example, necessitates a systematic process to evaluate potential interventions effectively. Agencies are often constrained by limited resources, and hence utilization of statistical methods is essential for prioritization and implementing safety interventions in an effective way. An ecologically based sketch planning system offers a data-driven methodology to the choice of effective interventions. An example of Chicago, utilizing a case study, illustrates how statistical methods can streamline decision-making processes and lead to improved safety reactions for vulnerable road users (Cottrill et al., 2010). Statistical methods are also an essential aspect of data measurement and analysis in transportation systems. Various organizations categorize information pertaining to fuels and oxidizers, assembling large datasets that inform engineering practice. Datasets contain thermophysical properties, combustion behavior, and safety protocols, which are essential to propulsion system design and optimization. Statistical analysis allows for the effective analysis of these data, setting the stage for advances in transportation technology. With the application of sophisticated statistical techniques, engineers can minimize risks associated with burning and consuming fuel, thus ensuring safety standards are maintained in the process of developing innovative transportation systems (Ludtke et al., 2025). Statistical techniques also aid in the measurement of efficiency and effectiveness of transportation systems, besides engineering performance and safety. The techniques enable transportation planners to study traffic flow, quantify congestion levels, and forecast future demand. By regression analysis and time-series forecasting, the planner is able to forecast trends and potential issues in advance so that pre-emptive measures can be implemented. Besides, statistical aggregation of several sources of information like traffic volumes and demographics enables us to have a comprehensive overview of what impacts transportation systems. Finally, the use of statistical techniques serves as the basis for evidence-based decision-making that ensures sustainable and efficient transport solutions that can be responsive to dynamic urban environments (Cottrill et al., 2010).

4.5.1. Traffic Management and Control

Successful traffic control and management are pivotal in optimizing transport systems, particularly in urban areas where congestion is the norm. Statistical and analytical methods make it possible for engineers to model and predict traffic flows so that they can make informed decisions. Application of methods such as traffic flow theory, demand forecasting, and simulation modeling facilitates identification of points of bottlenecks and peak usage times. By data collection from various sources, including sensors and cameras, researchers can have a real-time analysis of actual traffic conditions and learn from past experiences. This analysis is not only useful in reducing traffic congestion but also in enhancing safety features by detecting potential hotspots for accidents. Hence, the incorporation of these measures in traffic management systems is a regular necessity to adapt to evolving urban landscapes and varying transport needs. The incorporation of advanced technology into traffic management systems has revolutionized the manner in which cities manage their transport systems. Intelligent Transportation Systems (ITS) apply real-time information and advanced algorithms to dynamically monitor and manage traffic flows. Through the use of automatic signals and dynamic traffic management algorithms, ITS is able to re-tune signal timing based on existing conditions, effectively reducing waiting times and improving overall efficiency. Also, the analytics that these systems produce allow the creation of preemptive interventions to remedy traffic problems before they become enormous ones. For instance, the use of connected vehicle technology can provide driver behavior and traffic feedback, thereby leading to improved law enforcement as well as public campaigns educating people on how to reduce accident rates. Therefore, utilization of statistical analysis not only facilitates smoother working but also allows for safer roads for all. Furthermore, analytical comparisons of varying traffic management plans provide a forum for evaluating the effectiveness of solutions. Through measures such as average traffic speed, travel time reliability, and user satisfaction questionnaires, policymakers can review the impact of traffic interventions. This kind of analysis is essential in assigning the return on investment for specific projects and in facilitating rational transport planning at larger levels. As reiterated by current research, understanding the reliability and maintainability of such systems is of major importance in supporting public confidence in transport infrastructure (Bedford et al., 2005). In addition, the need for environmental perspectives within transport planning highlights the need for a systems approach to also consider socio-economic as well as environmental relationships (Balietti et al., 2011). Combining these analyses together helps stakeholders to come up with systematic traffic management that enhances the mobility of cities as a whole while fostering sustainable development (Figure 4.11).

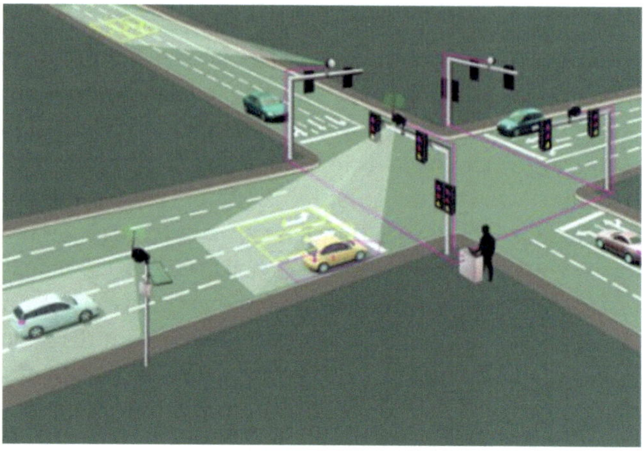

Figure 4.11. Traffic management and control systems.

Source: https://external-content.duckduckgo.com/iu/?u=https%3A% 2F%2 Fth-vnext.bing.com%2Fth%2Fid%2FOIP.xi9_DK6yRuFBwHBISGdqzgHaE8%3Fcb% 3Ducfimg2%26pid%3DApi%26ucfimg%3D1&f=1&ipt=d04e97b7b4ad589e1ec8abba27aedbb670350446ed1bc50199f789e3250a1167.

4.5.2. Infrastructure Planning and Design

One of the pillars of transportation engineering's planning and infrastructure design is efficiency and resiliency analysis for transit systems. With urban expansion continuing to grow, high-capacity infrastructure needs are more critical than ever. Statistical analysis and modeling can be mixed in the examination of critical infrastructures' performance, such as mass transit systems and bridges, across varied conditions. The research carried out in New Jersey endorses this approach, providing a comprehensive framework for the evaluation of resiliency in bridge infrastructure before and after disasters. Utilizing methods like structural health monitoring (SHM) and remote sensing, data are accessible and analyzable to inform improvements such that transit modes remain functional and attuned to environmental adversity (Bartin et al., 2017). This forward planning not only strengthens defense measures but also optimizes resource investment for long-term infrastructure management. Toll roads' positioning in determining regional traffic patterns is another applicable area of investigation under the topic of transportation infrastructure planning. By analysis of the effects of toll roads on traffic, engineers can infer insight into travel time and reliability, which are essential metrics for the understanding of system-wide performance. The Raleigh, North Carolina-based Triangle Expressway is a prime case study in which statistical measures were employed to determine variances in travel times over the years. It was reported that while the toll road became more consistent, parallel corridors were seriously affected in this region. These findings emphasize the need to utilize advanced travel time analysis when planning infrastructure to assist in decision-making regarding future projects and changes in operation to enhance the resilience and efficiency of the transport networks (Mathew et al., 2019). Infrastructure planning must also consider the long-term impacts of transport systems on urban development. Transit planning is complemented by the growth of a city and necessitates a high level of spatial and temporal understanding. It is essential to include both qualitative and quantitative elements to predict any problems due to urbanization. It ensures that new constructions not only meet current needs but are also flexible to grow for future development. Through the use of large data analysis through surveys and simulation, planners will be able to predict traffic flow and environmental impact more accurately. These analyses provide a framework for building sustainable transportation systems that foster economic activity with few adverse effects on urban environments. Successful infrastructure planning ultimately leads to improved quality of life and enables the long-term goals of sustainable urban development.

4.5.3. Environmental Impact Assessments

Environmental Impact Assessments (EIAs) are a crucial component of transportation engineering, providing a systematic approach to identifying the potential environmental impact of proposed projects. By combining statistical and analytical techniques, EIAs allow engineers to forecast potential impacts on ecological systems, air quality, and public health. These assessments use quantitative data and modeling approaches to predict the effects of construction activities on surrounding environments to ensure that transport projects are compatible with sustainability principles. Moreover, the evidence-based aspect of EIAs ensures informed decision-making through the identification of areas of concern before project implementation. Thereafter, EIAs go beyond ensuring statutory compliance to instill public confidence and community participation, with stakeholders being informed of potential risks and benefits of transportation developments. Assessment is an interdisciplinary process to evaluate in considerable detail the potential environmental impacts. The assessment process requires multidisciplinary teams to gather and analyze data relevant to proposed schemes, involving transportation engineers, urban planners, environmental scientists, and economists. Analytical tools such as Geographic Information Systems (GIS) and predictive modeling are key to this collaborative effort. To illustrate, they facilitate spatial analysis of natural resources, population dynamics, and transport flows, allowing for improved comprehension of the manner in which new infrastructure would affect existing conditions. Through the application of such techniques, transportation engineers are able to quantify probable effects, ensuring planned projects do less harm to natural landscapes and cities. The conclusions obtained through EIAs ultimately lead to a more sophisticated method of preventing adverse impacts, incorporating projects within the broader context of environmental protection and community development. Further, employing tools such as Life Cycle Sustainability Assessment (LCSA) in EIAs reinforces the integration of the broader consequences of transportation initiatives. LCSA offers an integrated approach by taking into account the economic, environmental, and social factors of road maintenance practices, particularly on road salt application during wintertime. The method takes into account not just the immediate impact of transport projects but also the significance of long-term sustainability practices. The interaction of these factors facilitates more intelligent decision-making through the inclusion of the life cycle of the materials and method employed in transport engineering. It draws attention to indirect transportation practice impacts, thereby enabling transportation agencies to balance competing demands while addressing public health and ecosystem integrity. Utilizing such combined analysis, the practice of transportation engineering can be established for coping with rising demands for sustainable development while maintaining efficient transportation systems (Figure 4.12).

Figure 4.12. Environmental impact assessments (EIAs).

Source: https://media.licdn.com/dms/image/v2/D4D12AQFt7takwso4qQ/article-cover_image-shrink_720_1280/article-cover_image-shrink_720_1280/0/1691141381860?e=2147483647&v=beta&t=I4HddnLkD4QZ5u5o3fsSwLYv1dewi4aGE5x50uWkqoY.

4.6. CHALLENGES IN STATISTICAL ANALYSIS

Statistical analysis is highly crucial in transportation engineering but has numerous challenges that may obscure the validity of findings. Among the biggest challenges is ensuring high-quality data, particularly for new projects such as high-speed rail systems. The California High-Speed Rail project, for example, recognizes major challenges including workforce development and education needs necessary for the execution of the project successfully (Estrada et al., 2012). If data are incomplete or drawn from unreliable sources, the statistical models built from such data can provide erroneous conclusions that undermine planners and policymakers' ability to make informed decisions about transport systems. It is one of the largest challenges that underscore the urgent need for devising effective data collection processes and training programs for those who are involved in transport analyses.

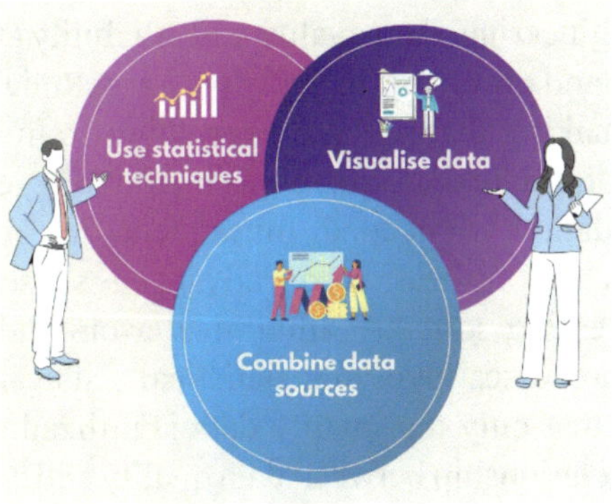

Figure 4.13. Challenges in statistical analysis.

Source: https://skillfine.com/wp-content/uploads/2023/03/4.jpg.

Integration of diverse data sources represents another major challenge in transportation statistical analysis. Transportation experts usually rely on crowdsourced information, which has gained more prevalence due to the advances in GPS technology. This kind of data, however, is prone to having missing observations that complicate statistical imputation (Henrickson et al., 2019). Integrating data from multiple sources requires sophisticated analysis techniques capable of accommodating spatial and temporal heterogeneities. With evolving analytical methods, robust statistical frameworks become necessary to ensure the quality of output. Enhancing the ability to properly integrate and analyze such diverse data becomes essential to create efficient transport solutions and understand traffic flow dynamics. Lastly, the interpretative stage of statistical analysis in transportation engineering poses its own problems. Even when data are collected and analyzed correctly, it becomes a difficult problem to interpret sophisticated statistical findings into actionable insights. Policymakers and city planners, especially, often lack the statistical expertise to comprehend in detail the intricacies of the models and their implications for transport systems (Estrada et al., 2012). Statistical findings may also be wrongly interpreted because of their complexity and lead to adverse effects on transport policies or project implementations. Sealing these communication loopholes through targeted training programs and clear-cut reporting procedures is vital as transportation projects increasingly rely on utilizing high-level statistical analysis. This multi-faceted effort will not only enable better interpretation of the data but also enhance the evidence-based foundation for transportation decision-making (Figure 4.13).

4.6.1. Data Quality and Reliability

For transportation engineering, data quality and reliability are crucial to ensuring safety, effectiveness, and efficiency in many projects. Reliable data are the basis for sound decision-making that touches everything from traffic simulation to infrastructure planning. Alike therefor, robust methodologies for data collection, processing, and validation are needed. Investments in robust data acquisition techniques, such as the utilization of advanced sensor technologies and automated data processing software, minimize errors and inaccuracies. Also, rigorous validation processes involving statistical tests can identify anomalies in collected data, so that only top-quality data is utilized to make engineering decisions. Thus, the relationship between data quality and engineering reliability cannot be overstated; poor data can lead to costly mistakes, reduced safety levels, and overall inefficiency within transport systems. To underscore the significance of data reliability in transport engineering, a formal data governance framework is necessary. This entails imposing stringent protocols that establish how data are acquired, stored, and shared among parties. Improved governance structures not only cultivate transparency and accountability but also ensure data integrity across the project life cycle. For example, the use of predictive analytics can shed light on prevailing patterns in high-quality data sets and, consequently, enable engineers to enhance resource utilization and identify possible infrastructural weaknesses. In addition, it is absolutely essential that all personnel handling data are trained to adhere to these standards as human mistake is still one of the main reasons behind data inconsistencies. Hence, a multi-pronged approach with both technological and human elements is needed to push the reliability and quality of data in transportation engineering to the next level. Moreover, the fluidity of transportation means there are regular evaluations of data quality, particularly with changing projects and externalities being reconfigured. The lessons of previous projects can be shared in future data management projects, promoting ongoing improvement. For example, safety case analysis may give insight into the best ways of maintaining data accuracy with time depending on various industry case studies (Bedford et al.,

2005). The pharmaceutical industry also provides insight into management of stocks and how data configurations influence process effectiveness (Andreu et al., 2017). Through comparisons with these industries, transportation engineers can replicate innovative methods for enhancing data quality and reliability so that systems are resilient to changing requirements. Efficient management of data therefore emerges as a critical determinant of long-term success in transportation engineering.

4.6.2. Complexity of Transportation Systems

Transportation systems complexity has increasingly been front and center in the topic of infrastructure and urban planning. The systems are not only facilities for the movement of individuals and goods but complex systems embedded with a host of socio-economic as well as environmental variables. A genuine ecological perspective is needed to comprehend the dynamic interplay of such complex systems, as stated. Effective transportation planning must consider the vast array of factors influencing traffic flow, environmental deterioration, and utilization of resources. To make this possible, researchers advocate the development of integrated design centers that will be able to simulate various scenarios, establish chains of causality, and identify adverse impacts of policy decisions prior to their occurrence. These forecasting abilities form a new paradigm within transport engineering where statistical and analytical methods play a pivotal role in identifying resilient and effective systems, hence meeting the demands of our contemporary society without causing any unforeseen effects (Balietti et al., 2011). The nonlinear dynamics that typify transportation systems also render their analysis and management even more complex. As opposed to the over-simplification inclinations of traditional linear models, complex systems theory would suggest that network dynamics are more inherently complex. For instance, traffic congestion is caused by a complex array of factors from weather, road surfaces, to human activity. By default, traditional approaches—linear datasets-based to predict traffic flows—cannot explain the emergent outcomes in these cases. As outlined in emerging literature, ever more there is recognition of the need for a paradigm shift to complexity frameworks to understand such dynamics. Such a transition entails the integration of heterogeneous causes that can render forecasting models stronger and enhance strategic decision-making in the transport sector (Baggio et al., 2007). Deep implications of adopting a complex systems approach exist for transport policymakers and engineers. By embracing statistical and analytical methods that acknowledge the sophistication of such systems, players can implement better solutions to problems of urban sprawl, climate change, and resource depletion. One such example is how transportation infrastructure and economic systems are interdependent, necessitating an integrated model towards the development of cities. This, as has been observed through various attempts at testing this interdependency, now becomes critical in integrating model initiatives with decision spaces. Decision-makers with data-rich information from several linked systems are better placed to craft policies that ensure sustainability while optimizing transport networks. Hence, the complexity of transport systems underscores the importance of innovation and adaptability in the field, affirming the use of advanced statistical and analytical methods in the resolution of tomorrow's and today's transportation problems (Balietti et al., 2011) (Figure 4.14).

Figure 4.14. Complexity of transportation systems.

Source: https://media.istockphoto.com/id/1524680694/vector/transportation-vehicles-icons.jpg?s=612x612&w=0&k=20&c=ifd_x_qJM5jiqp-6KeU1sG66WQKqrJnZ1a47YDpbD5g=.

4.6.3. Limitations of Current Analytical Tools

The analytical techniques employed in transport engineering at present exhibit significant flaws that do not allow them to be highly effective in addressing multidimensional engineering problems. One of the most significant concerns is related to the employment of historical information, which may not reflect future or present reality. Relying so much on past data could lead to ill-informed projections and suboptimal decision-making. Besides, while several organizations reduce essential information regarding fuels and oxidizers, a perceived lack of balance in data quality and coverage of available data can be witnessed. As witnessed while different organizations attempt to summarize, the data that they have listed is complicated, yet the set of these data is typically ignored in practical uses and consequently hinders detailed understanding required for effective planning of transportation as well as policy formulation (Ludtke et al., 2025). Apart from data quality considerations, current tools for analysis rarely succeed in embedding artificial intelligence fully into them. Although AI holds the promise of enhancing transportation analysis through improved data processing, incorporation has, in most cases, been incomplete. Experienced judgment remains an integral component of analysis as human experts provide invaluable insights that are not yet replicable by AI technology. Past AI development failures, particularly within the UK, have led to a reluctance to adopt such technologies. While unabated keen research elsewhere, transport systems' slow adoption of AI led to lost ground for innovation and heightened analytical capability (Bonsall et al., 1986). Hence, there continues to be more reliance on traditional methods that prevent the industry from being capable of taking advantage of technological advances in favor of greater analysis power. Finally, the shortcomings of current analysis tools are also supplemented by the inherent nature of the complexities of transportation systems themselves. As these systems involve numerous interdependent variables and extraneous variables, reductionist models fall short in proper analysis. Current tools make use of linear models to analyze inherently nonlinear issues, which simplify the interactions involved and yield false conclusions. This is especially important when transportation infrastructure investment is planned, which must consider multifaceted factors such as social, economic, and environmental issues. If newer, more sophisticated analytical approaches offering such degrees of complexity are not adopted, transportation engineering solutions become likely to be ineffective and short-sighted. Thus, the science must evolve to encompass more sophisticated analytical methods that not only bypass the limitations of existing tools but also better capture the dynamics of transportation systems.

SUMMARY

The culmination of everything explored in statistical and analytical methods applied in transportation engineering reveals significant achievements and challenges that have characterized the practice. The application of artificial intelligence in data analysis steps is a breakthrough step towards reducing the traditional requirement for human judgment by itself. Experts previously performed crucial analysis of transport data, but emerging technologies are making it easier for computers to learn from it, increasing the precision of analysis and decision-making. This change is important in addressing not only routine traffic issues but also complex scenarios such as congestion during high-turnout events. Bonsall et al. (1986) illustrates the importance of adapting to technological advancement, emphasizing the need for ongoing research and practice in the transport sector so as to be effective and competent in interpreting data. Case study research, particularly in the field of traffic control at mass events, showcases real-world usage of statistical techniques. Various research studies, including those utilizing emerging technologies such as Dynamic Bayesian Networks, illustrate how statistical techniques can contribute to improved traffic pattern and hotspot knowledge. For instance, research of travel behavior surrounding games of the Nebraska Cornhuskers provides insightful information on how some events exacerbate traffic such that intentional interventions can be directed towards these. Through pattern detection and parameters that affect traffic movement, professionals in the transportation engineering discipline particularly acquire their system of addressing infrastructure planning and real-time control. Ahsani et al. (2019) has a good argument regarding the necessity of applying rigorous analytical methods, with their role in making transportation infrastructure more effective and reliable. Ultimately, the intersection of statistical methods with advanced technological platforms puts transportation engineering on the verge of unrivaled potential. With advancing artificial intelligence, the promise of optimizing transportation systems rises exponentially. This not only enhances everyday activities but also prepares infrastructures to deal with such unique challenges, such as those brought about by mega events. The future of transportation engineering hinges on integrating such advanced analytical techniques into practice, so it is essential that educators, practitioners, and researchers provide their significance and priority in development and adoption. The meeting of human understanding and computational analysis is set to redefine transportation planning, transforming the manner in which experts operate and innovate within this critical field. The findings obtained through application of these processes will definitely make transportation systems safer, more efficient, and responsive.

Statistical and analytical methodologies hold a pivotal place in the discipline of transportation engineering in informing and maximizing traffic movement and traffic system operations. The secret to this analysis is dealing with heterogeneous data, from basic traffic volumes to sophisticated algorithms that try to predict traffic movement under various scenarios. Such methods facilitate engineers to effectively assess the impact of external factors, including weather conditions and urban sprawl, on transportation systems. Moreover, such analytical models provide information facilitating policymaking, thus ensuring effective utilization of resources in managing congestion and safety in a comprehensive manner. Integration of evidence-based approaches not only enhances the effectiveness of operations but also contributes to the formulation of strategic plans that assist in future forecasting of challenges within transport networks, resulting in a better, more reliable, and easier transport system (Hutchinson et al., 2013). Statistic methods are one of the primary applications, where traffic patterns during events are predicted, which create tremendous disturbance to normal traffic movement. A fine example is traffic management on highways associated with massive sporting events, where unexpected jams can happen since thousands of spectators are channeled into event grounds. For a five-year study of Nebraska Cornhuskers football games' contribution to traffic, INRIX data were utilized to compare game-day travel behavior with normal days' travel behavior. Such a comprehensive study revealed major hotspots—specific areas with unusually high traffic congestion during events. Such data is also crucial to the formulation of applicable traffic management policies, thereby enabling municipalities to implement customized interventions aimed at alleviating congestion as well as commuter safety (Ahsani et al., 2019). Furthermore, new methods such as Multi-EigenSpot and Dynamic Bayesian Networks (DBN) have revolutionized the study of traffic dynamics. The Multi-EigenSpot algorithm effortlessly identifies and analyzes several locations of traffic congestion, revealing spatiotemporal data patterns required for optimized traffic flow. Conversely, the DBN approach also performed impressively in terms of prediction ability for forecasting potential congestion conditions due to event timing and opposition popularity, with a phenomenal weighted mean absolute percentage error (WMAPE) of just 13.8%. These new statistical approaches not only provide instant traffic knowledge but also allow transport engineers to design adaptive management systems that are demand sensitive. Collectively, these advances in methodology emphasize the importance of employing statistical methods to augment transport engineering processes with a perspective to making transport systems more effective and reliable in various circumstances (Hutchinson et al., 2013).

REFERENCES

1. Ahsani, V., & Sharma, A. (2019). Assessing the impact of game day schedule and opponents on travel patterns and route choice using big data analytics. https://core.ac.uk/download/286729950.pdf (accessed on 21st October 2025).
2. Anthoine, S., Chen, B., Chen, K., Fitch, D., et al. (2020). Research synthesis for the California Zero Traffic Fatalities Task Force. https://core.ac.uk/download/323082492.pdf (accessed on 21st October 2025).
3. Baggio, R. (2007). Symptoms of complexity in a tourism system. https://core.ac.uk/download/44162476.pdf (accessed on 21st October 2025).
4. Balietti, S., & Helbing, D. (2011). From social simulation to integrative system design. http://arxiv.org/abs/1011.3970 (accessed on 21st October 2025).
5. Bartin, B., Deka, D., El-Awar, A., Lou, Y., et al. (2017). Performance measures to assess resiliency and efficiency of transit systems. https://core.ac.uk/download/129532928.pdf (accessed on 21st October 2025).
6. Bedford, T. J., Revie, M., & Walls, L. A. (2005). The safety case and the lessons learned for the reliability and maintainability case. https://core.ac.uk/download/9021796.pdf (accessed on 21st October 2025).
7. Bhatt, M., & Wallgruen, J. O. (2013). Geospatial narratives and their spatio-temporal dynamics: Commonsense reasoning for high-level analyses in geographic information systems. http://arxiv.org/abs/1307.2541 (accessed on 21st October 2025).
8. Bonsall, P. W., Kirby, H. R., & Kwan, R. S. (1986). Applications of expert systems in transport. https://core.ac.uk/download/53713.pdf (accessed on 21st October 2025).
9. Brudka, C., Bąkowska, E., Czepkiewicz, M., Jankowski, P., et al. (2016). Public participation GIS for sustainable urban mobility planning: Methods, applications and challenges. https://core.ac.uk/download/154444688.pdf (accessed on 21st October 2025).
10. Chang, S. K. J., De Backer, H., De Maeyer, P., Lauwers, D., et al. (2019). Integrating spatial and temporal approaches for explaining bicycle crashes in high-risk areas in Antwerp (Belgium). https://core.ac.uk/download/225020216.pdf (accessed on 21st October 2025).
11. Chatterjee, A., Manohar, M., & Ramadurai, G. (2016). Statistical analysis of bus networks in India. http://arxiv.org/abs/1509.04554 (accessed on 21st October 2025).
12. Clark, S. D., & McKimm, J. (2005). Estimating confidence intervals for

transport mode share. https://core.ac.uk/download/60261.pdf (accessed on 21st October 2025).

13. Cools, M., Farooq, B., Saadi, I., Teller, J., et al. (2017). An investigation into machine learning approaches for forecasting spatio-temporal demand in ride-hailing service. http://arxiv.org/abs/1703.02433 (accessed on 21st October 2025).

14. Cottrill, C., Thakuriah, P., Thomas, N., Vaughn, S., et al. (2010). A sketch planning methodology for determining interventions for bicycle and pedestrian crashes: An ecological approach. https://core.ac.uk/download/9652228.pdf (accessed on 21st October 2025).

15. Cui, N., Shi, X., & Xie, N. (2016). A framework for life cycle sustainability assessment of road salt used in winter maintenance operations. https://core.ac.uk/download/286999905.pdf (accessed on 21st October 2025).

16. Deptula, D. A., Ransone, R. K., & Yorke, G. G. (2025). Integrated mass transportation system study/definition/implementation program definition. https://core.ac.uk/download/pdf/42885483.pdf (accessed on 21st October 2025).

17. Estrada, K., Haas, P. J., & Hernandez, P. (2012). Estimating workforce development needs for high-speed rail in California, Research Report 11-16. https://core.ac.uk/download/70410031.pdf (accessed on 21st October 2025).

18. Gayah, V. V., Wood, J. S., & Yu, Z. (2016). Estimating uncertainty of bus arrival times and passenger occupancies. https://core.ac.uk/download/70426385.pdf (accessed on 21st October 2025).

19. Henrickson, K., Pereira, F. C., & Rodrigues, F. (2019). Multi-output Gaussian processes for crowdsourced traffic data imputation. http://arxiv.org/abs/1812.08739 (accessed on 21st October 2025).

20. Hutchinson, J., Kwiatkowski, J., & Metzger, A. T. (2013). Characterizing the load environment of ferry landings for Washington State Ferries and the Alaska Marine Highway System. https://core.ac.uk/download/286607575.pdf (accessed on 21st October 2025).

21. Jacobson, I. D., & Kuhlthau, A. R. (2025). The development of a model for predicting passenger acceptance of short-haul air transportation systems. https://core.ac.uk/download/pdf/42877258.pdf (accessed on 21st October 2025).

22. Kuhlthau, A. R., & Vermuri, R. S. (2025). Survey of air cargo forecasting techniques. https://core.ac.uk/download/pdf/42873889.pdf (accessed on 21st October 2025).

23. Ludtke, P. R. (2025). Register of specialized sources for information on selected fuels and oxidizers. https://core.ac.uk/download/pdf/42885811.pdf (accessed on 21st October 2025).

24. Magdalena, R. (2017). Financial performance analysis of location, rental rate and parking revenue on shopping centers. https://core.ac.uk/download/201756360.pdf (accessed on 21st October 2025).

25. Masvidal Andreu, C. (2017). Inventory drivers in a pharmaceutical supply chain. https://core.ac.uk/download/157811883.pdf (accessed on 21st October 2025).

26. Mathew, S., & Sulugurtha, S. S. (2019). How effective are toll roads in improving operational performance? https://core.ac.uk/download/289097252.pdf (accessed on 21st October 2025).

27. Mazzella, A., Pinna, F., & Piras, C. (2011). Use of kriging technique to study roundabout performance. https://core.ac.uk/download/54583426.pdf (accessed on 21st October 2025).

28. McGee, G. (2018). Modeling commodity flow as a statistical function of lock unavailability and usage. https://core.ac.uk/download/158200240.pdf (accessed on 21st October 2025).

29. N/A. (1990). Graduate Catalog, 1990–1991. https://core.ac.uk/download/144553218.pdf (accessed on 21st October 2025).

30. N/A. (2017). SciTech News Volume 71, No. 1 (2017). https://core.ac.uk/download/83049835.pdf (accessed on 21st October 2025).

31. Park, T., & Pyeon, J. H. (2010). Improving transportation construction project performance: Development of a model to support the decision-making process for incentive/disincentive construction projects, MTI Report 09-07. https://core.ac.uk/download/70409929.pdf (accessed on 21st October 2025).

32. Ślusarczyk, B., & Kot, S. (2025). Process simulation in supply chain using LOGWARE software. https://core.ac.uk/download/pdf/6481403.pdf (accessed on 21st October 2025).

CHAPTER 5

MACHINE LEARNING APPLICATIONS IN TRANSPORTATION

Contents

INTRODUCTION

The application of machine learning to the transport sector marks a significant shift toward increased efficiency and safety. With road mobility increasingly under threat from jams and environmental damage, machine learning offers innovative solutions to these vital problems. Through sophisticated algorithms, the transport sector can analyze considerable amounts of data gathered from various sources, such as traffic flows and car telemetry. This provides room for the development of forecasting models that continue to enhance decision-making and operational efficiencies, paving the way for intelligent transportation systems. This transition not only reduces the operational cost but also toward improved delivery of services, thus fundamentally affecting user experience in cities. Moreover, machine learning in application transcends various modes of transport, ranging from public transport to ride-sharing, as well as traffic management systems. For instance, bike-sharing programs benefit through machine learning by predictive modeling that analyzes user behavior and optimizes supply of service.

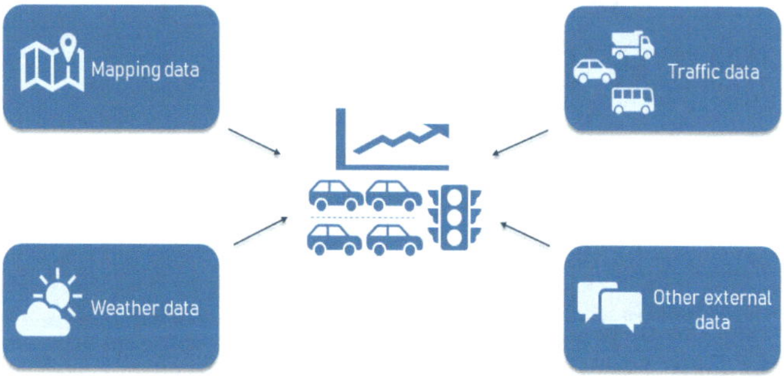

Figure 5.1. Machine learning applications in transportation.

Source: https://www.altexsoft.com/media/2022/01/word-image-17.png.

There has been research suggesting the capability of machine learning technologies to incorporate predictive elements for enabling better access to system real-time status for end-users, hence improving overall city mobility (Bacciu et al., 2015). Such types of strategies allow operators to anticipate variation in demand, optimize resource allocation, and ultimately increase end-user satisfaction. This shift towards data-driven planning is critical to cities hoping to balance growing populations with lower environmental footprints. In addition to user-focused applications, machine learning is transforming the incident detection and traffic management paradigm.

Traditional methods rely generally on stationary sensors and human feedback, leading to delinquent reaction times to traffic incidents. With social media data through high-level natural language processing techniques, machine learning can be applied to extract relevant traffic information autonomously, thereby improving incident response time (Ghazzai et al., 2020). In addition to enhancing real-time reporting of traffic conditions, this new approach helps navigation systems provide timely warning to users. As such technologies evolve, the transportation sector stands at the threshold of transformation characterized by enhanced security, productivity, as well as user engagement, ultimately leading to wiser cities (Figure 5.1).

Definition of Machine Learning

Machine learning (ML), the key subfield of artificial intelligence (AI), is broadly speaking all about designing algorithms capable of enabling computers to learn from experience. Learning typically involves building mathematical models capable of detecting patterns or predicting things from input data. In transportation, applications of ML techniques have been transformative, influencing everything from traffic control systems to autonomous vehicle routing. By harnessing large volumes of data, ML models can analyze routes for optimization, anticipate maintenance requirements for vehicles, and improve overall safety. With the transportation industry becoming more data-driven, the meaning and extent of machine learning will only grow, highlighting its far-reaching implications in terms of operational efficiency and user experience within this sector. Besides, the application of machine learning in transport systems not only enhances performance but also assists in addressing issues like congestion and safety that are relevant. For instance, advanced ML algorithms can make congestion forecast predictions based on real-time traffic information, thereby enabling adaptive traffic signals and best routing for drivers. Furthermore, in autonomous driving, ML is used to process sensor data and enable navigation safely through complex environments.

Application of ML technologies, particularly coupled with high-definition mapping, has already begun to revolutionize navigation systems by permitting the vehicle to adapt dynamically to road conditions. These innovations are the general potential of machine learning to revamp traditional transport models into smarter, more efficient systems that prioritize safety and efficiency. Central to machine learning is the paradigm approach that enables its applications. Machine learning encompasses several learning paradigms, including supervised, unsupervised, and reinforcement learning, each with specific roles based on the kind of data and desired outcomes. In transportation, these paradigms facilitate diverse applications such as predictive maintenance, traffic flow modeling, and modeling of driver and pedestrian behavior. For example, reinforcement learning has been particularly promising in optimizing vehicular communication protocols, wherein control algorithms learn to optimize performance over time adaptively. As highlighted in recent research, such developments greatly enhance vehicular network performance by lowering latency and raising throughput, a significant consideration of achieving seamless communication for transport systems (Redondo et al., 2024). Overall, the ongoing development of machine learning increasingly redefines its applications in transport, paving the way for developments that ensure a safer and more efficient future (Bhagwan et al., 2024) (Figure 5.2).

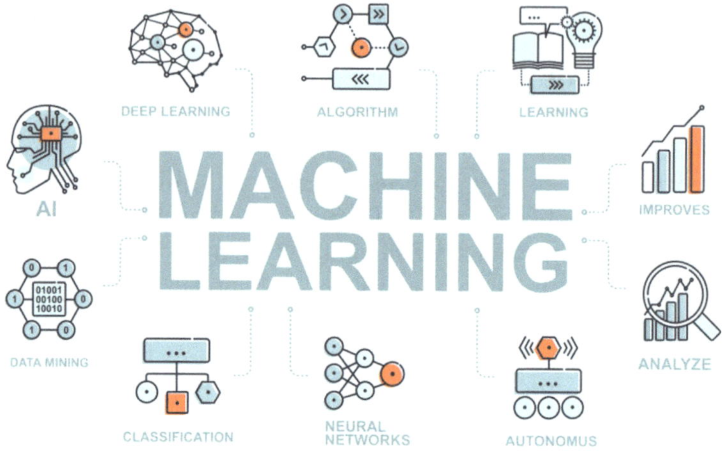

Figure 5.2. Machine learning (ML).

Source: https://www.fsm.ac.in/blog/wp-content/uploads/2022/08/ml-e1610553826718.jpg.

Importance of Transportation in Modern Society

The contribution of transport to modern society cannot be overstated as it does not only enable the movement of people but also facilitates efficient movement of services and commodities. Transport infrastructure is vital to the growth of the economy as it ensures access to markets, reduced costs of operation, and heightened productivity. Second, enhanced transportation technologies are assisting in developing smart cities, where machine learning and big data integration play a vital role. For instance, predictive analytics in traffic control facilitates dynamic routing and optimized public transportation timetables, which ultimately leads to reduced congestion and smaller environmental footprints. As explicated in Doctor et al. (2020), where computational intelligence and big data intersect, there are possibilities for intelligent transportation systems, which are revolutionizing urban mobility and contributing to a healthy economy. Transport, beyond its economic contributions, has profound social and environmental significance. Functioning transportation promotes accessibility as well as mobility, making it possible for individuals to access education, healthcare, and jobs. This enhances overall quality of life as well as encouraging social equality, particularly for disenfranchised groups who may otherwise be excluded from access to secure transport modalities. Furthermore, machine learning-based solutions in transport can also address pressing environmental concerns. By directing commercial transport vehicles to most efficient routes and employing smart traffic control systems, greenhouse gas emissions can be significantly lessened. Abtahi et al. (2018) highlights the way in which the implementation of computational intelligence techniques will facilitate data-driven decision-making towards more sustainable conduct, thereby achieving both societal imperatives and ecological stewardship. The use of machine learning in transport not only promotes innovative solutions but also ethical and legal concerns that have to be taken into account. With transport increasingly dependent on artificial intelligence, privacy, security, and accountability concerns become more significant. In addition, the ability to predict driver intention, as outlined in Abtahi et al. (2018), challenges the impact of automated decision-making on social norms and individual freedom. Autonomous vehicle deployment calls for robust regulatory systems to deliver public security and assurance. While society progresses further towards more automated forms of transport, it is imperative that we continue questioning these ethical sides, ensuring the potential of technology aligns with public benefit. How this balance is achieved will determine the way towards the future of transport in modern society, impacting not only systems on which we rely but also cultural values we enshrine.

Overview of Machine Learning's Role in Transportation

The application of machine learning (ML) technology in the transport sector is a transformative step that enhances efficiency, safety, and sustainability in operations. As cities face increasing pressure for effective mobility solutions, ML models are applied to optimize traffic management, predict mobility patterns, and enhance operations. For instance, predictive analytics can search for real-time traffic and chart congestion patterns and suggest alternative routes, which reduces travel time and emissions. Furthermore, the ability of ML to handle large volumes of data also enables modeling complex transport networks, enabling urban planners to have a visual of potential outcomes and implementing decisions that are data-driven. Such technologies not only improve day-to-day commuting but also enable long-term sustainability goals by reducing fossil fuel reliance, which is a mandatory requirement in transport systems innovation. In addition, the integration of machine learning in self-driving vehicles is another landmark in transportation technology history. Self-driving vehicles utilize deep learning to scan sensor data, reading their surroundings in order to drive themselves. Self-driving vehicles learn continuously from real-world environmental information and enhance their performance through experience. The ability of ML to facilitate advanced driver-assistance systems (ADAS) is similarly central to the avoidance of accidents and road safety improvement. Through detection of potential hazards and adaptation to diverse driving conditions, such systems help prevent human mistakes that are responsible for a significant percentage of traffic accidents. With the continuous evolution of the technology, it is imperative to manage regulatory and ethical issues surrounding autonomous driving to offer safety as well as tap the potential of the innovation. Apart from improving the utilization of personal cars and autonomous vehicles, machine learning has a key position in the operation of public transit systems. Sophisticated ML models are able to forecast ridership patterns as well as modify routes and schedules based on demand forecasting, thereby improving service efficiency. This is more emphasis on the intelligent management of resources, which will enhance the overall travel experience of users. With the integration of ML into transport networks, there will be an ability to implement dynamic pricing regimes and load balancing techniques, adapting to demand fluctuations in real time. Besides, with cities going for sustainability, ML has the potential to facilitate the development of low-carbon modes of transport, adapting to the environmental concerns related to the use of private cars. Through harnessing of findings from ML, transport authorities are able to create systems that respond to user requirements and wider societal aspirations, thereby opening the way for more intelligent and environmentally friendly urban spaces (Figure 5.3).

Figure 5.3. Overview of machine learning's role in transportation systems.

Source: https://stl.tech/wp-content/uploads/2022/12/machine-learning.jpg.

5.1. AUTONOMOUS VEHICLES

The development of driverless cars marks a significant breakthrough in the development of transportation technology. Machine learning is at the heart of this innovation, with the ability to interpret reams of sensory data collected from the surrounding environment. Equipped with technologies such as radar, lidar, and computer vision, autonomous cars are capable of navigating complex environments without any human intervention. The support for processing gray-scale images in detecting roads is an indication of this convergence of machine learning and real-world solutions.

The development of a new annotated dataset for urban road scenes has been suggested, which enhances training a machine learning model to efficiently identify road patterns, roadblocks, and traffic lights (Alkhorshid et al., 2016). This capability not only contributes to security but also makes the efficiency of urban transport networks better, illustrating the revolutionary potential of driverless cars using machine learning. Moreover, the automation of transport networks offers long-term benefits across many areas by saving money and increasing operating efficiency. The transition from human-driven vehicles to driverless vehicles can curtail human error immensely, which is a key source of accidents on the roads. As identified in the broader context of automation, control levels of operations can vary pretty widely depending upon the need and complexity of the environment in which these systems must operate (Kopardekar et al., 2025). By a focus on improving machine learning algorithms that inform decision-making for driverless cars, the industry aims to achieve higher levels of automation. This change facilitates real-time responses to varying traffic scenarios, eventually leading to not only individual safety but also total transport system efficiency. Machine learning applications also have implications that extend beyond the vehicles themselves to entire transportation systems. Connected vehicles can move intelligent transportation systems (ITS) forward, which facilitate improved traffic flow management and decreased congestion in urban centers.

The interconnectivity of autonomous cars enables network features that enable them to communicate with each other and infrastructure elements and therefore create a synchronized transportation system. With research continually advancing the processes of automating advanced systems, advances in autonomous vehicle technologies serve as an example for other industries, particularly high-risk contexts such as aviation where speed in decision-making is critical (Kopardekar et al., 2025). Thus, the synergistic relationship between autonomous vehicles and machine learning not only transforms personal transport but also lays the groundwork for more intelligent, responsive cityscapes.

5.1.1. Overview of Self-Driving Technology

The integration of autonomous technology is a revolutionary idea within the transport sector, initiated through the convergence of artificial intelligence and machine learning technologies. As the focal point of the technology is the ability to process vast amounts of data that support core functionalities like perception, decision, and navigation. These systems must read environmental information in real-time from a variety of sensors, including LiDAR and cameras, in an effort to create sophisticated maps and make real-time driving decisions. Progress in this field has promised to make transport networks safer and more efficient. However, there remain the issues, particularly with the capacity of the systems to operate under complicated conditions, such as in the event of uncertain human behavior on the road, hence the need for continued research and development in enhancing the resilience of automated driving systems. Automated driving systems (ADSs) have introduced a new paradigm for automobile efficiency and safety, but autonomous vehicle crashes have raised inherent issues pertaining to their reliability.

Data of rising fatalities that are associated with vehicles fitted with ADS confirm the necessity to transcend technical inadequacies. Profound insight into problems of current systems, including localization, mapping, and human-machine interfaces, is paramount in developing autonomous technologies that are safer. The resolution to such issues demands a multi-perspective solution that marries sophisticated algorithms and high-order system designs. With experts exploring such technological problems, they are also working towards embracing new approaches that hold the potential to close the divide between current practices and realizing end-to-end autonomous vehicles capable of navigating complex terrain easily. The autonomous technology's revolution is not merely about straightforward automation; it can redefine urban space and lifestyles. As autonomous cars become mainstream, they would inevitably influence industries like logistics, ride-sharing, and public transport systems. Also, with the incorporation of machine learning in such technologies, they not only get efficient but also bring forth regulatory and ethical concerns that must be addressed. They include ensuring the security of private data, fostering trust in the public domain, and addressing advanced complexities of liability in the event of an accident. As the distinction between human-driven and autonomous driving experiences disappears, it becomes ever more critical to establish parameters governing the embedding of this technology into current infrastructures of transportation. These parameters will be instrumental in insuring that the transportation revolution brought about by self-driving technology increases overall societal well-being and not add to risks or challenges (Figure 5.4).

Figure 5.4. Overview of self-driving technology.

Source: https://www.cprime.com/wp-content/uploads/2022/12/testing-self-driving-cars-5-best-practices.jpg.

5.1.2. Machine Learning Algorithms Used in Navigation

In navigation, machine learning algorithms have emerged as prime movers in enhancing the accuracy and efficiency of transport networks. These algorithms leverage humongous amounts of data from various sources, including GPS signals, vehicle sensors, and even social media comments, to facilitate real-time route optimization. For instance, supervised learning approaches, including decision trees and regression models, have a central role in predicting traffic flow through historical data, making it possible for navigation systems to direct users on best routes. These algorithms not only conserve time but also lower fuel consumption and exhaust emissions, thus ensuring urban transportation sustainability. By incorporating predictive analytics into navigation systems, the transport industry can take advantage of machine learning benefits and drive innovation that makes it easier to travel. In addition, application of deep learning approaches, particularly neural networks, has revolutionized navigation systems by making them more competent to process unstructured information. Such advanced models can detect subtle patterns in large data to pick up traffic patterns, road hazards, and even dynamic weather patterns. For example, convolutional neural networks (CNNs) are increasingly being utilized to scan images from cameras fitted on cars to detect road hazards and markings. The capability of real-time analysis not only aids in autonomous vehicle travel but also inputs into traditional navigation systems that require rapid alteration of route suggestions. Secondly, as city expansion increases, the incorporation of these deep learning algorithms is necessary in developing smart transportation systems to seamlessly evolve within different patterns of traffic, which aids in safety and efficiency. The union of machine learning and natural language processing (NLP) is also bringing about sophisticated navigation solutions. NLP-based automatic traffic alert systems can sift through vast amounts of real-time social media updates and other data to detect traffic-related occurrences and warn users in real time. For instance, applying models such as Bidirectional Encoder Representations from Transformers (BERT) allows for the projection of relevant traffic information, including accident reports or traffic congestion warnings, into navigation applications. Not only does this increase the user experience but also underscores the potential of social media as a useful tool for managing traffic. With such machine learning solutions continuing to advance, they also seek to enhance navigation systems further, evidence of the significant impact technological advancements have on transport (Kopardekar et al., 2025; Ghazzai et al., 2020).

5.1.3. Safety and Regulatory Challenges

The introduction of machine learning technology in transport systems has the potential to revolutionize safety features but also brings huge regulatory challenges. With advancements in self-driving technologies, such as self-driving cars, the regulatory environment is lagging behind addressing the rapid innovation in this field. These vehicles operate on a paradigm that demands rigorous scrutiny of engineering standards and ethical concerns that are crucial for public security. Extensive policies that not only address technical specifications but also social implications are of top importance. As suggested in Dodig-Crnkovic et al. (2018), there is a need for a global applied engineering ethical framework in order to develop autonomous technology. The framework should evaluate practical problems, balancing innovation and user trust and safety in order to guarantee that autonomous systems do not jeopardize public confidence or introduce unforeseen danger into the transportation system. Moreover, regulatory frameworks must cater to the complexities of big data application in transport. The levels upon which machine learning models are drawn from vast databases raise issues regarding data privacy, security, and ethical utilization. As Doctor et al. (2020) depicts, the intersection between big data and intelligent transport systems identifies the necessity for policies that govern the gathering, protecting, and commodification of data. Not only is this a privacy concern for the person, but without proper management, it can lead to biased algorithms that drive decisions within autonomous systems and potentially injure lives. Ensuring machine learning models are trained on representative and ethically sourced data is important to deploying them safely. Consequently, regulatory authorities must enhance current frameworks in a way that incorporates guidelines that address these big data challenges head-on, ensuring accountability and transparency for the sector. Further, the ethical considerations surrounding machine learning application in transportation often converge into severe dilemmas, such as the trolley problem. As Dodig-Crnkovic et al. (2018) argues, such dilemmas reveal the complexity entailed in coding decision-making algorithms for autonomous vehicles. Within the framework of this development there lies a dilemma: the necessity to innovate technological leaps with adherence to ethical standards. The fact that these vehicles must react in real-time according to risky human action necessitates a high level of understanding regarding moral decision making within algorithmic frameworks. Any regulatory body would do well to establish parameters to push forward innovative solutions that incorporate ethical considerations directly into machine learning development in transportation. As supportive regulation emerges, the focus will be on cooperative collaborations among technologists, ethicists, and lawmakers to address the safety concerns posed by such dynamic technologies.

5.2. TRAFFIC MANAGEMENT SYSTEMS

Machine Learning (ML) integration of Traffic Management Systems (TMS) is revolutionizing the efficiency and effectiveness of city roads transport networks. Algorithms developed using ML are able to scan significant amounts of data gathered from various sources, such as traffic cameras, sensors, and GPS, and, in the process, facilitate real-time traffic monitoring along with predictive analysis. This allows TMS to respond dynamically to dynamic patterns of traffic flow, mitigating congestion as well as public safety. Further, through the use of sophisticated data modeling techniques, such as Hierarchical Spatial-Temporal State Machine (HSTSM), traffic managers can identify patterns and make forecasts regarding traffic movement and significantly enhance overall transport efficiency in the context of smart cities. With increasing complexity in cities, the role of ML in TMS is essential to addressing modern issues of urban mobility, engineering more responsive and flexible transportation systems that respond to commuters' needs and promote sustainable city planning strategies (Doctor et al., 2020). Further, the application of ML in TMS gets its extension in wireless technology, which constitutes the vital backbone for coordinating connected infrastructure of smart cities. Following the onset of heterogeneous networks, the demand for adaptive learning in traffic management has been increasing. These networks require effective communication technologies that can support seamless information transfer between segments like vehicles, traffic lights, and pedestrian systems. The use of ML supports accurate parameter estimation and decision-making processes that adapt according to real-time and user demand. Through use of these intelligent systems, traffic light times can be optimized, routes for public transit optimized, and the general user experience on transportation networks enhanced, overall resulting in reduced travel times and reduced emissions (Chen et al., 2019). Beyond reducing traffic flow, ML solutions within Traffic Management Systems are responsible for enhancing safety and emergency response operations. Through analysis of historical traffic accident and vehicle movement patterns, ML models are able to provide insights on dangerous areas and predict potential incidents. Therefore, TMS can better deploy resources, making response to accidents and natural disasters quicker. Predictive analytics can also guide engineering countermeasures, such as traffic calming devices or improved signs, to avoid accidents from occurring. The coming together of ML, urban traffic management, and emergency preparedness is a great step towards creating safer road conditions. As urban centers expand, the use of such technologies will be unavoidable in meeting the complexity of modern transport systems while ensuring safety and health for their inhabitants (Doctor et al., 2020) (Figure 5.5).

Figure 5.5. Traffic management systems.

Source: https://zh.digi.com/getattachment/Blog/post/Introduction-to-Smart-Transportation-Benefits/GettyImages-1207768535-1280x720.jpg?lang=en-US.

5.2.1. Real-Time Traffic Prediction Models

As urbanization accelerates, efficient transportation management has never been more vital. Real-time traffic prediction models provide the vital solution through the use of advanced machine learning techniques to forecast traffic flow and congestion. Through large datasets, such as historical traffic patterns, weather patterns, and special events, these models provide proactive decision-making for traffic control and infrastructure planning. Real-time use of data facilitates dynamic traffic signal control and routing planning. This increases the mobility of cities and reduces congestion, leading to the overall efficiency of transport systems. With increased growth in traffic volumes, the application of real-time forecasts will be vital in constructing smart cities with the capability of effectively managing the prevailing challenges of transport. One of the major breakthroughs in traffic prediction is the development of the Long-term Traffic Prediction Network (LTPN). The special-purpose model employs a combination of convolutional and recurrent neural network layers to predict accurate real-time traffic. Research indicates that the LTPN performs better than the conventional predictive models under all instances, with reduced error rates on both short-term and long-term traffic predictions (Ha et al., 2024). Its ability to perform well despite changing traffic conditions is proof of the potential of machine learning in real-world implementation. By being able to accurately forecast, LTPN helps the goals of intelligent transportation systems (ITS) be more efficient and responsive. With cities continuing to grapple with increasing urban density and traffic complexity, the application of such sophisticated models will be instrumental in improving traffic management initiatives and meeting the demands of city mobility. Moreover, efficient coping with the distinctive challenges of dynamic traffic is crucial to the success of real-time traffic forecasting models. A significant impediment in this field is the issue of class imbalance and distribution drift, which has an adverse effect on the performance of machine learning models in real-world situations. Methods like the Test-Time Training (TTT) mechanism enhance the capability of the models to learn during test time, and thus they work efficiently under variable traffic situations (Peng et al., 2025). Employing new components like Adaptive Memory Layers and Class-Balanced Attention enables the models to address such challenges by tuning the parameters and the classification rate. Therefore, these advancements point towards the potential for enhanced adaptability and resilience in intelligent transport systems, paving the way for smarter and more effective traffic management innovations within the context of evolving urban environments.

5.2.2. Optimization of Traffic Flow Using Machine Learning

Traffic flow optimization is rapidly becoming an essential element of urban planning and transport engineering as cities grow more congested. Machine learning techniques are revolutionizing the handling and utilization of traffic data, enabling transport agencies to develop plans for decreased congestion and improved mobility. With the application of big data from traffic sensors, GPS, and social media, machine learning models can identify patterns and perform traffic predictions with high accuracy. Dynamic prediction allows anticipatory control of traffic networks and implementation of optimized routing tactics, which in turn generates smoother traffic and reduced travel times. The integration of these technologies not just gives temporary relief to congestion but also paves the way for enhanced urban transportation systems in the future. Further, the application of machine learning to maximize traffic is more than just mere data mining.

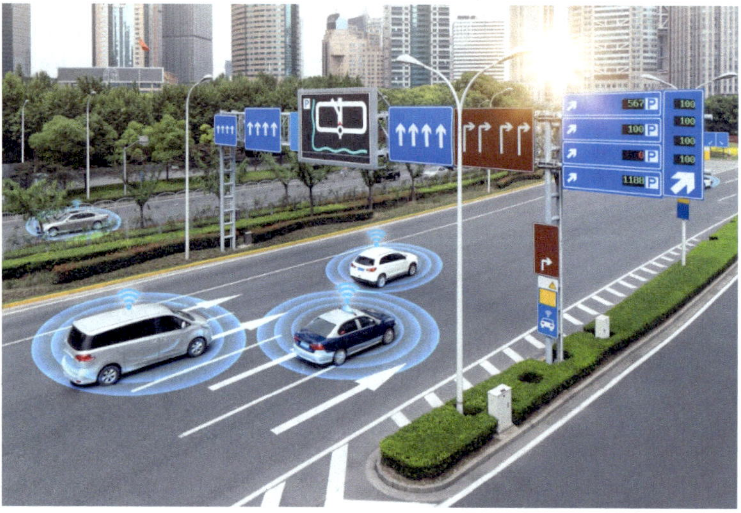

Figure 5.6. Optimization of traffic flow using machine learning and transportation.

Source: https://media.licdn.com/dms/image/v2/D5612AQEBCpo9lbRKKA/article-cover_image-shrink_720_1280/article-cover_image-shrink_720_1280/0/1721125594435?e=2147483647&v=beta&t=zfSCWDU6a10BJ2U1XmQp3APrbDS0V5ix1ysVZf_ques.

The Hierarchical Spatial-Temporal State Machine (HSTSM) is one of these models that enables planners to simulate a vast array of traffic situations and estimate potential effects on flow. This computer-based data modeling is a sign of the growing emphasis on computer intelligence in smart city models. In attempting to achieve efficiency and sustainability, cities' capacity to learn how to optimize traffic flow through machine learning results in reduced emissions and optimal public resource usage. As part of broader smart city strategies, employing such advanced methods can benefit significantly in enhancing city living by crafting impeccable transport systems that serve citizens as well as visitors (Doctor et al., 2020) best. Moreover, data-driven techniques like inverse optimization and variational inequalities facilitate the ongoing enhancement of traffic equilibrium models. By drawing inferences about utility functions from what is observed of traffic behavior, machine learning facilitates enhanced representations of user activity and congestion within road networks. By virtue of being a data-driven approach, it not only facilitates the easier creation of robust traffic models but also adaptive traffic management systems that can react in real-time to changing conditions. Innovative technologies resolving the complexities of city traffic behavior enhance predictability, enabling transportation networks to adjust to dynamic demands. These technologies underscore the critical role played by machine learning in refining traffic movement, eventually producing safer and more efficient cities (Bertsimas et al., 2014) (Figure 5.6).

5.2.3. Case Studies of Successful Implementations

The use of machine learning in transportation is illustrated in a recent case study of optimizing bridge design. Historically, engineers have relied on their experience and knowledge to select structural systems, leading to suboptimal designs due to time and financial constraints. However, a new approach with supervised machine learning was developed to enhance the structural system selection process. Considering data on over 600,000 bridges in the National Bridge Inventory, researchers established key characteristics influencing bridge design. Bayesian Networks and Decision Trees were two of the machine learning models employed. Utilizing seismic data and state-level variability in material costs was found to improve model performance significantly with more than 80% average recall and precision. This case study illustrates how machine learning can potentially transform traditional engineering practices into data-driven ones, for the best possible optimized design decisions and reduced long-term maintenance costs (Jootoo et al., 2017). An even more compelling example comes from transport planning's application of AI, where advancements have made better analyses of transport systems possible. The long-standing reliance on human expertise to comprehend complex data sets is being transformed by the synergy of artificial intelligence and traditional methods to enable placing expert knowledge in machine learning models. This paradigm shift has its lineage through global AI research initiatives, from the Japan Fifth Generation Computer project to Europe's ESPRIT program, that created new technologies relevant to many industries, including transportation. The interplay between these research efforts has yielded innovative ways of studying traffic trends, predicting congestion, and planning public transport. Consequently, the transport industry gains from more guided decision-making processes through massive data analyses, reducing over-reliance on intuition and optimizing operational efficiency (Bonsall et al., 1986). Lastly, the incorporation of machine learning in autonomous vehicle technology highlights further the transformative potential of such tools in transport. The need to make complex real-time decisions amidst dynamic traffic conditions necessitates sophisticated data processing techniques. Machine learning algorithms are crucial to make sense of excessive sensory input from the environment, such as the detection of pedestrians, reading road signs, and reacting to varied weather conditions. Industry players have been capable of utilizing these technologies efficiently, taking advantage of the predictive capability of machine learning to drive safety and mobility improvement. The continuous innovations in these uses not only increase the efficiency of the vehicles but are also making transportation strategies as a whole more capable of alleviating congestion and decreasing environmental impact. As case studies indicate, integration with machine learning is not a passing fancy but a significant shift in the way transport systems operate and can respond to needs of modern society (Jootoo et al., 2017; Bonsall et al., 1986).

5.3. PUBLIC TRANSPORTATION ENHANCEMENTS

Improvement in public transport is becoming increasingly important in city planning, particularly in light of new emerging technologies like machine learning. Effective public transit is not only crucial to economic sustainability but also in reducing congestion and decreasing ecological footprints. Through advanced data analysis, intelligent transportation systems (ITS) enable transportation agencies to develop safer, more efficient, and convenient public transit systems. Public transport mobility prediction has been identified as a significant challenge, and to address this, location and communication devices that offer operational information have been developed. By employing tailored machine learning, patterns in such data can be detected to help streamline operational schedules and the efficiency of public transportation in general, thereby providing a foundation for robust transit policies that are user experience-based and responsive to logistical concerns in real-time (Thiagarajan et al., 2025). Machine learning application in public transportation is also very promising in terms of dynamic resource allocation. Specifically, real-time prediction of passenger demand has been a dominant theme, using algorithms like DBSCAN clustering and the SARIMA Algorithm to support a more dynamic approach to route planning and scheduling buses. This dynamism allows transit authorities to pre-emptively react to changes in demand, which optimizes service frequency and reduces waiting time for passengers. These forecast models can especially enhance customer satisfaction and system performance and eventually encourage more public transport use. With changing local travel behavior, it is critical to have a sound transport demand forecasting to improve the provision of services and ensure that public transport systems have the capacity to satisfy varying users' needs (Thiagarajan et al., 2025). Besides, ITS complementarity with artificial intelligence is full of prospects for sustainable urban development. Case studies of various world cities prove the catalytic ability of AI to enhance traffic flow, security, and overall transit efficiency. For instance, Austin and New York cities have been able to implement AI-based solutions successfully to counteract traffic issues, thus promoting not only efficient commuting but also sustainable city practice. While such advances pose monumental benefits, issues continue as data quality, real-time processing, and issues of public acceptance that need addressing. Ultimately, the capability of machine learning technologies to improve public transportation systems' efficiency underscores the imperative need for continued innovation and planning in metropolitan transit environments (Figure 5.7).

Figure 5.7. Public transportation enhancements.

Source: https://www.silenthillglow.com/images/6cd3ed91.png.

5.3.1. Predictive Analytics for Scheduling

The advent of predictive analytics for transit schedules has revolutionized the work of transit systems. With enormous amounts of traffic data, the agencies can effectively forecast demand, optimize vehicle deployment, and reduce waiting times for passengers. For instance, machine learning software tracks traffic patterns and past trends, allowing scheduling systems to forecast peak travel times and adjust service frequency in response. These forecasting models, aside from enhancing operation effectiveness and customer satisfaction, also ensure environmental sustainability by eliminating unnecessary emissions in the form of idling vehicles. In addition, this approach enables the generation of intelligent traffic management systems, whereby timetables get dynamically updated in real-time based on current conditions, thus building a more adaptive transport system (Palla, 2021).

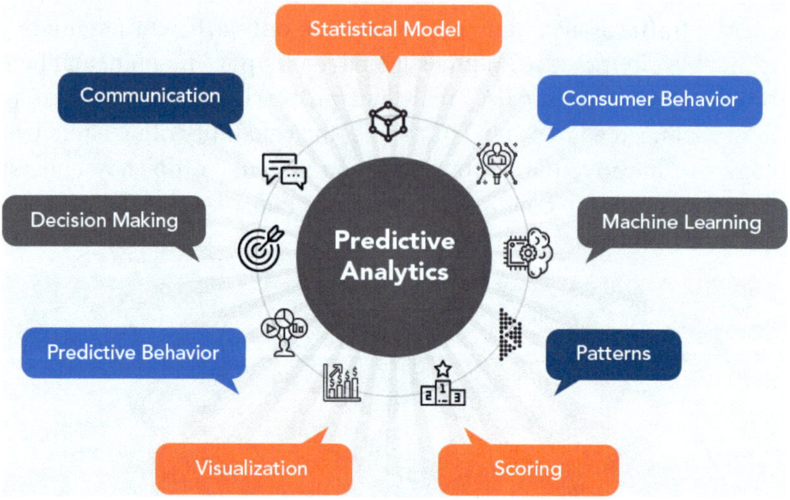

Figure 5.8. Predictive analytics for scheduling.

Source: https://www.queppelin.com/wp-content/uploads/2020/08/Predictive-7.png.

Adding machine learning techniques further enhances predictive analytics for scheduling by further improving traffic flow forecasts' precision. Tests have demonstrated that integration of multiple sets of data—such as GPS tracking by cars, real-time traffic camera images, and weather data—into forecast models delivers spectacular improvements in forecasting accuracy. For example, models incorporating weather conditions into traffic information have been discovered to have larger predictive ability in measuring levels of traffic congestion, offering insight into the complex relationship between outer conditions and travel behavior. This capability allows transportation planners to make informed decisions on scheduling and service deployment while keeping public transport as dependable and sensitive to the needs of society. With cities continuing to embrace smart technologies, predictive analytics will be key to creating more efficient urban mobility systems (Xia et al., 2018). Beyond legacy public transit systems, predictive analytics touches on new ride-hailing services.

As the popularity of on-demand transportation modes such as Uber and Lyft grows, real-time data analysis is crucial to dynamic pricing and optimal routing. By expecting areas of peak demand and price points accordingly, these services can best manage supply, maximize customer satisfaction, and minimize waiting times. Predictive models can also predict traffic conditions to recommend the most efficient routes, which permit timely arrivals without clogging road traffic. This multidimensional strategy makes available the disruptive potential of predictive analytics in modern transportation, its significance across numerous modalities in the sector (Palla, 2021) (Figure 5.8).

5.3.2. Demand Forecasting for Transit Services

Transit demand forecasting is a critical component of effective urban transport planning as it directly affects the allocation of resources and the overall performance of the public transport system. As city populations keep growing and mobility demands increase, accurate forecasting is ever more vital. The imperative task requires sophisticated methods to analyze vast volumes of data from various sources like ticket sale data, real-time passenger flows, and socio-economic factors. The addition of machine learning models to these forecasting models has shown good results, enabling transit agencies to make service adjustments for adapting to shifting patterns of demand. Through the application of big data and advanced analytics, agencies can optimize routes, increase the service frequency, and the rider experience, ultimately resulting in more sustainable urban mobility solutions (Thiagarajan et al., 2025). Aside from this, Intelligent Transportation Systems (ITS) have emerged as a focal point for enhancing transit service demand forecasting reliability. With the use of ITS applications, transit agencies are able to harvest and analyze real-time service operational information for the purposes of proactively adjusting service delivery. Dynamic management of transit is particularly useful in coping with peak hour demand and unexpected disruptions. As an example, machine learning techniques such as predictive modeling and clustering are able to capture passenger behavior patterns by way of travel time, route, and even weather insights. Using new machine learning algorithms such as DBSCAN clustering and SARIMA algorithms is a case of how the agencies can improve demand forecasting techniques so that they can do robust predictions that boost operational efficiency (Thiagarajan, 2023). In total, the development of demand forecasting of transit service through the use of machine learning not only addresses short-term urban mobility demands but also sets the stage for long-term strategic planning. With the information gained from advanced demand forecasting models, transit agencies are able to make informed decisions that will improve the level of service and rider satisfaction. The resolution of different forecasting models, such as the Prophet Model versus conventional approaches like ARIMA and SARIMA, emphasizes the need to continually improve analytical methods in terms of being able to better forecast passenger demand. As cities transform and the complexity of transit systems increases, the dependency on reliable demand forecasting based on machine learning continues to be more and more critical in designing responsive and sustainable transport systems (Thiagarajan, 2023; Thiagarajan et al., 2025).

5.3.3. User Experience Improvements Through Data Analysis

In the transport industry, user experience (UX) has more and more been linked to the application of data analysis and machine learning (ML) techniques. By leveraging huge volumes of data collected from various sources including user feedback, navigation behavior, and trip behavior, the transport systems can achieve radical improvements in service quality and user satisfaction. Data analysis enables transportation planners to identify user experience trends and pain points, which lead to tailored solutions that optimize overall usability. This focus on user experience design not only ensures services that meet the requirements of travelers but also ensures an improved user-friendly and efficient transportation system, demonstrating the role of data analysis in determining user experience improvements. With advancements in machine learning technologies, their applications in transport are also perpetually evolving, giving rise to new means of enhancing users' experience. Predictive analytics, for instance, can predict users' needs and desires through the study of past patterns, thereby enabling transport services to see ahead of problems and respond to them even before they materialize. Customized routing recommendations and real-time information on probable delays can significantly enhance the overall efficiency and user satisfaction level. With the emphasis placed on making travel hassle-free, the ability to utilize ML algorithms to make data-driven decisions is a great tool in optimizing transportation networks. These innovations underscore the importance of integrating the improvement of user experience with the functionality provided by data analysis. Further, the future transportation user experience will benefit from improvements in wireless communications systems, particularly from the development of 5G and 6G systems. The new technologies will provide high-speed and low-latency communications to facilitate real-time information exchange that is critical in optimizing user experience. With increased connectivity, it becomes possible to implement sophisticated ML models in transportation applications to facilitate enhanced decision-making and service delivery. As prophesied in newer research, zero-touch network optimization through the use of ML will be a key remedy to address issues within the transportation sector, with promise for new services that are designed wholly to user wants and needs (Chen et al., 2019; Abbas et al., 2020). Ultimately, this harmonious union of enhancing user experience and data analysis brings the transportation network more responsive and smarter and prepared for future demands.

5.4. LOGISTICS AND SUPPLY CHAIN OPTIMIZATION

The integration of machine learning into supply chain and logistics optimization is a paradigm shift in how companies are addressing operational effectiveness. Through predictive analytics and data-driven forecasting, organizations can anticipate patterns within demand, inventory, and delivery schedules. This leaves room for more accurate planning and deployment of resources, generating less waste and less expense. Algorithmic advances in the recent past enable processing large data sets, assisting companies in forecasting demand fluctuations and, in turn, optimizing the supply chain. As logistics operations become more complex with time, employing machine learning becomes essential to retain competitive advantage in a globalized market environment. Through better decision-making procedures, businesses not only optimize the logistics activities but also deliver greater customer satisfaction and responsiveness to the shift in the market. Moreover, the use of simheuristics—a cutting-edge optimization technique that is a combination of simulation and heuristics—is highly advantageous in logistics and supply chain problems where things are uncertain. It aids in finding solutions to intricate problems by employing simulation of various consequences based on probabilistic models, thus making the optimization part easier through a viable mechanism of managing uncertainty. For instance, if an organization is facing volatile demand or mixed delivery schedules, simheuristics can lead to more effective and well-informed decisions about resource usage and service quality. Agent-based simheuristics can also enhance cooperation and information sharing along the supply chain by modeling interaction between different agents, i.e., suppliers and distributors. The application of these sophisticated techniques puts firms in an improved position to address stochastic combinatorial optimization problems, thereby operations improvement as well as logistics strategy innovation stimulation (Estrada-Moreno et al., 2018). In addition to predictive modeling, the inclusion of soft constraints is important when addressing logistics optimization problems. The majority of real-world applications introduce flexibility into operational constraints, which may be crucial in achieving optimal performance. For example, in freight transportation, firms will adjust fleet sizes or delay certain deliveries in reaction to varying conditions to be more attuned to market demands. By treating such constraints as soft ones, which are breakable at a penalty, firms can seek creative solutions to overcome logistics management complexities. It focuses on the need for advanced methodologies, such as biased-randomized algorithms, to solve NP-hard and non-smooth optimization issues within supply chain scenarios effectively. Such algorithms are refined forms of traditional heuristics, which incorporate nonuniform randomization to explore favorable areas of the solution space more precisely. By doing this, they assist in creating adaptive and resilient supply chains that can perform optimally under uncertain environments (Biosca et al., 2020) (Figure 5.9).

Figure 5.9. Logistics and supply chain optimization.

Source: https://throughput-1cdea.kxcdn.com/wp-content/uploads/2024/05/08-Supply-Chain-Optimization-Infographics.png.

5.4.1. Route Optimization Algorithms

Optimization route algorithms form the cornerstone of modern transport networks, contributing significantly to ensuring maximum efficiency in route planning. The basis of the algorithm is still the modeling of the path choice behavior of network users, otherwise referred to as the route choice problem. Transportation science has been interested in this problem across a broad scope, studying the intricate dynamics of how users select their paths under various conditions. Discrete choice models are one of the chief points of prediction for these behaviors, yielding useful models that can be developed further with more complex techniques such as recursive modeling. The significance of such models is great, since they not only offer a valid foundation of user preference, but they also form the basis of future endeavor in optimizing travel routes over diverse networks of transport (Frejinger et al., 2020). Application of machine learning techniques adds to the extensibility of route optimization algorithms to update in real-time according to fluctuating traffic conditions and user behavior. Smart cities' real-time traffic sensors gather information necessary for route guidance; however, sensor failures render such a system unreliable. Faulty data caused by these malfunctions has the potential to produce suboptimal route decisions, and it is thus imperative that the data collection processes are reliable. The monitoring mechanisms need to be effective to detect and eliminate the impacts of faulty sensors. The advanced detection algorithms, such as those based on Gaussian Processes, provide a method of early detection and correction of these issues, hence preserving a higher degree of precision in route optimization processes (Dubey et al., 2017). Not only is the use of machine learning in route optimization involved in improving the algorithms but also in fine-tuning the quality of the input data. By being able to effectively identify critical sensors whose fault would directly affect route planning, transport networks can enhance overall efficiency. Anticipating this, early interventions can be made with least disruption and improved user experience. Putting these novel methodologies into practice and testing them using actual data provides us with real-world insights on how they apply. Overall, the ongoing improvement of route optimization algorithms is the revolutionary potential of machine learning in transport, leading to intelligent and responsive urban mobility solutions (Frejinger et al., 2020; Dubey et al., 2017) (Figure 5.10).

Figure 5.10. Route optimization algorithms in modern transportation systems.

Source: https://s7.easternpeak.com/wp-content/uploads/2021/03/Frame-63.png.

5.4.2. Inventory Management Using Machine Learning

Inventory control is one of the critical supply chain functions in the transport sector, demanding efficient techniques for resource optimization. The use of machine learning (ML) techniques has emerged as a revolutionary approach to tackling the complexities of inventory management. With data-driven algorithms, companies are able to properly plan for demand, prevent stockouts, and reduce excess inventory, hence minimizing waste and maximizing productivity. Machine learning software can review past sales data and identify patterns that can result in insights that may be overlooked by traditional methods. As a result, businesses can get more efficiency and responsiveness and attain a competitive advantage with the dynamic nature of transport. The most popular application of ML in stock management is the development of optimal classification trees (OCTs), through which more interpretable decision-making is feasible. OCTs redefine optimization problems in inventory contexts as multiclass classification problems.

Figure 5.11. Inventory management using machine learning.

Source: https://citrusbug.com/wp-content/uploads/Key-Benefits-of-Machine-Learning-in-Inventory-Management-Systems.webp.

Managers are provided with clarity on how key parameters influence their inventory strategies through the application of these trees, essentially making what was initially considered a black box a comprehensible paradigm. Interpretability is most crucial as it enables logistics operators to identify the underlying realities that determine optimal inventory quantities and replenishment cycles. Furthermore, the application of OCTs, as described in Bertsimas et al. (2020), is a compelling avenue for enhancing decision-making for intricate mixed-integer inventory optimization problems. Another interesting innovation in inventory management is the application of gradient-free algorithms, which are the key to solutions of convex resource allocation problems without needing strict convexity or differentiability in the applied models. These algorithms are particularly effective in variable demand and constrained resource systems, actually offering near-optimal solutions in a timely manner even for big problem sizes, as noted in Gribel et al. (2018). The ability to handle problems involving big variables—typically a million or so, as examined under experiments—makes the algorithms essential elements in transportation inventory management systems. By enhanced precision and speed of inventory optimization activities, machine learning ultimately enables an even more responsive and flexible supply chain, with a direct impact on operational efficiency and customer satisfaction levels (Figure 5.11).

5.4.3. Impact on Delivery Times and Costs

The addition of machine learning in transportation significantly enhances delivery time and reduces expenses, which address significant logistical issues for today's supply chains. With rich data sets, machine learning algorithms are capable of simplifying routing and scheduling, resulting in more efficient processes. For instance, predictive analytics is able to anticipate demand patterns to enable companies to make sound resource allocation and reduce waiting time. From recent studies, Big Data analytics in transport contributes to making operational tactics more intelligent, increasing overall standards of service delivery (Doctor et al., 2020). Therefore, organizations can respond to altering demand with agility, enhancing customers' satisfaction and retaining their position in the marketplace. Besides, machine learning applications extend beyond routes to cover vehicle predictive maintenance that also costs delivery less. Advanced algorithms can monitor real-time vehicle sensor data to predict potential failure prior to occurring and thus minimize downtime as well as unexpected repair expenses. This innovative approach not only streamlines operations but also optimizes logistics networks by ensuring a more stable fleet. With the information obtained from new machine learning techniques such as optimal classification trees, decision makers will have improved decision-making on fleet management and resource allocation, hence cost-effectiveness (Bertsimas et al., 2020). The contribution of such technologies to operational sustainability and delivery efficiency, therefore, cannot be overstated in the transport sector. In addition, with the development of machine learning, effects on cost and delivery time will also become more important. Improved computational intelligence techniques provide logistics professionals with a better insight into what is happening and enable them to optimize processes to their optimal performance levels. For example, the combination of Big Data and machine learning allows for the development of sophisticated models that can simulate diverse traffic scenarios and potential disruptions, in order to make anticipatory changes in delivery schedules. As noted in a case study of intelligent transportation systems, such advanced methods significantly facilitate the capacity of smart cities to manage transportation assets efficiently (Doctor et al., 2020). Hence, those organizations that embrace such technology will stand to benefit from enhanced delivery performance and reduced operating expenses, making the argument of machine learnings pivotal role towards transforming the transport industry.

5.5. MAINTENANCE AND SAFETY MONITORING

The integration of machine learning (ML) into safety monitoring and maintenance in transportation systems increases the operational efficiency and reliability much greater. This is particularly so in the rail industry, where application of robotic and autonomous systems (RAS) has transformed traditional maintenance systems. Through automation of processes such as rolling-stock inspections and track inspections, ML-based systems enable predictive maintenance, which avoids failures from taking place by forecasting them. In this way, rail companies can reduce downtime and maintenance costs without compromising round-the-clock operations reliability and safety. The competitive nature of the industry also propels growing interest in RAS, as the industry demands higher operating practices and cost-control standards, which ML solutions naturally counteract (Ackerman et al., 2019).

Safety monitoring is a focal point in the rail industry, where occurrences at railway stations have severe consequences. In order to minimize accident risks, ML technologies are more and more utilized to analyze previous accidents and develop efficient safety protocols. Using complicated algorithms like decision trees, the sector can classify accidents and predict the type of individuals who are responsible for causing them, thus intensifying preventative measures. Such technologies not only raise the level of understanding of accident dynamics but also increase rail passengers' and infrastructure's overall safety. Successful application of ML in this scenario serves as proof of the potential for employing innovative technologies to make conventional safety measures proactive, data-driven processes (Alawad et al., 2019). As machine learning progresses, the potential for improved maintenance and surveillance of safety in transport advances considerably. It establishes a culture where information is utilized to make decisions, and consequently, results in efficient resource allocation and risk management. The shift from reactive to proactive maintenance techniques is particularly significant in environments where safety is paramount, such as railways. Leaning on algorithms capable of filtering through amounts of operation data, organizations are no longer simply capable of predicting when maintenance is necessary but also identify underlying flaws that may result in safety hazards. The integration of machine learning with traditional safety regimes results in an integrated strategy with the capability to reduce accidents and provide more reliable services, ultimately leading to a safer transport environment as it responds to the growing demands of modern society (Ackerman et al., 2019; Alawad et al., 2019) (Figure 5.12).

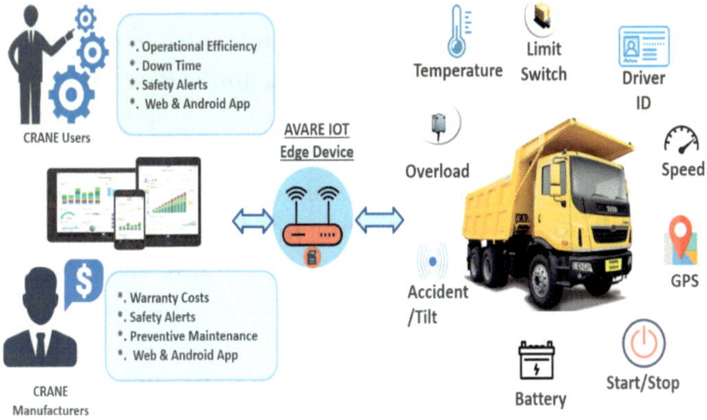

Figure 5.12. Maintenance and safety monitoring within transportation systems.

Source: https://www.vervetronics.com/wp-content/uploads/2019/10/Screenshot-2019-10-07-13.38.10.png.

5.5.1. Predictive Maintenance Techniques

In the context of predictive maintenance techniques, machine learning (ML) plays a major role in ensuring optimal transportation system reliability and efficiency. These techniques employ algorithms that analyze information from a variety of sources so that forthcoming system breakdowns may be detected prior to their occurrence. Based on historical performance data, and current sensor data from sensors installed on transportation infrastructure and vehicles, ML systems can predict when maintenance is required. This transition from predictive to reactive maintenance decreases unplanned downtimes, which ultimately results in significant cost reductions and improved security within transport networks. For instance, with railway transport, ML technology can improve safety analysis by automatically reviewing incident data, solving problems that have caused accidents in stations before (Alawad et al., 2019). Therefore, through the use of ML, apart from enhancing the efficiency of operations, it increases system safety and reliability of service in transport. Moreover, the use of predictive maintenance techniques in transport employs a variety of artificial intelligence (AI) techniques. Genetic algorithms, for example, and neural networks are among the AI techniques applied in optimizing maintenance schedules and improving fault diagnostic systems. For instance, an AI applications survey in operations management is a manifestation of the key technological advancements surrounding predictive maintenance (Kobbacy et al., 2007). Such a development allows for the integration of fuzzy logic and hybrid approaches that also result in improved wear and tear insights on transportation assets. With the application of these sophisticated approaches, transport companies can schedule maintenance interventions based on actual usage rates and forecasting analyses, rather than relying on isolated checks. This adaptive maintenance strategy conserves resources while enhancing the overall performance and safety of transport systems. Apart from sustaining operating efficiency, predictive maintenance techniques have a direct influence on sustainability within the transport sector. As the vehicles and infrastructure are maintained proactively, their longevity is enhanced, therefore reducing environment-damaging wastage due to premature replacement. Moreover, by optimizing maintenance schedules through machine learning, transport organizations can reduce emissions associated with idling or inefficient operation. Ongoing adoption of predictive maintenance technologies can enable transition to cleaner transport modes, ultimately contributing to global sustainability efforts. By doing this, predictive maintenance becomes not only a driver for enhancing operational efficiency but also a facilitator of broader environmental goals, a perfect example of the critical nexus between technological innovation and environmentalism in transport.

5.5.2. Anomaly Detection in Transportation Systems

Transportation system anomalies are typically caused by sensor faults, which can negatively affect efficiency in operation. Sensor faults typically produce traffic data that has a negative effect on some of the applications like real-time route planning. Negatively affected sensor accuracy can lead to increased travel times and frustrate commuters. To tackle these problems, one needs strong anomaly detection methods with the capability of identifying faulty sensors quickly and correctly. Traditional detection algorithms are often challenged by false positives and false negatives, and this complicates the detection process and can lead to amplified inefficiencies.

Therefore, advanced detection systems involving predictive models such as Gaussian Processes have been known to effectively maximize the detection process by minimizing the possibility of misclassifications in traffic data analysis. In the machine learning for applications context, development of sophisticated anomaly detection techniques is critical towards fostering the reliability of transport systems. Anomaly-based intrusion detection systems (IDS), which are extensively used in the context of cybersecurity, provide valuable insight into how exactly the same method can be applied to transportation industry needs. These models take historical data and use it to train their detection mechanisms, in essence detecting malicious activity or system anomalies. Hybrid feature selection methods, such as the application of particle swarm optimization or genetic algorithms, can also enhance such systems' classification efficiency. With fine-tuning of the datasets utilized, transportation systems are able to benefit from better detection rates and reduce operational halts. It is critical that such methodologies be rigorously tested, including statistical significance testing, for their effectiveness to be verified in actual environments. In addition, defining the critical sensors in the transportation infrastructure is crucial to lessening the influence of anomalies.

Defining sensors whose failure would cause the greatest operational threats enables the monitoring of these sensors individually and immediate action to take place. Machine learning algorithms can be formulated to assess sensors' performance indicators on an ongoing basis, detecting those that fluctuate from typical norms. This anticipatory approach not only lowers the likelihood of disruptions caused by sensor failure but also assists in forecasting future anomalies prior to their transformation into enormous systemic failures. Utilizing enormous detection methods, transport networks can develop their robustness and maintain ideal operating effectiveness. Lastly, these advancements put into perspective the criticality of anomaly detection towards enhancing the performance of modern transport systems as part of broader machine learning uses in the industry (Figure 5.13).

Figure 5.13. Anomaly detection in transportation systems.

Source: https://www.researchgate.net/profile/Alessandro-Podda/publication/ 348-390012/figure/fig7/AS:978788615720964@1610372799211/Some-examples-of-correct-anomaly-detection-when-the-trained-Faster-R-CNN-is-executed-on.ppm.

5.5.3. Case Studies of Machine Learning in Safety Monitoring

In today's world of transportation, the application of machine learning (ML) in monitoring safety has shown to have revolutionary potential. It is especially relevant in industries like railways and industrial settings, where safety tops all else. For example, railway companies increasingly use ML algorithms to forecast maintenance requirements so as to avoid unexpected breakdowns that can jeopardize safety. Through huge volumes of data collected by sensors planted in rolling stock and in tracks, such systems analyze patterns to make probable malfunctions predictions effectively. The study shows that machine learning not only enhances operation effectiveness but, and more crucially, significantly reduces chances of human mistakes and thus assists in making the transport system safer as per requirements of modernization in the infrastructure and safety systems.

Furthermore, as connectivity of industrial control systems advances, safety monitoring software using ML will be required to overcome new cybersecurity challenges. Various case studies indicate that the traditional approach of compartmentalizing such systems is rapidly losing its significance with the introduction of IoT concepts into industrial landscapes. For example, interoperability between various management systems is possible and can make production processes simplified but result in windows of vulnerability to be exploited by cyber-attacks. To address these threats, researchers are exploring potential solutions that leverage ML for real-time anomaly monitoring and adaptive security capabilities. Convergence of safety and cybersecurity calls attention to the need for robust machine learning implementations that concurrently provide operational safety while also protecting critical infrastructure against evolving digital threats (Johnson et al., 2016).

Lastly, the other case studies outside rail transport illustrate the two-role uses of automated systems in monitoring for safety reasons across various modes of transport. Of particular note here is the focus placed on maintenance in the rail transport industry, where RAS adoption is on the rise in popularity. Indications point to RAS specifically being created for rolling stock maintenance, and then track maintenance, as a strategic focus on enhancing safety through mechanization. The demand for new solutions stems from heightened competition and the need to keep lowering the cost of maintenance, resulting in organizations needing to implement machine learning technologies. As pointed out in recent literature, this trend marks a heightened regard for the central role of ML in guaranteeing not only efficiency but also safety protocols in the transportation sector, promising a future where technology forms the foundation of safe transit operations (Ackerman et al., 2019).

5.6. ENVIRONMENTAL IMPACT REDUCTION

The application of machine learning in transportation significantly enhances the capacity to reduce environmental impacts, particularly the field of infrastructure planning and maintenance. A proper use is the structural selection procedure in bridges. Using supervised machine learning, engineers can analyze big datasets, for instance, the National Bridge Inventory, to discover the most sustainable and efficient bridge structures as a function of a variety of variables like seismic intensity and material cost. This increases design standardization and optimizes the allocation of resources, ultimately leading to reduced maintenance costs and lighter environmental impacts on multiple infrastructures. Better decision-making through predictive modelling not only supports best structural decisions but also contributes to the overall goal of minimizing environmental degradation in engineering use (Jootoo et al., 2017).

Figure 5.14. Environmental impact reduction in transportation.

Source: https://sustainablyforward.com/wp-content/uploads/DALL%C2%B7E-2024-01-31-10.26.10-A-horizontal-image-depicting-various-modes-of-sustainable-transportation-like-bicycles-electric-vehicles-and-public-transit-in-an-urban-setting.-The-900x515.jpg.

Furthermore, machine learnings contribution extends into the field of automotive design and operational efficiency. In fact, consideration of low-carbon motor vehicles, such as hybrids and electric cars, necessitates a full understanding of how users engage with vehicle technology. By applying ergonomic principles together with machine learning, designers are able to create interfaces that promote energy-efficient driving and simultaneously minimize distraction for drivers. This ergonomic approach acknowledges challenges that new car technologies such as range anxiety and increased cognitive load present. Through triggering driver behavior change through improved interface design, the transport sector can promote a culture of sustainability, which will result in significant greenhouse gas savings from private vehicle usage (Harvey et al., 2013). The consequences of applying machine learning to transport extend even further to traffic management and urban planning. Through the analysis of real-time data across multiple modes of transportation, machine learning can maximize traffic flow, cutting congestion and excess idle time that generates excess emissions. For example, adaptive traffic lights can regulate their timing according to prevailing traffic, providing smoother travel and less fuel use. Also, machine learning can assist in public transportation planning by forecasting passenger needs, enabling cities to better deploy resources. This integrated approach not only improves the efficiency of current transport systems but also promotes a transition towards cleaner options, significantly contributing to the mitigation of the overall environmental cost related to city transport networks (Figure 5.14).

5.6.1. Fuel Efficiency Improvements Through Machine Learning

New machine learning advancements have significantly enhanced fuel efficiency in transportation systems by way of data-driven decision-making processes. Huge amounts of car and environmental data are processed by machine learning models to create predictive models that optimize fuel utilization. Predicative maintenance, for instance, can identify oncoming mechanical errors that would lead to increased fuel usage and maintain the cars in their optimal condition of performance. Moreover, algorithms can keep track of driving patterns and suggest changes to improve fuel efficiency. This union of machine learning not only addresses the urgent requirement of lowering greenhouse gas emissions but also aligns with the broader vision for sustainable transportation systems (Harvey et al., 2013). Machine learning is not just applied to individual vehicles but to influence entire transport systems. By integrating machine learning in traffic management, cities can optimize traffic flow and hence reduce idle time and overall fuel consumption. Data analysis in real-time allows for adaptive traffic signals and dynamic routing algorithms that take into account evolving traffic conditions. These systems are able to efficiently reduce the fuel consumption associated with stop-and-go traffic, a common issue faced by cities. Furthermore, the use of machine learning for route optimization reduces operational expenses for logistics companies without growing their carbon footprint (Belsley et al., 2025). The power of machine learning to transform fuel efficiency is also achieved in the design and operation of hybrid and electric vehicles. The vehicles employ advanced algorithms that optimize energy consumption and distribution efficiently based on driving situations, battery life, and driver action. Through predictive energy management via machine learning, producers can create cars that beat even current efficiency standards. The car and transport industries alike benefit from this ongoing innovation as machine learning promotes a culture of ongoing improvement in fuel efficiency, driving the grand vision of environmental minimization while maximizing system productivity (Harvey et al., 2013).

5.6.2. Emission Reduction Strategies

Greater urgency in addressing climate change has led to greater emphasis on reducing emissions policies throughout the transportation sector. Laying at the core of such measures is the integration of machine learning technologies offering new ways of greenhouse gas emissions reduction. Through algorithms capable of handling vast volumes of data, transportation specialists can identify inefficiencies in traffic flow, vehicle usage, and energy use. This data-driven approach allows real-time optimization that can realize significant emission reductions. For example, predictive analytics can optimize traffic lights, thereby minimize idling, and facilitate smoother car flow. Such optimizations not only promote environmental sustainability but also enhance mobility in cities and demonstrate how emission reductions are possible along with improved infrastructure and public transportation. Second, design and operation of the vehicle also play a crucial part in minimizing emissions. Low-carbon vehicles, such as hybrids and electric vehicles, are essential in mitigating fuel dependence on fossil energy. Machine learning can be applied in ergonomic vehicle design to facilitate easy-to-use operations and driving practices that promote increased fuel economy. As literature discussed, the driver behavior and vehicle design interaction can be a fuel economy determinant. For instance, Ergonomics environments emphasize discussions on how user-friendliness can promote energy-saving driving habits without distractions and cognitive load. By structuring in-car technologies to an optimized form through a design process like Ecological Interface Design, designers can encourage sustainable driving practices that are in sync with overall emissions reduction strategies, thereby supporting a shift towards a cleaner transport landscape (Harvey et al., 2013). Furthermore, gaining public acceptance and awareness of new technology is key to successful emission reductions. The rise of social robotics is an opportunity in a lifetime to engage and educate citizens on sustainable driving habits and mobility options. Using social robots with the capacity to automatically send energy-efficient driving behaviors to drivers will allow us to promote behavior changes resulting in emission savings. As an example, robots can provide interactive remarks on driving habits and fuel consumption in a bid to get users adopt green practices. The creation of low-cost, adaptable platforms like Ono is a vision for the future in developing technologies that are not only affordable but also well-tailored. By facilitating open participation in these projects, it is easy to integrate machine learning technologies into routine transportation decision-making, leading to enhanced environmental control and the promotion of sustainable transport systems (Saldien et al., 2014).

5.6.3. Sustainable Transportation Solutions

The search for sustainable transportation solutions increasingly counts on machine learning to optimize efficiency with the least environmental impact. A compelling one is a combination of electric vehicles (EVs) and advanced infrastructure advancements. Research indicates that deploying a scalable and safe charging architecture is essential to fully leverage the benefits of EVs, particularly as self-driving vehicles become popular. There can be inadequate existing manual charging processes; thus, new technologies like automated charging using blockchain technology have been introduced. These innovations not only ease the process of charging but also facilitate real-time data analysis and machine learning solutions that are capable of revolutionizing the energy distribution and use throughout transport networks (Hussain et al., 2018). As there is greater demand for cleaner transport, these technologies are central to promoting sustainable behavior in the industry. Data-driven solutions in the context of smart cities are central to achieving transportation network sustainability. Big Data coupled with machine learning allows the development of smart systems that can improve mobility in cities. By leveraging vast quantities of transport data, cities are better able to manage traffic, reduce traffic, and lower emissions. By way of a case study on an advanced modeling technique—Hierarchical Spatial-Temporal State Machine (HSTSM)—it illustrates how computational intelligence can provide insights required for effective transport planning and management (Doctor et al., 2020). Furthermore, real-time data analytics enable transportation networks to react in real time to changing patterns, to best utilize resources while minimizing the carbon footprint. By doing so, sustainable transportation options are not only reactive but proactive and aligned with greater sustainability goals for urban growth. As cities evolve further and grow, machine learning applications in transport will be key to constructing effective and sustainable systems. Traffic flow analysis, the use of public transport, and environmental surveys can be used to inform data-driven policy that reduces carbon footprint. Machine learning algorithms have the capability of identifying wasteful practices and recommending optimal routes for transport operators, thus lowering costs and enhancing delivery. Moreover, involving the general public through data analysis encourages civic ownership and responsiveness, which is critical in sustainable works (Doctor et al., 2020). By putting a holistic approach to transportation management that balances technological advancement with community participation ahead, urban planners can develop integrated solutions that address immediate transport needs as well as longer-term environmental goals, facilitating resilient and sustainable urban transport systems (Figure 5.15).

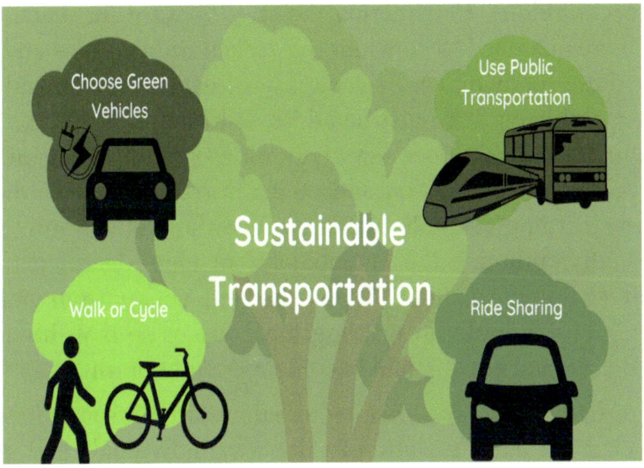

Figure 5.15. Sustainable transportation solutions.

Source: https://blog.techstork.sg/imgDB/Sustainable%20Transportation.png.

SUMMARY

In summarizing the greatest advances brought about by machine learning to transportation, it is clear that integration of intelligent systems has revolutionized traffic control and vehicle communication. They not only enhance operational efficiency but also safety and customer experience. As indicated in applications like TensiStrength, which effectively analyzes social media stress-related inputs, machine learning surpasses the traditional approach through the offering of better response to varying conditions. This attribute demonstrates the ability of machine learning to utilize real-time information to enhance decision-making mechanisms in transport networks, demonstrating its potential as an integral part of future development in the industry (Thelwall et al., 2016). In addition, the development of advanced traffic monitoring and vehicle classification systems is a promising avenue toward smart city infrastructure. Through the application of advanced algorithms, such systems provide real-time and precise performance measures feeding intelligent traffic control systems. The emphasis on precise sensor data capture to facilitate dynamic control methods indicates shifting dynamics of handling transport (Al-Askary et al., 2018). These capabilities enable cities to implement schemes maximizing the flow of traffic and reducing congestion, and resulting in improved environmental performance and general public satisfaction. Machine learning, therefore, not only offers real-time analytical capabilities but is also a template for planning sustainable urban transport solutions. Considering the way forward for machine learning in transport, there is a need to consider the ethical implications and challenges presented by these innovations. Privacy issues, data integrity, and fairness in algorithms are problems that must be addressed to ensure that the integration of these intelligent systems does not undermine the confidence of the public. Balance in technology innovation and ethics is required in building a future where machine learning technology can grow without undermining the values of society. Finally, although the potential for machine learning to transform transportation is immense, it needs coordinated action from policymakers, technologists, and the public in tackling these issues so that the dividends of such use are distributed equally and become a popular norm.

The integration of machine learning into the transport sector has introduced far-reaching changes that revolutionize the way individuals and goods move around in cities. One of the primary applications is in traffic control systems, which utilize real-time information and predictive analytics to control traffic. The systems are able to process massive amounts of data from various sources such as sensors and mobile applications and react within time. This ability to comprehend and analyze Big Data has not only increased efficiency but also decreased congestion levels, thus emissions being less and urban sustainability better. However, data privacy issues have turned into a timely issue, wherein excessive data collection tends to lead to user distrust. Governments and technology companies must navigate these ethical landscapes with care so as not to allow the machine learning virtues at the expense of human rights (Ishimwe et al., 2020). A further field where machine learning proves its usefulness is where public security and accident prevention are concerned. Through advanced algorithms, transportation systems can analyze previous accident records for identifying and preventing future threats. For example, risky places can be identified through machine learning models by mapping patterns and trends that will not readily be apparent through traditional data investigation methods. This is preventive in the sense that it allows for strategic positioning of security measures, such as improved lighting or signs, thereby improving road safety. Additionally, through urban transfer learning models, cities have the ability to learn from one another, adopt effective strategies from pioneering smart cities into developing their own transportation systems efficiently. Not only does it encourage collaboration among city planners, but it also accelerates the process of building more secure transportation ecosystems across various regions (Guo et al., 2018). Finally, machine learning's role in optimizing logistics and supply chain operations only amplifies its revolutionizing influence in transport. Being able to predict demand patterns and even direct delivery, businesses can reduce their operating expenses as well as increase customer satisfaction. Machine learning algorithms can decode past events to predict future events, and hence companies can optimize the allocation of their resources. Such optimization not only allows for streamlining processes but also contributes significantly to the abolition of the carbon footprint generated in logistics. Taking advantage of these advanced technologies, businesses can make informed decisions consistent with sustainable practices. But the balance between effectiveness and ethical use of data continues to remain supreme because organizations need to place high stakes on user privacy and transparency in their operations to build trust and gain a longer-lasting future in transport (Ishimwe et al., 2020).

REFERENCES

1. Abbas, R., Ahmad, I., Ali, S., Amuru, S., et al. (2020). 6G white paper on machine learning in wireless communication networks. http://arxiv.org/abs/2004.13875 (accessed on 21st October 2025).
2. Abtahi, A., Grzywaczewski, A., Agrawal, S., Al-Sultan, M., Asimov, B., Bezdek, P., et al. (2018). A novel big data analytics and intelligent technique to predict driver's intent. https://core.ac.uk/download/228154021.pdf (accessed on 21st October 2025).
3. Ackerman, J., Guler, M., Li, R., McNulty, J., Márquez, M., Reason, B., & Smorenburg, T. (2019). Importance and applications of robotic and autonomous systems (RAS) in railway maintenance sector: A review. https://core.ac.uk/download/293867868.pdf (accessed on 21st October 2025).
4. Al-Askary, M., Dorn, D., Haferkamp, M., Sliwa, B., et al. (2018). A radio-fingerprinting-based vehicle classification system for intelligent traffic control in smart cities. http://arxiv.org/abs/1801.03291 (accessed on 21st October 2025).
5. Alawad, H., An, M., & Kaewunruen, S. (2019). Learning from accidents: Machine learning for safety at railway stations. https://core.ac.uk/download/286269815.pdf (accessed on 21st October 2025).
6. Alkhorshid, Y., Aryafar, K., Bauer, S., Wanielik, G., et al. (2016). Road detection through supervised classification. http://arxiv.org/abs/1605.03150 (accessed on 21st October 2025).
7. Bacciu, D., Gnesi, S., & Semini, L. (2015). Using a machine learning approach to implement and evaluate product line features. https://core.ac.uk/download/pdf/80264816.pdf (accessed on 21st October 2025).
8. Belsley, S. E., Rosenberg, N., & Thompson, A. (2025). Technological change and productivity growth in the air transport industry. https://core.ac.uk/download/pdf/42871908.pdf (accessed on 21st October 2025).
9. Bertsimas, D., & Stellato, B. (2020). The voice of optimization. http://arxiv.org/abs/1812.09991 (accessed on 21st October 2025).
10. Bertsimas, D., Bertsimas, D., Gupta, V., & Gupta, V. (2014). Data-driven estimation in equilibrium using inverse optimization. https://core.ac.uk/download/142070194.pdf (accessed on 21st October 2025).
11. Bonsall, P. W., Kirby, H. R., & Kwan, R. S. (1986). Applications of expert systems in transport. https://core.ac.uk/download/53713.pdf (accessed on 21st October 2025).

12. Chen, K. C., Hanzo, L., Jiang, C., Ren, J., et al. (2019). Thirty years of machine learning: The road to Pareto-optimal wireless networks. http://arxiv.org/abs/1902.01946 (accessed on 21st October 2025).

13. Comuzzi, M., Rhee, K. H., & Tama, B. A. (2019). TSE-IDS: A two-stage classifier ensemble for intelligent anomaly-based intrusion detection system. https://core.ac.uk/download/225164258.pdf (accessed on 21st October 2025).

14. Doctor, F., Iqbal, R., Mahmud, S., More, A., et al. (2020). Big data analytics: Computational intelligence techniques and application areas. https://core.ac.uk/download/228156656.pdf (accessed on 21st October 2025).

15. Dodig-Crnkovic, G., Holstein, T., & Pelliccione, P. (2018). Ethical and social aspects of self-driving cars. http://arxiv.org/abs/1802.04103 (accessed on 21st October 2025).

16. Dubey, A., Ghafouri, A., Koutsoukos, X., Laszka, A., et al. (2017). Optimal detection of faulty traffic sensors used in route planning. http://arxiv.org/abs/1702.02628 (accessed on 21st October 2025).

17. Dwivedi, Y. K., Hughes, L., Baabdullah, A. M., Ribeiro-Navarrete, S., Giannakis, M., Al-Debei, M. M., Dennehy, D., et al. (2022). Metaverse beyond the hype: Multidisciplinary perspectives on emerging challenges, opportunities, and agenda for research, practice and policy. International Journal of Information Management, 66, 102542–102542. https://doi.org/10.1016/j.ijinfomgt.2022.102542 (accessed on 21st October 2025).

18. Estrada-Moreno, A., Freixes, A., Juan Pérez, Á. A., Panadero, J., et al. (2018). Current trends in simheuristics: From smart transportation to agent-based simheuristics. https://core.ac.uk/download/187768826.pdf (accessed on 21st October 2025).

19. Ferrer Biosca, A., Gunes Corlu, C., Juan Pérez, Á. A., Tordecilla Madera, D., et al. (2020). On the use of biased-randomized algorithms for solving non-smooth optimization problems. https://core.ac.uk/download/286776300.pdf (accessed on 21st October 2025).

20. Frejinger, E., & Zimmermann, M. (2020). A tutorial on recursive models for analyzing and predicting path choice behavior. http://arxiv.org/abs/1905.00883 (accessed on 21st October 2025).

21. Ghazzai, H., Massoud, Y., & Wan, X. (2020). Leveraging personal navigation assistant systems using automated social media traffic reporting. http://arxiv.org/abs/2004.13823 (accessed on 21st October 2025).

22. Gribel, D., Jaillet, P., & Vidal, T. (2018). Separable convex optimization with nested lower and upper constraints. http://arxiv.org/abs/1703.01484 (accessed on 21st October 2025).

23. Guo, B., Wang, L., & Yang, Q. (2018). Smart city development with urban transfer learning. http://arxiv.org/abs/1808.01552 (accessed on 21st October 2025).

24. Ha, V. S., Nguyen, H., & Thi, B. (2024). Machine learning models for real-time traffic prediction: A case study in urban traffic management. Journal of Science and Transport Technology. https://www.semanticscholar.org/paper/c046952040cd49d594bdab204ef0d89b1dff19ed (accessed on 21st October 2025).

25. Harvey, C., McIlroy, R., Robertson, D., Stanton, N., et al. (2013). Sustainability, transport and design: Reviewing the prospects for safely encouraging eco-driving.

26. Hussain, R., Lee, J., Mazzara, M., Mustafin, R., et al. (2018). On M2M micropayments: A case study of electric autonomous vehicles. https://core.ac.uk/download/492594510.pdf (accessed on 21st October 2025).

27. Ishimwe, A. T. (2020). Identifying privacy policy in service terms using natural language processing. https://core.ac.uk/download/322958978.pdf (accessed on 21st October 2025).

28. Johnson, C. (2016). Securing the participation of safety-critical SCADA systems in the industrial internet of things. https://core.ac.uk/download/46664188.pdf (accessed on 21st October 2025).

29. Jootoo, A., & Lattanzi, D. (2017). Bridge type classification: Supervised learning on a modified NBI dataset. http://arxiv.org/abs/1803.04478 (accessed on 21st October 2025).

30. Khakre, V. B., & Kadam, S. (2024). A comprehensive study on artificial intelligence and machine learning. International Journal of Advanced Research in Science, Communication and Technology. https://www.semanticscholar.org/paper/8c3e98ea7b3e6e062c6f41efdf265c118e9bbe79 (accessed on 21st October 2025).

31. Kobbacy, K. A. H., Rasmy, M. H., & Vadera, S. (2007). AI and OR in management of operations: History and trends. https://core.ac.uk/download/104325.pdf (accessed on 21st October 2025).

32. Kopardekar, P. H., & Sridhar, B. (2025). Towards autonomous aviation operations: What can we learn from other areas of automation? https://core.ac.uk/download/pdf/286393131.pdf (accessed on 21st October 2025).

33. Palla, T. (2021). Intelligent traffic management using big data analytics and IoT. International Journal for Research in Applied Science and Engineering Technology. https://www.semanticscholar.org/paper/888b834fba61b9690c2aa7fd6f38f4bcd39f9694 (accessed on 21st October 2025).

34. Peng, D., & Yan, W. (2025). Test-time training with adaptive memory for traffic accident severity prediction. Computers. https://www.semanticscholar.org/paper/85638d0e687361896d6845b33d86fa5a8e86d436 (accessed on 21st October 2025).
35. Redondo, J., Yuan, Z., Aslam, N., & Zhang, J. (2024). Cross-layer adaptable contentionwindowforhigh-definitionmapQoSenhancement.2024International Wireless Communications and Mobile Computing (IWCMC). https://www.semanticscholar.org/paper/f0414dc48cb87ad6e6cead15ca1ccdc2b2bfe2e8 (accessed on 21st October 2025).
36. Saldien, J., & Vandevelde, C. (2014). Ono: An open platform for social robotics. https://core.ac.uk/download/55690833.pdf (accessed on 21st October 2025).
37. Thelwall, M. (2016). TensiStrength: Stress and relaxation magnitude detection for social media texts. https://core.ac.uk/download/42606702.pdf (accessed on 21st October 2025).
38. Thiagarajan, R. (2023). Performance of public transport appraisal using machine learning. International Journal on Recent and Innovation Trends in Computing and Communication. https://www.semanticscholar.org/paper/2a0777fac483945ad8c0a2ad84b8b36fbf767f26 (accessed on 21st October 2025).
39. Thiagarajan, R., & Kumar, S. P. (2025). Performance of public transport appraisal using machine learning. The Scientific Temper. https://www.semanticscholar.org/paper/b02825a2544defdd96ea0a4f53e38c02e61d4c7e (accessed on 21st October 2025).
40. Xia, F., Li, J., So-In, C., & Rodrigues, J. (2018). Guest editorial: Big traffic data analysis and mining. IET Intelligent Transport Systems. https://www.semanticscholar.org/paper/28a049a0102b61076688976cd6161270f3d5c672 (accessed on 21st October 2025).
41. Yurtsever, E., Lambert, J., Carballo, A., & Takeda, K. (2020). A survey of autonomous driving: Common practices and emerging technologies. IEEE Access, 8, 58443–58469. https://doi.org/10.1109/access.2020.2983149 (accessed on 21st October 2025).

CHAPTER 6

INTELLIGENT TRANSPORTATION SYSTEMS (ITS)

Contents

INTRODUCTION

The technological growth in transport led to the emergence of Intelligent Transportation Systems (ITS), a system that makes use of advanced information and communication technology to enhance road safety and efficiency. ITS brings about systems with the capability to monitor, manage, and optimize traffic flows in real-time in response to the intricacy of today's transport networks. The basic premise of ITS is to reduce accidents, relieve congestion, and improve travel times through the use of various autonomous and co-operative systems. Governmental agencies, research establishments, and business groups are turning their focus towards ITS because it has huge potential for transforming traditional transport modes into more intelligent, networked infrastructures (Mol et al., 2008).

Figure 6.1. Intelligent transportation systems (ITS).

Source: https://carmanah.com/wp-content/uploads/intelligent-transportation-systems-featured-image-750x750-1.jpg.

It is significant to know about ITS because it signifies two different classes: autonomous systems, which are vehicle-based, and cooperative systems, which are vehicle-to-vehicle and vehicle-to-infrastructure communication-based. Autonomous systems are exemplified by features such as radar-based blind-spot detection and dynamic traffic control by variable signs. Cooperative systems leverage this idea by enabling vehicles to share critical information with one another, such as about traffic situations or danger, enhancing situation awareness for drivers (Mol et al., 2008). Generally, these systems represent a paradigm shift in solving transport problems, emphasizing the integration of technology into the daily travel of individuals, which has the potential to significantly improve safety and efficiency for all users of roads. As ITS technologies continue to develop, the potential for their impact on transport networks means that they need a complete assessment of their effectiveness and deployment tactics. The assessment of the vast array of applications for ITS not only involves technical considerations but also the broader society implications, including public acceptability and regulatory policies. The ongoing research and development that are continuing to bring the spotlight on these organizations' reports emphasize the need for harmonized policy-making that will allow adoption of the technologies. In general, as cities expand and more traffic is built up, the applicability of ITS continues to build, laying the groundwork for revolutionary solutions that have the potential to revolutionize the mobility of persons and goods in and between cities (Figure 6.1).

Definition of Intelligent Transportation Systems

Intelligent Transportation Systems (ITS) is a giant step ahead in optimizing and managing vehicle and transportation infrastructure networks. In effect, ITS applies information technology and electronic communications to road networks and even vehicles, enabling life-saving capabilities and reducing traffic congestion while maximizing overall transportation efficiency. ITS represents a wide spectrum of applications, such as traffic surveillance and routing optimization, and safety improvement. To achieve ITS effectiveness, one must differentiate between its two broad categories: autonomous systems and cooperative systems, both with their unique roles in the evolution of roadway safety and operational efficiency. Further development of these systems is not only meant to enhance user experience but also to solve fundamental safety problems pertaining to modern transport. Diving deeper into ITS categories, autonomous systems too are independent and do not rely on vehicle-to-infrastructure or vehicle-to-vehicle communication. Automatic emergency calls initiated by a crash, electronic stability control, and dynamic traffic management systems utilizing variable message signs to send appropriate information to drivers are some of the examples. On the other hand, cooperative systems depend on communication and cooperation to facilitate real-time alerts on hazardous road conditions or traffic congestion between infrastructure and vehicles. That interconnection heightens driver alertness and stimulates an active approach to roadway safety, ultimately resulting in reduced accident levels and improved traffic flow. The contrast between these systems highlights the intricate character of ITS and underlines the necessity of sophisticated technologies in establishing safer transportation environments. With ITS building pace globally, standards and guidelines are being established to ensure cohesiveness and interoperability of the technologies used. Global collaboration among stakeholders like government, industry, and academia is critical to effective deployment and implementation of such systems. The establishment of standards by organizations such as the ISO has played a key role in this direction, allowing the development of a robust communication framework needed for ITS to function efficiently. Moreover, as the automotive sector moves towards more automated modes, the quest for harmonic convergence between traditional automotive systems and networked infrastructure will require draconian systems engineering paradigms to preclude possible safety problems and ably ensure system response. Thus, the ongoing progression of ITS is not only a technological challenge but an interdisciplinary problem requiring inter-transdisciplinary solutions.

Importance of ITS in Modern Transportation

The introduction of Intelligent Transportation Systems (ITS) has revolutionized modern transport by optimizing the efficacy and security of mobility services in urban and rural environments. ITS is one of the most significant aspects of being compatible with Big Data analytics, which has significantly transformed building intelligent cities. By applying real-time data from various modes of transportation, traffic control can be made more efficient, less congested, and more efficient public transportation service provision. Application of computational intelligence techniques offers new solutions to advanced transport issues, and facilitating forecasting and prevention of traffic jams even before it gets intensified. The intelligent transportation case study also shows how technologies like these could create more responsive transport systems, eventually leading to higher convenience and satisfaction among commuters. Therefore, ITS not only addresses current transport problems but also speculates future innovations in mobility that can contribute to environmentally friendly urban expansion (Doctor et al., 2020). Secondly, the security of transport infrastructure has also gained greater significance in the current world, particularly with growing global security concerns. ITS comes straight to the forefront in terms of improved safety through improved risk assessment and surveillance systems. This is especially remarkable in sea transport, where there are compliance demands that are shifting due to global security norms. Through the use of emerging technology such as wireless communications and radiofrequency identification, seaports are able to maximize operations while simultaneously maximizing safety measures. This technology helps to better manage the handling of goods, thereby maximizing cargo flow as well as augmenting security surveillance aspects. As such, ports that employ ITS technologies are better placed to compete in ever-evolving global markets, thus underlining the necessity of such systems in empowering transport infrastructures to respond to security compliance and operational excellence (Borri et al., 2025). Integration of ITS represents a broader trend of sustainability and efficiency in transportation according to global actions of reducing carbon footprint emissions and promoting ecological responsibility. ITS promotes environmentally friendly practices through reducing routes and minimizing idling times, which lower fuel usage as well as emissions. Moreover, research has indicated that the use of evidence-based solutions leads to more intelligent multimodal transport schemes and logistics that enhance mobility efficiency in general. With rising populations and transport demands, ITS is a key component to ensure transport schemes will be capable of meeting future growth without over-stretching environmental resources. So, ITS importance extends beyond the immediate benefits of operation in that it is likewise brought into harmony with long-term goals of sustainability, which become provided with a great priority in urban planning and policymaking activities (Figure 6.2).

Figure 6.2. Intelligent transportation systems (ITS) has transformed modern.

Source: https://www.marketsandmarkets.com/Images/autonomous-vehicles-and-its.jpg.

6.1. HISTORICAL DEVELOPMENT OF ITS

Historical development of Intelligent Transportation Systems (ITS) dates back to the initial use of information and communications technology in city planning and traffic management. During the latter part of the 20th century, tremendous advances in telecommunications became the backbone of the intelligent systems present today. This period was characterized by the introduction of technologies such as vehicle detection systems and automatic traffic lights, which were designed to maximize traffic flow and safety. These early uses were designed to counteract mounting congestion and were not as advanced compared to today's systems. The evolving nature of city planning necessitated the smarter solutions to transportation problems effectively. As city populations grew, requirements for systems that could provide real-time information and decision authority grew, thus creating a turning point in the development of transport management towards more integrated and responsive systems. The early 21st century was a decisive time in the evolution of ITS through the incorporation of big data analytics and computational intelligence. As data from various sources—mobile phones and GPS—became increasingly ubiquitous, transportation networks evolved to make use of this data in order to enhance operational efficiency and ease of use for users. Smart cities became a top theme, with cities focused on an integrated approach for managing cities combining transportation, economy, and socio-cultural aspects. The innovative utilization of big data not only improved traffic control but also helped in the development of traffic forecasting tools and route optimization. Through this groundbreaking extension, ITS played a key role in addressing urban problems of the present day and, therefore, improving the overall quality of urban life in cities where it was applied (Doctor et al., 2020). With ITS always evolving, ubiquitous eco-cities development necessitated the implementation of large-scale urban management systems that are backed by high-level telecommunications. Enhanced connectivity between various technologies created a seamless flow of information among transport systems and city infrastructure. Telecommunications development allowed for the creation of real-time monitoring systems that would enhance decision-making schemes involving urban mobility. This convergence of telecommunications and information technologies provided a framework for the smart management of cities, improving the travel experience and enabling sustainability interventions. The evolutionary trajectory of ITS thus moved away from isolated innovations into an overarching system, consistent with the general goals of smart urbanism and improved public service provision (Yigitcanlar et al., 2009).

6.1.1. Early transportation Technologies

Early transport technologies made history in human movement and logistics, paving the way for Intelligent Transportation Systems today. Transport initially involved rudimentary processes such as walking, chariots with animals, and primitive boats, which limited the scope and scale of movement. The invention of the wheel around 3500 BCE spurred advanced means of transport, allowing heavier loads and longer distances. The advent of rail transport in the early 19th century represented another monumental leap, incorporating steam engines that enabled efficient movement of goods and people at unprecedented speeds. This transition from mechanized-to-mechanized transport not only enhanced operating effectiveness but also profoundly altered social arrangements, and in due course, urbanization came to thrive with increased connectivity that ultimately laid the groundwork for contemporary ITS patterns that leverage automated, integrated technologies to improve transport networks. As technology evolved, early motor vehicles appeared in the late 19th century, further revolutionizing transport practices. The advent of the internal combustion engine introduced the age of personal and commercial car travel, and this consequent explosion in mobility was accompanied by the building of enormous road networks. All this, with the addition of traffic lights and signs in the early 20th century, witnessed the transition away from primitive transport systems toward more systematic traffic management strategies.

Figure 6.3. Early transportation technologies Systems.

Source: https://geospatialmedia.s3.amazonaws.com/wp-content/uploads/2018/05/hyperq.jpg.

These initial advancements in transport emphasized the need for systematic responses to address increasingly crowded roads, anticipating today's ITS technologies that draw upon revolutionary telecommunication advancements. These developments, as described by (Yigitcanlar et al., 2009), brought with them the bringing together of various systems to facilitate better control of the urban area as well as a more general quality of life within the urban space, moving towards the idea of smart and connected transport networks. Advances in transportation automation and robotics in the second half of the 20th century further enhanced the development of modern ITS platforms. Examples of some of the first automated systems were basic cruise control in vehicles and navigation system development, which significantly improved route optimization. Incremental embedding of sensors and data-centric technologies into transport infrastructure platforms marked a milestone for intelligent system design. Not only did these technologies optimize motor performance and safety, but they also facilitated real-time decision-making processes critical in modern traffic management. As Kopardekar et al. (2025) argue, this evolution from basic automated systems to complex autonomous networks predicts a critical move towards higher levels of automation in transportation, paving the way for the modern ITS that seeks to maximize operating efficiency and safety within a modernizing transportation landscape (Figure 6.3).

6.1.2. Evolution of ITS From the 20th Century

The evolution of Intelligent Transportation Systems (ITS) during the 20th century is defined by a striking shift in technological abilities and approaches to managing cities. Towards the end of the 20th century, the development of advanced information and telecommunications technologies created the basis for ever more integrated systems of city management. Technologies were developed at this time that had the potential to monitor and manage transport networks through wired as well as wireless technologies. The incorporation of these technologies not only made these services accessible but also enhanced the overall quality of urban living. With improvement in sophistication in these systems, they began to incorporate real-time information to feed decision-making, which enabled the cities to respond more reactively towards traffic patterns and accidents, ultimately paving the way for contemporary ITS solutions based on efficiency and sustainability (Yigitcanlar et al., 2009). On entering the 21st century, ITS development gained increasing priority, driven by advances in technology as well as mounting urbanization problems. The merging of sensors and communication technologies revolutionized roadways, enabling them to have features like vehicle-to-infrastructure (V2I) communication. What this shift of paradigm implies is not just that vehicles talk to each other, but that they also interact with the infrastructure around them, making transportation safer and more efficient to operate. ITS platform flexibility was achieved through municipalities seeking tailored solutions to support increasingly advanced transportation systems. In addition, the application of intelligent urban management systems was recognized as a key strategy in the creation of Ubiquitous Eco Cities, where technology would be instrumental in promoting sustainable city living and carefree mobility (Cantisani et al., 2017). Additionally, increased focus on roadside safety in the ITS context was a critical turning point during this period. Since some of the fundamental technologies in roadside barriers remained static for decades, new approaches began introducing new materials and design philosophies with the goal of improving safety performance. These developments acknowledged the dynamic interaction between vehicles and roadside infrastructure, particularly with the introduction of autonomous vehicles. The use of embedded sensors in safety barriers is a classic case in point, enabling real-time sensing and analysis to inform real-time responses to an accident or congestion. As ITS advances further, the integration of such advanced safety features reflects commitment to leveraging technology in managing the complexity of transportation systems today (Cantisani et al., 2017).

6.1.3. Key Milestones in ITS Implementation

The history of Intelligent Transportation Systems (ITS) is marked by a series of major milestones that have basically improved transportation safety and efficiency. One of the earliest of these milestones was the development of Advanced Traffic Management Systems (ATMS) in the latter half of the 20th century, which used real-time data collection and control of traffic signals to reduce congestion. This was a major departure from the old traffic flow strategies to data-intensive ones that would be capable of managing congestion ahead of time. Integrating multiple communications technologies also accelerated ITS evolution, enabling better coordination between vehicles and infrastructure elements. Such early applications laid the foundation for subsequent systems that placed heavy emphasis on the importance of glitch-free connectivity among infrastructure elements and vehicles. The turn of the century saw utter innovation in car technology bringing about the concept of Vehicle-to-Everything (V2X) communication, which took the discussion further than just vehicle-to-vehicle (V2V). The historic mile broke the path for enhanced security features and traffic management features where cars could communicate with roads and other road users. The development of these systems was a direct reaction to the increased complexity of the transport network, which required scalable means of managing the soaring numbers of vehicles on the roads. In addition, these communication technologies had to redefine design and operations in the auto sector, hence changing the traditional methods and clearing the way for the creation of more self-governing vehicle solutions. With the evolution of ITS technology, the launch of Autonomous Vehicles (AVs) was a significant milestone towards the development of safer and more efficient transportation systems. Implementation of Hierarchical Emergent Behaviors (HEB) in AV communication strategies signifies a significant achievement in overcoming the complexities and scalability issues in autonomous driving. By applying simple rules for local regulation, HEB promotes adaptive vehicle behavior that is capable of navigating surprising circumstances. The paradigm shift here is to depart from the rigid programming and approach more adaptive, flexible systems, both improving safety and efficiency on roads. Thus, ITS implementation landmarks reflect not just technological but strategic advancements which reshape the larger transport landscape, positioning it well for future evolution (Ang et al., 2017; Milito et al., 2020) (Figure 6.4).

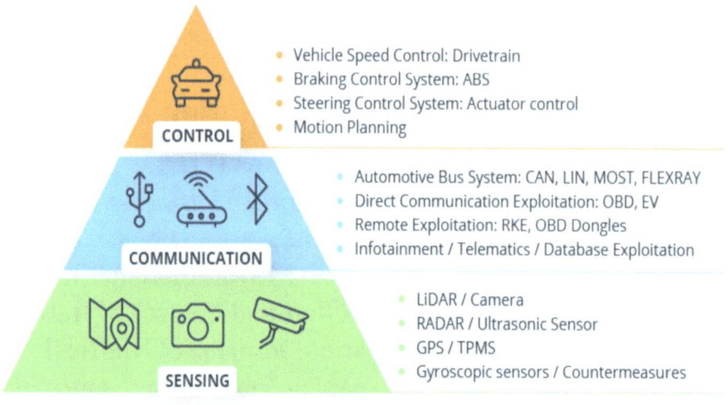

Figure 6.4. Key milestones in ITS implementation.

Source: https://d17ocfn2f5o4rl.cloudfront.net/wp-content/uploads/2021/09/bp-intelligent-tms-grapics-1.png.

6.2. COMPONENTS OF INTELLIGENT TRANSPORTATION SYSTEMS

A critical component of Intelligent Transportation Systems (ITS) is the integration of autonomous and cooperative systems, each taking on a particular function in ensuring traffic safety and efficiency. Autonomous systems, such as electronic stability control and dynamic traffic management technologies, act independently to facilitate safer driving without requiring interaction with other vehicles or infrastructure. Such systems are critical to monitoring personalized car performance and efficiently managing traffic flows, thereby ensuring fewer accidents and congestion. As pointed out in literature, emphasis on such autonomous systems is key, particularly with government and industry regulators looking to witnessing significant improvements in road safety as well as operating efficiency through their adoption (Mol et al., 2008). Cooperative systems represent another key component of ITS, with the aim of vehicle-infrastructure communication to further improve traffic management. Cooperative systems are distinguished from autonomous systems in that data exchange is utilized to notify drivers of possible dangers, such as weather or traffic conditions. Cooperative action has the potential to significantly enhance situational awareness and pre-emptive measures, reducing the likelihood of crashes. For instance, vehicle-to-vehicle (V2V) or vehicle-to-infrastructure (V2I) communication technologies enable real-time reporting and warnings that can guide drivers to adjust their behavior in response to changing traffic conditions (Mol et al., 2008). The potential returns of such systems in the future are humongous, with research estimating that they would cut accidents by as much as 50%, putting their critical role in the traffic management programs of the future in perspective. Apart from these fundamental components, a robust simulation and modeling platform improves the functioning of ITS. Due to the complexity inherent in actual traffic networks, replicating several traffic scenarios provides valuable information without the interfering impact of physical changes. By employing cloud-based systems and adhering to standards such as IEEE's High-Level Architecture (HLA), stakeholders are able to facilitate interoperability between various simulation tools to be able to work together when analyzing transportation systems. This multi-resolution approach empowers experts to examine transport dynamics comprehensively, and eventually make more informed choices regarding traffic policy and infrastructure design. Through the utilization of complex simulation techniques, stakeholders are able to leverage the maximum performance and safety of intelligent transport solutions in tangible, real-world terms (Figure 6.5).

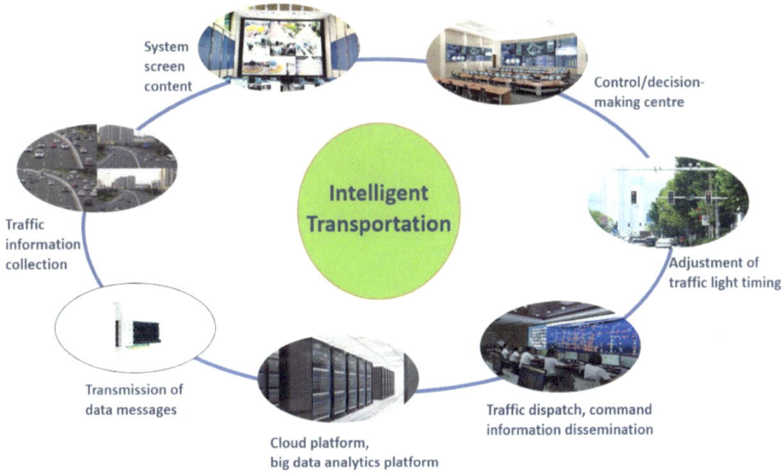

Figure 6.5. Components of intelligent transportation systems.

Source: https://www.lr-link.com/attached/image/20231023/20231023202128522852.jpg.

6.2.1. Traffic Management Systems

The evolution of Traffic Management Systems (TMS) is a landmark achievement in Intelligent Transportation Systems (ITS), which allows for enhanced vehicular mobility and road safety. By assimilating real-time data from disparate sources such as traffic cameras, sensors, and road sensors, TMS can monitor congestion levels dynamically. This can allow proactive control over traffic signals and road usage to alleviate bottlenecks, facilitating overall mobility. Most importantly, such systems can provide connectivity between vehicles and infrastructure, a function that becomes increasingly important with the advent of connected and autonomous vehicles (CAVs). Integrating CAVs in TMS increases operational efficiency and facilitates eco-mobility in cities, all leading to lower travel times and energy consumption. One of the most promising applications of TMS is the optimization of merging operations, particularly for situations involving traffic flowing off on-ramps onto highways. Traditional merging policies typically fail to take into account vehicle dynamics and state changes at the moment of merging, leading to inefficiencies as well as increased risks.

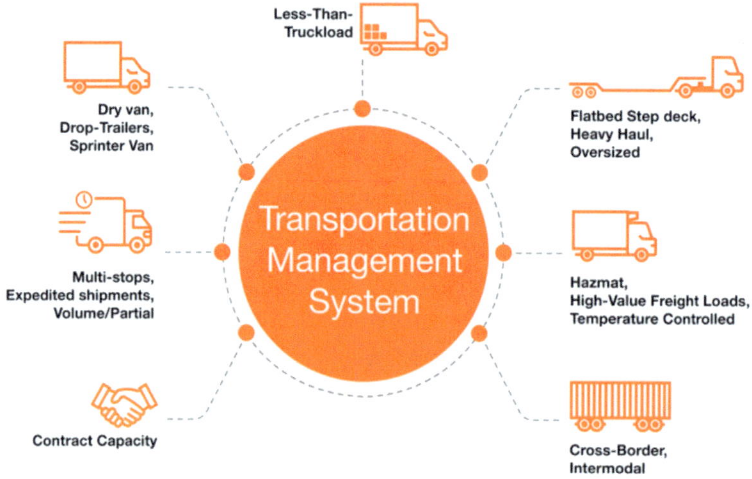

Figure 6.6. Traffic management systems (TMS).

Source: https://www.gw-world.com/fileadmin/_processed_/0/1/csm_TMS_Icons_2187017ee5.png.

However, advances in TMS can offer smoother merging procedures through cooperative maneuvers offered by CAVs. According to research, the hierarchical ramp merge system enhances merging zone traffic flow through the regulation of ramp inflow rates and system mobility. Beyond smoothing the merge, this preemptive system suppresses congestion-induced delay significantly, sustaining the merit of real-time traffic management in urban areas, as established under extreme simulations with a corresponding boost in average speed. Further, traffic management and energy efficiency overlap is an area of growing importance with respect to connected and autonomous electric vehicles

(EVs). A data-driven decision-making facilitated TMS has the capability to optimize speed profiles of vehicles, minimizing stop-and-go driving and highly optimizing energy efficiency. Optimized traffic management has the potential to save energy for CAVs by intelligently adjusting speed limits based on real-time conditions, according to research. With vehicle-to-vehicle communication and on-board sensors, CAVs are able to fine-tune their movements to make traffic flow smoother and lower energy usage. Lastly, the implications of introducing cutting-edge TMS into the transport network extend beyond traffic efficiency; they are an all-embracing approach to emission reduction, environmentally friendly urban mobility, and ushering in possibilities for future development in Intelligent Transportation Systems (Figure 6.6).

6.2.2. Vehicle-to-Everything (V2X) Communication

The concept of Vehicle-to-everything (V2X) communication is a foundation stone of Intelligent Transportation Systems (ITS), and it increases communication between vehicles and the external world significantly. V2X is employed for encompassing all various forms of communication like Vehicle-to-Vehicle (V2V), Vehicle-to-Infrastructure (V2I), and Vehicle-to-Pedestrian (V2P) so that real-time information can be exchanged to increase traffic control and safety. Through the use of V2X technologies, vehicles can share critical information on road conditions, traffic, and hazards so that decisions can be made in real-time to avoid accidents and enhance the overall mobility experience. Besides, V2X deployment in ITS systems complements global visions toward the development of safer, cleaner, and more efficient transport systems, and requesting analysis and field testing of these communication protocols to determine they are optimally suited for development, as emphasized by Cartolano et al. (2008). To obtain the high goals designated by ITS programs, there is a need to appreciate the global ramifications of V2X communication on traffic and environmental effects. Road safety, smooth traffic movement, and controlling pollution by ensuring the best utilization of fuel are the challenges that need to be met by automobile manufacturers at present (Denis et al., 2014). Leveraging connected car technologies for enabling V2X interaction allows automobile manufacturers to move their attention away from traditional means of transport to smarter systems for monitoring and optimizing driving habits. This shift changes not only how autos operate but how drivers conceptualize driving life as well, understanding automobiles as a continuation of virtual life—similar to smartphones. While these developments are occurring, a significant aspect is to fill the communication gap between autos and road facilities so that innovations become user-centric while addressing society's problems. V2X communication deployment involves extensive testing and verification processes to guarantee the maximum capability of the technology. Specifically, the establishment of legitimate testbeds and operation models is required for conducting field tests that can comment on the real performance of V2X applications for real traffic conditions. Such tests enable researchers and practitioners to identify the merits and demerits of wireless vehicular cooperative systems in managing traffic (Cartolano et al., 2008). As the industry evolves, the successful implementation of V2X in ITS has to be taken with a holistic approach that deals with the technical, regulatory, and social dimensions of transport innovation. Ultimately, by encouraging collaboration between stakeholders like vehicle makers, infrastructure providers, and technology companies, V2X communication can be leveraged to transform the future of intelligent transportation systems into one that is safer and greener (Figure 6.7).

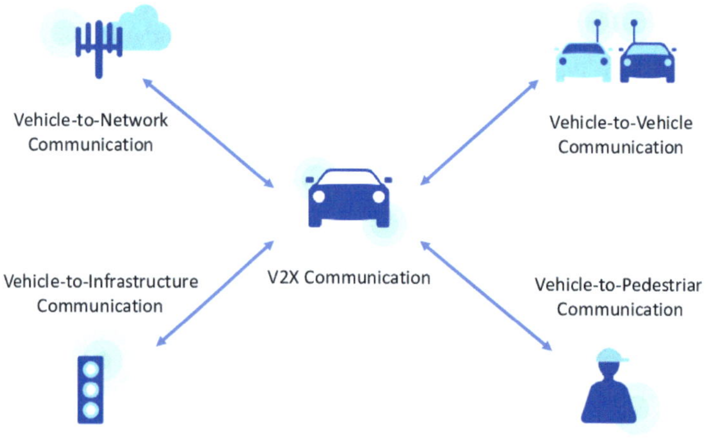

Figure 6.7. Vehicle-to-everything (V2X) communication.

Source: https://emizentech.com/blog/wp-content/uploads/sites/2/2021/05/Towards-seamless-ubiquitous-vehicle-to-everything-V2X-communication.png.

6.2.3. Data Collection and Analysis Technologies

The successful functioning of Intelligent Transportation Systems (ITS) relies heavily on advanced data acquisition systems which facilitate traffic management and policy development. Traditionally, data for transport has been created through fixed sensors and manual counts, which had limited analysis scope. However, with the advent of GPS-enabled devices and smart technologies, real-time data acquisition is now achievable through individual mobile devices and vehicles. This crowd-sourced data upturns traditional approaches by providing a more comprehensive and dynamic representation of traffic. As in the case of transport simulation platforms, communication between sources of incompatible information is paramount for seamless analysis. With the provision of cloud-based solutions and commonplace frameworks, transport experts are able to collaborate in investigating complex matters, facilitating responsive and knowledge-informed decision-making in urban planning and traffic management (Azevedo et al., 2015).

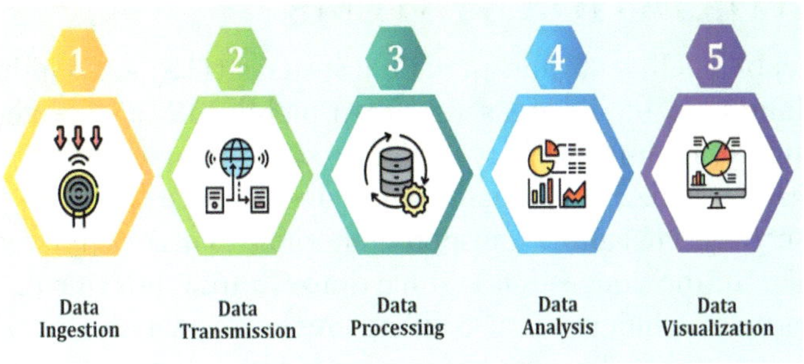

Figure 6.8. Data collection and analysis technologies.

Source: https://psiborg.in/wp-content/uploads/2024/04/IoT-Data-Collection-2.webp.

Besides, the quality of collected data matters most, particularly traffic speed data as it is subject to significant vulnerabilities from gaps and inconsistencies in crowd-sourced measures. It has been established through research that traffic speed data imputation plays a critical role in enhancing data-driven analyses in transport to be more effective. The adoption of multi-output Gaussian processes (GPs) provides a pragmatic solution, as it can capture complicated spatial and temporal trends in the data within crowdsourced traffic streams. This approach addresses uncertainty in missing observations effectively, enabling practitioners to draw more trustworthy conclusions from the data. As indicated in studies of probe-vehicle data in cities, the enhanced imputation methods are superior to traditional ones, promising improved decision support use for traffic management (Henrickson et al., 2019). The application of such advanced data collection and analysis technologies in ITS is uncovering the door to more effective and cleaner transport solutions. Through real-time monitoring and adaptive management, intelligent systems can react in advance to traffic and congestion levels, thus boosting overall efficiency. In addition, the cooperative nature of these technologies makes it possible to have a unified approach to urban mobility, with participation from government, academia, and private industry stakeholders. As cities expand and traffic complexity increases, ongoing innovation and deployment of new data technologies will be fundamental to the future development of transportation systems. Enhanced data integrity and integration among different data sources will ultimately enhance the versatility and effectiveness of ITS in addressing challenges for tomorrow's cities (Figure 6.8).

6.3. BENEFITS OF INTELLIGENT TRANSPORTATION SYSTEMS

The benefits of Intelligent Transportation Systems (ITS) are multifaceted and extend to long-term implications for urban mobility. At the center of ITS is the integration of high-technology devices such as computer vision and sensor networks that maximize traffic management and enhance road safety. Real-time data from such systems allow transport authorities to make informed decisions that minimize traffic congestion and maximize traffic movement. The use of machine learning techniques is also aiding predictive analytics, in which traffic patterns may be analyzed to anticipate and avoid jams even before they escalate into a traffic problem. This pre-emotive management benefits commuters by saving them time but also from the perspective of fuel efficiency and reduced emissions, thereby projecting the environmental benefits of ITS installation (Kamble, 2024). In addition to maximizing traffic efficiency, ITS initiatives also advance public security. Due to new technologies such as autonomous vehicles and intelligent traffic lights, the probability of accidents can be greatly reduced. To illustrate, real-time alerts on road conditions, accidents, or roadwork provide drivers with critical information that may influence their choices on the road. Besides, the employment of ITS has the potential to make passage through cities safer and more efficient, hence decreasing the risk of accidents. Improved coordination of emergency responses is also possible, as ITS optimizes routes for emergency services to respond as fast as can be expected in critical cases. This coordinated approach towards safety not only guards individual members of the transport system but also fosters the health of communities as a whole (Azevedo et al., 2015). Also impressive is the economic benefit of implementing Intelligent Transportation Systems. ITS systems provide cost benefits for the private sector and the public sector through decreased delay and enhanced traffic flow. Time saved in travel is equivalent to added productivity for the private sector as well as commuters. Furthermore, the operation of ITS can bring about reduced vehicle operating costs resulting from decreased fuel consumption, together with decreased vehicle wear and tear. Such benefits to the economy apply at the societal level as well, where improved mobility can stimulate local economies by enhanced access to employment and services. As metropolitan areas contend with the reality of urbanization, the economic rationale for the adoption of ITS becomes increasingly irresistible, marking a significant step on the way to sustainable city development (Kamble, 2024) (Figure 6.9).

Figure 6.9. Benefits of intelligent transportation systems.

Source: https://www.loginextsolutions.com/blog/wp-content/uploads/2025/04/image-3-2.jpg.

6.3.1. Improved Traffic Flow and Reduced Congestion

Incorporation of Intelligent Transportation Systems (ITS) has played an important role in enhancing traffic movement and minimizing congestion in urban areas. Through the application of data-centric technologies, including real-time traffic monitoring and adaptive control of traffic signals, cities can better control vehicular movement. These technologies can examine traffic flow and adjust signal timing accordingly, leading to minimized waiting times at intersections and consistent traffic movement along congested corridors. These developments not only improve travel dependability but also reduce fuel usage and emissions, attesting to the environmental advantages of ITS. The prospects for future research are vital to the further development of these technologies; for instance, the evaluation of congestion control measures illustrates the use of state space control theory to optimize traffic signal performance at peak times, providing a guide to future advances in urban mobility (Shepherd et al., 1992). In addition, the use of ITS is not limited to automobile control but also encompasses multi-modal transportation solutions, thus incorporating the use of public transit solutions that reduce congestion on the roads. The use of smart transit systems, which offer real-time information and location tracking, promotes the use of buses and trains. This transition from private cars to public transport not only minimizes the number of vehicles on the road but also creates a sense of community about different modes of travel. Such cities show notable decreases in peak hour traffic congestion and better access for all commuters.

This congestion management method serves to demonstrate how ITS can revolutionize urban transport through support of a multi-mode system that integrates different modes of travel, which will eventually improve the entire transportation network's efficiency while taking care of environmental issues. Novel technologies linked with ITS also lead to further advancements in transport capabilities on the way forward, further mitigating congestion on a systemic scale. As described in current research, new aircraft technologies are capable of preventing air traffic congestion and are comparable to advancements in ground-based transportation systems. The study emphasizes enhancing the production of new vehicles by tapping the existing aeronautics technology to maximize efficiency, safety, and environmental friendliness (Albers et al., 2025). Investing in such new technologies can make both air and ground transportation systems develop an integrated system that effectively counteracts congestion problems. This integration of various forms of transportation is critical to developing economic growth and improving the competitive advantage of industries that are dependent on effective logistics and transportation operations. Briefly, the ongoing development and coordination of ITS present an integrated solution to traffic congestion problems in contemporary cities.

6.3.2. Enhanced Safety for Drivers and Pedestrians

The introduction of Intelligent Transportation Systems (ITS) has immensely enhanced the safety of drivers as well as pedestrians by virtue of advanced technology. One of the most significant advancements is the utilization of demand-responsive transverse rumble strips, which enhance vehicle awareness without posing noise pollution. They respond to pedestrian presence or changes in traffic patterns to warn drivers on time without offering continuous interruption like traditional rumble strips. Research indicates that these strips effectively generate in-vehicle audible noise and vibration, helping to warn drivers to hazards downstream without generating an uncomfortably high level of noise for home settings. It is hence the application of traffic calming interventions for community needs that represents a real safety improvement, achieving a more amiable pedestrian setting without sacrificing vehicular perception (Hossen et al., 2019). Besides, image processing technology innovations bring thrilling possibilities for detecting dangerous driving behaviors, enhancing further road safety. For instance, algorithms to detect careless driving through wide-angle camera sensors offer an innovative answer to a centuries-old problem long underserved by research. By applying image processing techniques, such systems can analyze driving behavior in real-time and alert authorities to drivers whose record might otherwise go unnoticed. This is particularly critical in densely populated areas, where reckless driving constitutes a tangible danger to drivers as well as the public at large. Successfully understanding and diverting these behaviors with ITS can lead to a drastic reduction in accidents, more appropriately empowering law enforcement and drivers with the tools to promote prudent driving practices across urban landscapes (Haloi et al., 2014). In addition, the combined application of ITS technologies not only increases safety for road users but also supports a holistic traffic management strategy. Through the combination of different elements, including vehicle-to-infrastructure messaging and real-time data analysis, cities are able to create systems that respond to varying traffic patterns and improve pedestrian crossing. By adjusting traffic lights dynamically based on pedestrian usage intensity, cities can also protect vulnerable users while improving overall traffic flow. This anticipatory approach not only maximizes safety but also improves a sustainable harmony of vehicular and pedestrian flow. Those advances create a strong argument for ongoing ITS investment, demonstrating that higher safety features not only save lives but also increase overall efficiency in transport systems (Figure 6.10).

Figure 6.10. Enhanced safety for drivers and pedestrians.

Source: https://www.researchgate.net/profile/Hassan-Masoomi/publication/329233150/figure/fig1/AS:745577381916672@1554770908089/nteraction-between-AVs-and-pedestrian-AVs-and-driver-Vehicles-to-Vehicles-photo.jpg.

6.3.3. Environmental Benefits and Reduced Emissions

Intelligent Transportation Systems (ITS) is a critical means for maximizing environmental benefits and reducing transport-related emissions. ITS employs new technologies such as real-time aggregation and analysis of information to optimize traffic flows to the maximum, hence reducing idling times and fuel usage. Through optimizing traffic flow, ITS not only maximizes the efficiency of vehicles but also achieves an enormous reduction of greenhouse gas emissions. Moreover, the inclusion of smart traffic systems will enable public transport to become more effective in timetables, making it easier for individuals to make a transition from personal to public transport.

This transition also decreases traffic and leads to reduced overall emissions. The synergy of these technologies testifies to the potential of ITS in creating a green transport system that balances environmental health with growing needs for urban mobility. In addition, ITS facilitates the implementation of measures that support environmentally sustainable transport modes. Recognized in several studies, tax incentives in the form of tax cuts targeted at supporting public transport modes such as car sharing and pooling are effective instruments for reducing emissions.

These incentives can make a notable difference in travel decisions by making alternatives to driving alone more attractive, hence encouraging improved use of resources. Besides, enhancing network connectivity and access through ITS not only makes public transport more attractive but also helps bring about the cultural transformation of using environmentally friendly modes of travel. Where fuel efficiency and emission-reducing measures are combined, measurable benefits are achieved, bringing on large greenhouse gas emissions reductions from the transportation sector (N/A, 2009). Public perception and realization of those benefits are central to developing effective environmental policy for transportation. Growth in greenhouse gas emissions from the transportation system is a matter of critical concern, particularly as urbanization rates continue to rise globally. This growth has prompted scientists and legislators to look for methods effective in stopping emissions, with ITS being a valid contender.

The rapid growth of the industry highlights the need for embracing new transport policies that accommodate diverse reactions from experts to environmental impacts (Certa et al., 20). Using analytic hierarchy processes, decision-makers can determine the best way of promoting sustainable transport, thus improving the energy efficiency, and reducing the carbon footprint of urban mobility systems. The collaboration of stakeholders in the use of ITS solutions not only addresses existing environmental challenges but also facilitates the accomplishment of long-term climate ambitions, offering support in the restructuring of transportation policies to be economically viable and environmentally friendly.

6.4. CHALLENGES IN IMPLEMENTING ITS

The implementation of Intelligent Transportation Systems (ITS) involves numerous challenges, especially in the integration of interlinked technologies. As ITS relies heavily on cloud platforms and edge computing, integrating these technologies is a great challenge. More specifically, developers are faced with compatibility, functionality, and performance issues in an interlinked environment. Another vital aspect of this integration deals with ensuring effective communication between various devices and systems, something that is vital in an effort to achieve the anticipated efficiency of ITS. Without addressing these integration problems, the systems will not deliver the traffic management and safety enhancements they are meant to deliver, ultimately hindering the overall efficiency of ITS projects in urban centers (Alhagagi et al., 2018).

Figure 6.11. Challenges in implementing intelligent transportation systems (ITS).

Source: https://cdn.continental.com/fileadmin/__imported/sites/corporate/usa/english/hubpages/10_20press/10_20press_20releases/20_20local/2021/12_27april_2021_continental_and_iteris/intelligent_infrastructure.jpg.

Moreover, cybersecurity is a pressing matter in ITS construction. Increasing connectivity between vehicles and infrastructure makes them susceptible to hacking by those who can have devastating impacts on public safety. As noted by research, security issues ought to be considered through ITS design and deployment processes, so that traditional approaches may prove insufficient (Alhagagi et al., 2018). Identification of high-risk areas with poor security measures is also critical in an effort to reduce risks. Without end-to-end cybersecurity, ITS deployment may be severely compromised, deterring investment in the systems by investors and ultimately hindering the progress of intelligent transportation technology. Another challenge that comes with the deployment of ITS is the need for robust application development platforms that sustain CV applications. The applicability of such apps, addressing safety, mobility, and environmental concerns, relies on their ability to be developed, tested, and debugged within real-time domains. The development of an industrial-grade, scalable, and secure platform such as CVDeP is the key to enabling developers to offer assurance on functional requirements such as latency and throughput. But setting them up is very time and resource-intensive, as well as requiring ongoing testing and verification to ensure their effectiveness. Without resolving these basic concerns, the promise of ITS to make transportation more efficient and safer can be left unrealized, creating another obstacle to widespread acceptance and implementation (Figure 6.11).

6.4.1. High Costs of Technology and Infrastructure

The implementation of Intelligent Transportation Systems (ITS) is accompanied by huge capital expenses, largely because of the costs of working on the technology and modernizing infrastructure. As municipalities attempt to integrate sophisticated technologies such as automated traffic control systems and vehicle-to-infrastructure communications networks, the initial cost can be astronomically high. Apart from funding the different elements of investments, including sensors and communication systems, constructing and reconfiguring existing road infrastructure, there is a risk of a lack of public funds for local and national authorities to give top priority to ITS investment because of the enormous list of applications that it can be employed for Mol et al. (2008). The necessity of the public sectors and private industries to work together increases the complexity of the funding process, but another such high-demanding field for a collective strategy in controlling these high costs. Aside from the initial investment in direct capital, the on-going operational costs and maintenance required with ITS technology are further added challenges. They need to be constantly updated and supported with technology to keep them working effectively and serving the public, and this can become costly over time. Further, as newer technologies become available, there is a certain expectation that older systems will conform, which can mean unnecessary costs for upgrades that may not bring short-term value (Gray et al., 2006). While cities grapple with the demands of environmentally friendly modes of transportation, the financial outlays of ongoing maintenance can divert funds away from other critical aspects of public service and thus impact the greater needs of communities. Policymakers must thus consider not only their own short-term budgets but also the long-term cost of ITS maintenance on finances. Reduction of the financial burden of ITS is paramount to leveraging the maximum capacity of such a system towards improved traffic flow and safety. Strategic cooperation among government, the private sector, and academia can be employed to reduce costs by pooled resources and expertise (Mol et al., 2008). Further, emphasizing public education and involvement in the planning process can lead to an informed citizenry willing to fund projects. Studies showing successful ITS adoption in other locations can serve as effective case studies to gain grassroots backing for technology investments. Ultimately, success in balancing the high technologies and infrastructure will require innovative strategies that invest heavily in cooperation while ensuring residents understand the long-term values of implementing Intelligent Transportation Systems into daily life.

6.4.2. Data Privacy and Security Concerns

The arrival of Intelligent Transportation Systems (ITS) has transformed city mobility at its heart, bringing with it an abundance of benefits. However, these advancements are accompanied by robust data security and privacy concerns. As ITS rely heavily on networked devices and Internet of Things (IoT), they create vast volumes of data that contain personal and operational information. This interconnectivity not only enhances the efficiency of transport services but also exposes the sensitive information to unauthorized usage, data loss, and cyber-attacks. The severity of data management in ITS demands efficient privacy protection policies for individuals and maintaining the operational integrity of the systems, and it is therefore crucial that researchers and practitioners address these threats holistically through innovative security mechanisms and frameworks (Kotian et al., 2025). With enhanced concerns regarding data privacy, centralized architecture that is typically employed in ITS amplifies these issues particularly with user data spread across multiple platforms. Relaxedness in adequate privacy rules typically results in spilling out of user details, and this incites ethical issues regarding consent as well as data ownership. Thus, alternative models need to be pursued that give equal priority to user privacy as with data availability. Among the promising strategies is the development of Blockchain-based systems, facilitating decentralized data management. Such systems offer greater security in terms of unchangeable ledgers and smart contracts, ensuring that data exchanges are authenticated without compromising user anonymity. ITS can achieve a balance between compulsory use of data for operational optimization and protection of personal privacy via the adoption of blockchain technology (Qureshi et al., 2023). Apart from decentralization through blockchain, there are numerous methodologies that ensure privacy and security in ITS. Technologies such as differential privacy and homomorphic encryption present feasible alternatives by making data analytics possible without revealing sensitive user information. These methods enable transportation agencies to gain insights from aggregated data while keeping individuals anonymous. Also, employing multilayer protection systems that have been tailored to specific transportation environments can minimize risks associated with cyber-attacks. Consequently, an arrayed approach for protecting data in ITS incorporating cutting-edge encryption techniques, robust access control, and proactive threat detection techniques will be crucial towards mitigating natural data privacy and security issues in this rapidly emerging sector (Kotian et al., 2025) (Figure 6.12).

Figure 6.12. Data privacy and security concerns of intelligent transportation systems (ITS).

Source: https://www.nexcom.com/UploadFiles/FCKeditor/image/Product_News/170914_ITSWC2017/ITSWCboothgraphics.jpg.

6.4.3. Integration with Existing Transportation Systems

The integration of Intelligent Transportation Systems (ITS) with existing transport infrastructure is essential to enhance urban mobility and sustainability. With cities growing at an exponential rate, there is a need for efficient management of transport, necessitating the integration of advanced technologies. By integrating ITS into existing infrastructure, cities will have the ability to utilize real-time information in managing traffic flow, minimizing congestion, and maximizing security. This integration creates a harmonious linkage among different modes of transport, such as public transport, private vehicles, and non-motorized transport, to build an inclusive transportation system. Additionally, simulation methods can facilitate this integration and help policymakers analyze the performance of different scenarios without experiencing the cost and dislocation of physical alterations to the transport system (Azevedo et al., 2015). So, effective integration is aiding in strengthening the resilience and adaptability of urban transport systems in the face of rising pressures. Beyond facilitating streamlined transportation, ITS integration into ongoing systems supports sustainable urban development through environmental stewardship.

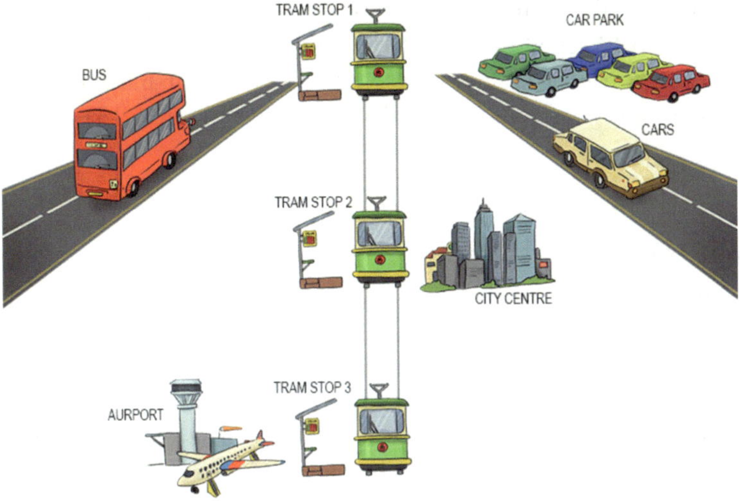

Figure 6.13. Integration with existing transportation systems.

Source: https://mammothmemory.net/images/user/base/geography/Urban%20Issues/integrated-transport-systems-in-urban-issues-geography.e945fad.jpg.

The combination of new technology with traditional transport can assist in the introduction of low-emission vehicles, encourage carpooling, or enable smart public transport systems. For example, vehicle-to-infrastructure (V2I) communication can enhance the interaction of individual vehicles with the transport network as a whole, enhancing routing effectiveness and overall emissions reduction. Moreover, integrating telecommunication technology into urban systems, as identified in recent studies, emphasizes the importance of a comprehensive system that integrates economic, socio-cultural, and environmental factors (Yigitcanlar et al., 2009). Thus, its incorporation within existing transport systems not only addresses immediate traffic issues but also long-term urban resilience and environmental sustainability. Besides, effective incorporation will require appropriate planning and coordination among a number of players including government, private institutions, and nongovernmental bodies. The character of urban transport systems necessitates interdisciplinary measures where technologies and approaches supplement one another. This collaboration enables the establishment of user needs and putting in place data-exchange policies that are optimized for interoperability across modes of transport. In addition, promoting an open decision-making environment can help ensure that the deployment of ITS is responsive to the priorities and issues of a range of communities. Through cooperation between professionals handling transport complexities, new possibilities exist, and this can mean the actualization of a better integrated, responsive, and robust urban transport system (Azevedo et al., 2015). Ultimately, integration into existing systems is necessary for ITS effectiveness maximization and for guaranteeing ITS contribution towards enhanced urban mobility (Figure 6.13).

6.5. CASE STUDIES OF SUCCESSFUL ITS IMPLEMENTATION

The use of Intelligent Transportation Systems (ITS) has been impacted to a significant extent by a series of successful case studies proving the value of such advanced technologies. One particularly persuasive example involves the development of better map-matching algorithms meant to resolve spatial ambiguity in transportation. In these cases, traditional map-matching algorithms break down when spatial data collected using GPS are incorrect or improperly integrated into digital maps. Development of a novel topological decision-rule algorithm provides an important solution, permitting data points to be accurately matched with roadway centerlines, thereby improving navigation and reducing error rates in real-time applications. As evidenced by experimentation in several geographical settings, such improvements demonstrated by this algorithm have been critical in optimizing the overall performance of ITS, facilitating easier operation within metropolitan traffic systems and maximizing the usability of data in traffic control (Blazquez et al., 2012). Another excellent example of successful ITS implementation is Californias Bus Rapid Transit (BRT) systems, which have been planned using the help of meticulous planning and coordination among various stakeholders. The California Department of Transportation (Caltrans), in association with the Mineta Transportation Institute (MTI), initiated a project to create a guidebook that would help transit agencies improve the design and operation of BRT systems on state facilities. This project took inspiration from detailed case studies of effective BRT schemes in major Californian cities such as San Diego, Los Angeles, and San Francisco. By collating opinion from municipal authorities, transit operators, and citizens, the program arrived at decisive legislative overhaul and regulation changes to ease BRT introduction. Therefore, the collaborative effort has not only enhanced the operational efficiency of public transit but has also boosted public engagement in transit planning (Gray et al., 2006). Furthermore, successful ITS implementations can significantly enhance urban mobility and alleviate congestion using innovative technologies and collaborative planning. One of the most significant examples is integrating intelligent traffic management systems utilizing real-time data to optimize traffic flow through urban corridors. By installing sensors and communication systems, cities are able to monitor traffic behavior and adjust signal timing in real time, consequently reducing bottlenecks and commuter waiting times. Their inclusion has been found to be especially helpful during rush hours and has been associated with higher public transit ridership as commuters enjoy more dependable travel times. Case studies featuring intelligent traffic management not only demonstrate the means by which ITS can transform urban transport but also reaffirm the importance of technology-based solutions to solving contemporary transport problems and thereby the need for continued investment and innovation in this area (Figure 6.14).

Figure 6.14. Case studies of successful ITS implementation.

Source: https://oil-gas.net/wp-content/uploads/2024/11/Understanding-Turbo-Expanders-in-LNG-Plants-scaled.jpeg.

6.5.1. Examples From Major Cities Worldwide

Cities everywhere are increasingly depending on Intelligent Transportation Systems (ITS) to address the challenges of urban mobility and maximize the efficiency of overall transport systems. Singapore's smart traffic management system, for instance, relies on real-time data analytics to optimize traffic flows and reduce congestion. With sensors and cameras, the system optimizes traffic signal timings based on real-time road conditions, cutting travel time drastically. This innovative urban transport management style has significantly increased the efficiency of existing infrastructure, demonstrating the profound role of technology in urban planning today. Similarly, cities like Barcelona have integrated ITS solutions that allow integration of public transportation, including bus and tram schedules, into a single useful application. Such converging systems make it easy to change between different transport modes, mimicking the direction towards converging multi-modal transport modes into a single converged system supported by new technologies. ITS functions are also being used by cities to get their transport system more environmentally friendly. In Amsterdam, for example, new GPS-tracked bicycle-sharing programs not only promote green commuting but also provide useful data to city planners. This data assists the policy-makers in attaining a better understanding of cycling trends and conceiving infrastructure modifications that consider the rising proportion of bicycles in the city's transport system. Besides, integrating electric vehicle charging stations operated on renewable energy into the ITS system expands the scope of sustainability to include cycling beyond it. This carbon-reducing footprint and improved urban mobility vision harmonizes with the Intelligent Community values, which promote double goals of sustainable economic development and environmental preservation (N/A, 2013). Through adopting such innovative approaches, cities across the world are setting new models for urban transport behavior. The implementation of ITS in metropolitan cities also has a tendency to address more intricate social issues, such as inclusiveness and accessibility. In Toronto, the City's Intelligent Traffic Management System prioritizes pedestrians and cyclists for a safe passage through urban settings. The provision of accessibility facilities like audible signals at intersections and enhanced bike lanes is a prime example of the emphasis on urban transit equity. Prioritizing the requirements of every citizen, these cities reflect how ITS is not only making transit more efficient but also improving the well-being of residents. other coordinated efforts to tackle transport-related challenges like increased demand for services and goods amid rising e-business can be witnessed in other metropolises as well. Metropolises must respond to these shifting demands by employing technology-driven methodologies that ensure maximum logistics and the optimal performance of transport networks, signaling the need to merge innovative systems that cater to the diversified needs of metropolitan inhabitants (Ferreira et al., 2002). From such instances, it can be observed that Intelligent Transportation Systems create long-term effects on city planning and residing in large cities of the world.

6.5.2. Analysis of Specific ITS Technologies Used

The deployment of Intelligent Transportation Systems (ITS) has driven impressive advances in traffic management, primarily through the application of specific technologies to real-time sensing and data processing. Traffic sensory systems have been among the most significant of these technologies. The Mobile Century experiment illustrates the efficacy of such systems in acquiring high-quality real-time traffic data through extensive preprocessing of the data. This is made possible through experimental roadmaps generation and the utilization of trajectory analysis in single vehicle tracking. The extraction of association rules from the dataset by utilizing the Apriori algorithm is another layer of enrichment for the systems analytical capabilities to comprehend traffic flow dynamics. These technologies ultimately render traffic management solutions more effective, to the shared goal of ITS to improve road efficiency and safety (Otesh et al., 2014). Merging with the sensory technologies are the Global Navigation Satellite System (GNSS) applications, which are increasingly vital in ITS. GNSS provides Position Velocity and Time (PVT) information that is essential for a variety of applications, including high-end driver support and fleet management systems.

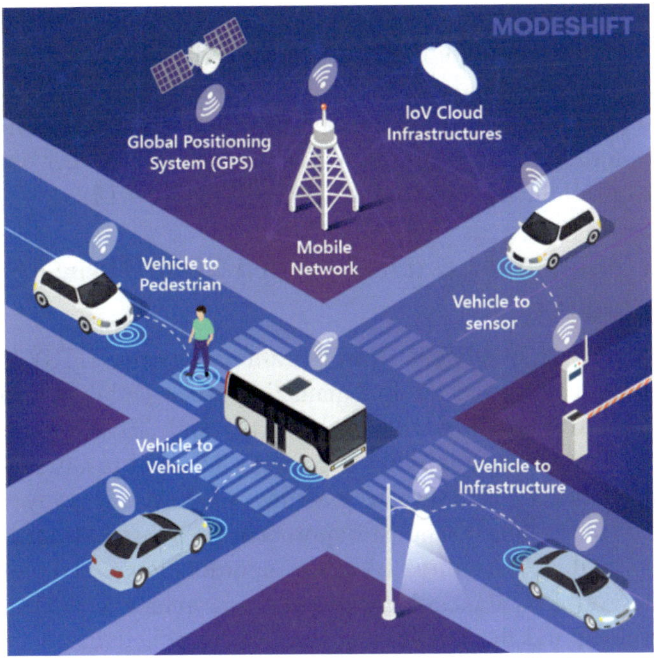

Figure 6.15. Analysis of specific technologies Intelligent Transportation Systems.

Source: https://www.modeshift.com/wp-content/uploads/2023/09/1_ITS_illustration.png.

The integration of GNSS technology has also proven beneficial in providing accurate vehicle localization as well as better navigation assistance.

Furthermore, the anticipated evolution of autonomous driving as well as Cooperative Intelligent Transport Systems (C-ITS) will largely be contingent on the resilience of such positioning technologies. Test campaigns demonstrating GNSS signal jamming and authentication have highlighted the necessity of security in GNSS systems. Experimental tests and studies of GNSS performance, particularly with regards to inertial mounted unit systems and other filtering techniques, are steps toward offering reliable navigation and safety procedures in ITS (Gianmarco et al., 2016). In addition, the sustainability requirements of the environment for ITS technologies cannot be overlooked since they contribute to better traffic regulation and reduced emissions. Traffic sensor technologies in particular facilitate real-time aggregation of data, which can be utilized to improve traffic flow and minimize congestion. This enhancement not only improves travel time but also minimizes fuel usage and greenhouse gas emissions, thus promoting global sustainability objectives. Moreover, the incorporation of GNSS technologies facilitates more precise routing and operational management of transportation fleets, adding another layer of optimization of transportation logistics. Basically, the analysis of these specific ITS technologies prove their diversified benefits, from improved safety and efficiency to environmental conservation. As technologies evolve, it is essential that these technologies be put to use in making smarter, greener transportation solutions that meet the needs of today's society (Figure 6.15).

6.5.3. Outcomes and Lessons Learned From These Implementations

The deployment of Intelligent Transportation Systems (ITS) has yielded significant takeaways in the field of managing urban mobility, particularly in the form of initiatives such as DELTA, EPOMM, and SEE-MMS. Financed by the European Union, these initiatives point to the need to integrate mobility management concepts into urban planning. A key takeaway from such deployments is the finding that holistic approaches in transportation can optimize sustainability and accessibility. Effective coordination and communication between the various stakeholders—e.g., government agencies, transport service operators, and the public—are needed for effectiveness. In response, lessons learned indicate that collaborative planning improves transportation effectiveness as well as offers an interface of community involvement, which is essential for mobility program adoption and sustainability (Sercaianu et al., 2025). Furthermore, recent ITS criticisms highlight the potential of employing telecommunications information to improve urban mobility plans. By pulling out traffic behavior and transactions, policymakers can gain insights into individuals' mobility patterns, thereby enabling more targeted and effective transport solutions. Nevertheless, the effectiveness of such data-led solutions lies in striking a delicate balance between making use of behavioral data and privacy. Impacts from such applications have brought about controversy over protection measures for data, highlighting the need to comply with legal guidelines without neglecting useful insights that can be derived from telecommunications data (Boman et al., 2013). Thus, the balance is imperative not only for public trust but also for the establishment of the overall effectiveness of ITS. Finally, experience from past ITS installations demonstrates that solutions are essential in addressing the diverse urban problems. The differences of every city require tailored strategies and not one universal one-size-fits-all system. For instance, implementing mobility management solutions in cities may be with different technologies and paradigms than for rural settings. The realization mandates the lesson that flexibility and the ability to adapt in the course of **ITS** design are essential in responding to dissimilar needs effectively. Moreover, constant research and development efforts should be aimed towards new, privacy-releasing techniques of deriving meaningful information from urban mobility trends. These efforts will additionally strengthen the capabilities of **ITS** to enhance urban living, ultimately promoting sustainable development goals and enhanced societal resilience (Sercaianu et al., 2025; Boman et al., 2013).

6.6. FUTURE TRENDS IN INTELLIGENT TRANSPORTATION SYSTEMS

With ongoing development of Intelligent Transportation Systems (ITS), future trends reveal an intensive interplay of advanced communication technologies that will revolutionize urban mobility. The interplay between information and telecommunication technologies has already begun to enhance the management of urban transport networks. Access to a variety of services through multifunctional devices like smartphones will become more important to successful real-time transportation management as the 21st century progresses. This transition not only facilitates improved decision-making but also facilitates a more sustainable city. Through the exploitation of these technological advancements, cities can install smart urban management systems that respond dynamically to traffic condition and user behavior. The positive effects anticipated are greater safety, reduced congestion, and an improved standard of living for the citizens, thereby locating the role of ITS as a prime factor in the design of tomorrow's city (Yigitcanlar et al., 2009). Another prominent trend in the ITS industry is the incorporation of advanced safety features through innovative roadway infrastructure. Current roadside systems work effectively but on outdated design principles that have received scant notable change over the past decades. As society continues to advance, it is crucial to embrace new materials and design philosophies for vehicle-barrier interaction. Vehicle restriction systems can be revolutionized in the future using smart technologies and sensors built into barriers, eventually giving way to far safer road users. Under the influence of growing autonomous vehicle integration, this has never been more critical. Such developments would enable enhanced interactions among vehicles and infrastructures, thus providing user safety and optimized traffic conditions. Accordingly, the future of ITS will not only be driven by efficiency but will also proactively advance safety through such revolutionary infrastructural technologies (Cantisani et al., 2017). Moreover, the shift towards autonomous cars is a vital trend in the context of ITS, transforming the urban mobility landscape. The arrival of autonomous vehicle technologies necessitates a reappraisal of all present-day transportation policy and infrastructure. With these technologies gaining momentum, the relationship between autonomous vehicles and city planning will need a single strategy to ensure seamless integration into cityscapes. Intelligent transport systems will play a critical role in enabling vehicle-to-everything (V2X) communication, which is vital for the safe running of autonomous vehicles. This will create a sophisticated network where cars communicate with each other and traffic control systems, ultimately leading to fewer accidents and smoother traffic flow. The city transportation environment is thus set for a full shake-up, courtesy of these anticipated ITS technology advancements (Figure 6.16).

Figure 6.16. Future trends in intelligent transportation systems.

Source: https://images.squarespace-cdn.com/content/6526a2cd1822263deb066af4/1708715186686-NKU6WZAY455BASVQYQY5/AdobeStock_616865920.jpeg?format=1500w&content-type=image%2Fjpeg.

6.6.1. Advancements in Autonomous Vehicles

The domain of autonomous vehicles has witnessed significant expansion through the incorporation of high-end computing systems and connectivity platforms. Vehicle-oriented Edge Intelligence (VEI) applications, supported by the Internet of Vehicles (IoV), are necessary for enhancing autonomous driving functions within Intelligent Transportation Systems (ITS). However, with such advancements providing improved navigation and automation, issues still remain to do with vehicular sensor capabilities that fall behind in effectively anticipating beyond near-future road conditions. This is most critical in regional transportation management, for which innovative means must be developed to merge real-time data with predictive analytics. Recent studies have advanced a holistic approach that combines task offloading and resource allocation into multi-access edge computing environments. This is done by focusing on the complexities of vehicular mobility and data dependency, thereby creating a dynamic and reactive environment for autonomous cars to function optimally in cities (Luo et al., 2025). Efficient simulation and testing are now a major requirement with the dynamic transport environment (Luo et al., 2025). The limitations of single simulation environments in terms of being able to have the whole picture of this technology suggest the necessity of co-simulation approaches in Intelligent Transportation Systems. These approaches provide end-to-end analysis by simulating the intricate interactions CAVs will experience in real environments. Various simulators, both open-source and commercial, have been grouped based on their functions in autonomous driving systems, emphasizing their interactive roles in addressing the unique challenge posed by autonomous driving tests. Simulators allow researchers to increase realism and fidelity of test environments as well as thorough evaluations of responsiveness in the system. The advancement in artificial intelligence and machine learning increases the strength of these co-simulation platforms, paving the way to more efficient and safer urban mobility solutions (Ashayer et al., 2025). The advancement in autonomous vehicles also necessitates a rethinking of urban planning and infrastructure development to release their true potential. Transition to smart transportation systems poses a challenge to conventional road design, traffic management, and public transport integration practices. These vehicles are more than high-tech features; they call for a systemic transformation of the manner in which cities are capable of embracing transport. Sensor-smart roadways have the ability to provide automobiles with real-time information, rendering them improved decision-makers in navigating. Besides, integration with public transit can enable a seamless multi-modal environment that promotes reduced congestion and enhanced accessibility. Policymakers and city planners will, as cities begin implementing these systems, have to weigh how best to maximize the benefits of autonomous vehicles to achieve sustainable urban mobility goals while also ensuring that infrastructure is compatible with advances in vehicle technology.

6.6.2. The Role of Artificial Intelligence in ITS

The application of artificial intelligence (AI) in Intelligent Transportation Systems (ITS) has revolutionized the operation of transport networks to become more efficient and secure. With the application of advanced computational techniques, AI offers the capability to analyze enormous amounts of information gathered by various transportation modes simultaneously. Such data, referred to as Big Data, include traffic movement, vehicle performance metrics, and weather conditions, which can be utilized to optimize the movement of traffic and reduce congestion. The role of AI in ITS is to apply machine learning algorithms to predict traffic conditions so that proactive management techniques can be implemented that can make overall transport networks far more dependable. Through these roles, AI transforms raw data into meaningful insights, fostering smarter urban environments and supporting the development of successful smart cities, as evidenced in the seminal Big Data-artificial intelligence convergence in the literature (Doctor et al., 2020). Further, AI advances the scope of inter-organization cooperation in transport systems, particularly through the application of Electronic Data Interchange (EDI). This system enables various organizations in the supply chain—suppliers, carriers, and customers—to easily share crucial information. Intelligent agents' integration with this system has the ability to automate processes by gathering data and facilitating decision-making that used to be manual and resource intensive. Agribusiness industries, for example, can optimize inventory control by maximizing stock levels according to real-time changes in demand with agent technology and EDI used together. Thus, the convergence of AI and EDI generates flexibility in transport logistics, allowing organizations to respond better to the environment, which could result in increased operational efficiencies and reducing logistical costs (Mangina et al., 2005). Finally, problems with the integration of AI in ITS need to be carefully weighed. While the benefits of improved traffic management and increased inter-organizational communication are enormous, data security and privacy concerns persist. The reliance on massive data gathering is the reason for concern for the security of sensitive data. Therefore, it is extremely important for policymakers to develop robust frameworks ensuring data security in addition to stimulating innovation in transportation technology. Furthermore, the moral implications of AI in making traffic management decisions need to be tackled to prevent bias and offer equal treatment to all sections of the population. Thus, the role of artificial intelligence in ITS is not merely a technological change but a multifaceted issue that is intertwined with social values and calls for far-reaching strategies for safe and equitable deployment (Figure 6.17).

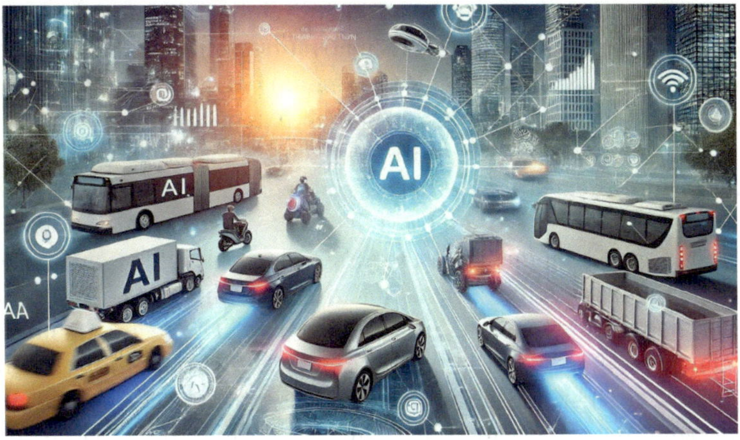

Figure 6.17. Role of artificial intelligence in intelligent transportation systems (ITS).

Source: https://miro.medium.com/v2/resize:fit:1200/1*RONCugY9Kw67fsfyoEhfew.jpeg.

6.6.3. Potential for Smart City Integration

The incorporation of smart cities within the scope of Intelligent Transportation Systems (ITS) presents enormous potential for restructuring urban mobility. Smart cities refer to integrated infrastructure whereby advanced technology and data analysis combine to ensure efficiency and sustainability. As populations within cities continue to expand, the strains on transport systems have grown, requiring innovative solutions that not only address the current inefficiencies but also anticipate future needs. ITS integration in smart city projects can optimize traffic management, reduce congestion, and overall improve safety. As argued in (Emeka et al., 2024), the intersection of big data and transport systems offers space for green urban development, whose interconnection of technology and data-driven decision-making forms a robust foundation for the evolution of smart cities. Examining the extent of integration in smart cities, one should also note the limitations and constraints of existing models. The utilization of smart mobility has been adversely criticized on numerous occasions for reflecting too much focus on technology, wherein innovations are being implemented without considering their social implications properly. As noted in (Lauwers et al., 2015), the consumerist tendencies of smart mobility sometimes reduce users to mere figures regardless of the broader context of city planning and government. This can lead to solutions that fail to address the diverse needs of individuals and may even exacerbate current inequalities in urban mobility and accessibility. Therefore, a broader and interdisciplinary integration model for smart cities is imperative, ensuring ITS innovations within the context of equity and inclusivity principles. Smart city integration potential also extends beyond traffic systems; it encompasses a set of urban functionalities that enhances the quality of life of citizens. Through fostering collaboration among stakeholders like government officials, private sector partners, and society, smart cities are able to create comprehensive solutions that address transport as well as social, environmental, and economic issues. In furthering the development of their cities towards higher connectivity, systems integration in the context of information sharing and mutual cooperation will be imperative. The focus should not only be on technological progress, but also on improving urban life through smart mobility solutions that engage citizens actively in the development process, consistent with the observation in Lauwers et al. (2015). Ultimately, the vision of smart transport systems in smart city initiatives has to prioritize a comprehensive and inclusive strategy that improves urban life.

SUMMARY

The foray into Intelligent Transportation Systems (ITS) provides a new and futuristic possibility for city traffic, necessitating innovative solutions to current transportation demands. As cities struggle with increasing congestion, pollution, and the necessity of free-flowing transportation, ITS provides an all-encompassing system where technology, data analysis, and user-centric policies merge. The application of intelligent technologies enables real-time traffic monitoring and management, effectively optimizing traffic flow and reducing delays. In addition, using information gathered from the urban experience, policymakers can formulate targeted policies to further develop cities as sustainable entities, thereby addressing the problems of modern urban living. This marriage between technology and urban planning is not simply a matter of perfecting transit but also of enhancing the value of urban living. Once more pointing to the success of ITS, a critical inclusion in the determination of the future of urban traffic control is that of precise key performance indicators (KPIs). These measurements, built from the dynamic interactions between street networks, serve as the foundation for intelligent traffic control systems. In such cases, sophisticated measurement mechanisms allow for thorough analysis of traffic flows, vehicle types, and system performance, allowing for further efficient decision-making. Solution-backed data like real-time vehicle typology and dynamic lane allocation enable adaptive responses from cities to emerging conditions on the ground. Further, putting forward solutions that highlight cost saving and protection of privacy enshrines the overarching strategy of creating inclusive, sustainable urban transport systems. The double-barrelled treatment of the priority aims at ensuring ITS development encourages equity in terms of taking advantage of sophisticated technology. Lastly, Intelligent Transportation Systems signal a paradigm shift with regard to how cities are to solve mobility challenges. Integration of cutting-edge data processing, adaptive policy-making, and technological innovation offers a holistic solution to the most pressing challenges of urban transport. Through the application of strategy grounded in existing research and practice—highlighted by the importance of dynamic control means in optimizing traffic flow and by the worth of urban positioning data for informed policy evaluation—ITS has the potential to pave the way for wiser, greener cities. As cities continue to evolve, the need for wise, data-driven transportation strategies remains front and center. Future advancements should not only focus on integrating technology but also on building sustainable practices that prioritize the well-being of urban residents, which is in line with the overall goal of making cities more mobile for all.

The development and implementation of Intelligent Transportation Systems (ITS) have emerged as core solutions to address ongoing problems in city mobility and road safety. Top highlights indicate that Intelligent Speed Adaptation (ISA) technology plays a significant role in managing car speeds in bottlenecks. In Sweden and the Netherlands, on-road experiments have demonstrated that ISA has the effect of actually inhibiting transient high speeds, thereby reducing speed variability, which is linked to improved safety results. In addition, simulations in the report show major benefits like a whopping eight percent reduction in the use of fuel when ISA is deployed to its maximum capacity, which shows that ITS can not only enhance safety but also ensure environmental sustainability through minimizing fuel consumption and emissions in urban traffic conditions, as evidenced by the research findings in Liu et al. (2004). The inclusion of connected cars within the ITS platform has redefined smart mobility. Today's developments in automotive tech focus more on transforming traditional means of transportation towards a more connected paradigm, where cars are not just vehicles for transportation but advanced, interactive devices. This is the paradigm that follows the increased need for cars to talk to one another, infrastructure, and users through IoT devices. This interconnectivity makes dynamic traffic management feasible, and thus innovative solutions to reduce congestion as well as improve the flow of traffic become viable. The literature presents classification frameworks dividing vehicle-centric, vehicle-network, and user-centric approaches as pointing to the fact that a coordinated ITS ecosystem must include multiple user interactions and technology evolutions in its design and implementation stages, as outlined in Denis et al. (2014). Besides, the social nature of ITS cannot be evaded because the systems must satisfy the needs and expectations of the users to be truly helpful. There is a wide gap between the users and their vehicles primarily because there is a mismatch between application interfaces and user interactions. Closing the gap is a challenge for future ITS design as it calls for an integrated approach combining technological innovation and satisfaction of the users. By focusing on smart vehicle designs that promote efficient interaction between user and his/her vehicle, future ITS advancements can make way for a more active driving experience which is both smart and efficient. Ultimately, such enhancements will not only enhance individual mobility but also meet broader social goals, including ecological compatibility and enhanced road safety, thus confirming the central role of ITS within modern transportation networks.

REFERENCES

1. Al-Askary, M., Dorn, D., Haferkamp, M., Sliwa, B., et al. (2018). A radio-fingerprinting-based vehicle classification system for intelligent traffic control in smart cities. http://arxiv.org/abs/1801.03291 (accessed on 21st October 2025).
2. Albers, J. A., & Zuk, J. (2025). Aircraft technology opportunities for the 21st century. https://core.ac.uk/download/pdf/42830883.pdf (accessed on 21st October 2025).
3. Alfeo, A. L., Cimino, M. G. C. A., Egidi, S., Lepri, G., et al. (2018). A stigmergy-based analysis of city hotspots to discover trends and anomalies in urban transportation usage. http://arxiv.org/abs/1804.05697 (accessed on 21st October 2025).
4. Alhagagi, H. A., Dianati, M., Lu, Y., Maple, C., et al. (2018). Analysis of cyber risk and associated concentration of research $(ACR)^2$ in the security of vehicular edge clouds. https://core.ac.uk/download/159995351.pdf (accessed on 21st October 2025).
5. Ang, J. H., Flores Saldivar, A. A., Goh, C., Li, Z., et al. (2017). Energy-efficient through-life smart design, manufacturing and operation of ships in an industry 4.0 environment. https://core.ac.uk/download/82893542.pdf (accessed on 21st October 2025).
6. Ashayer, S., Elhag, F., Sun, P., & Sartipi, M. (2025). Co-simulation in intelligent transportation systems: A survey targeting connected and autonomous vehicles. Transportation Research Record: Journal of the Transportation Research Board. https://www.semanticscholar.org/paper/c1c2bb698d96f22e9fa8f54603876a19544198cd (accessed on 21st October 2025).
7. Azevedo, T., Barbosa, J. G., & Rossetti, R. J. F. (2015). A state-of-the-art integrated transportation simulation platform. http://arxiv.org/abs/1601.08162 (accessed on 21st October 2025).
8. Baldini, G., Borio, D., Cano Pons, E., Gioia, C., & Giuliani, R. (2016). Robust GNSS services for road transportation: Analysis and studies to mitigate GNSS threats in the road transportation sector. https://core.ac.uk/download/81685442.pdf (accessed on 21st October 2025).
9. Banoth, K. (2014). A study on vehicle trajectory analysis. https://core.ac.uk/download/53190022.pdf (accessed on 21st October 2025).
10. Barth, M. J., Wang, Z., Wu, G., Zhao, Y., et al. (2020). Development of eco-friendly ramp control for connected and automated electric vehicles. https://core.ac.uk/download/286729285.pdf (accessed on 21st October 2025).

11. Becker, J., van Hillegersberg, J., & Verduijn, T. (2004). Agent technology in supply chains and networks: An exploration of high potential future applications. https://core.ac.uk/download/pdf/11481348.pdf (accessed on 21st October 2025).

12. Blazquez, C., Miranda, P., & Ponce, A. (2012). Performance of a new enhanced topological decision-rule map-matching algorithm for transportation applications. https://core.ac.uk/download/288915403.pdf (accessed on 21st October 2025).

13. Boman, M., Bylund, M., Hirsch, B., Sanches, P., et al. (2013). Knowing your population: Privacy-sensitive mining of massive data. http://arxiv.org/abs/1412.2247 (accessed on 21st October 2025).

14. Borri, D., Camarda, D., Circella, G., & Barletta, G. L. (2025). Smart Technologies for Environmental Safety and Knowledge Enhancement in Intermodal Transport. https://core.ac.uk/download/pdf/7035946.pdf (accessed on 21st October 2025).

15. Cantisani, G., Di Mascio, P., & Polidori, C. (2017). New research opportunities for roadside safety barriers improvement. https://core.ac.uk/download/153367220.pdf (accessed on 21st October 2025).

16. Cartolano, F., Filali, F., Gozalvez, J., Krajzewicz, D., et al. (2008). iTETRIS: An integrated wireless and traffic platform for real-time road traffic management solutions. https://core.ac.uk/download/11140413.pdf (accessed on 21st October 2025).

17. Certa, A., Enea, M., Berrittella, M., & Zito, P. (2025). An analytic hierarchy process for the evaluation of transport policies to reduce climate change impacts. https://core.ac.uk/download/pdf/6461214.pdf (accessed on 21st October 2025).

18. Chowdhury, M., Deka, L., Islam, M., Khan, R., et al. (2019). Development and performance evaluation of a connected vehicle application development platform (CVDeP). https://core.ac.uk/download/287585310.pdf (accessed on 21st October 2025).

19. De Mol, J., Van Leeuwen, T., Vandenberghe, W., Vlassenroot, S., et al. (2008). Intelligent transport systemin ITS en verkeersveiligheid. https://core.ac.uk/download/55814175.pdf (accessed on 21st October 2025).

20. Denis, R., Dillschneider, J., Lèbre, M.A., Le Mouël, F., et al. (2014). VANET applications: Hot use cases. https://core.ac.uk/download/49610995.pdf (accessed on 21st October 2025).

21. Doctor, F., Iqbal, R., Mahmud, S., More, A., et al. (2020). Big data analytics: Computational intelligence techniques and application areas. https://core.ac.uk/download/228156656.pdf (accessed on 21st October 2025).

22. Ernst, R., Nolte, M., & Schlatow, J. (2019). Controlling concurrent change–A Multiview approach toward updatable vehicle automation systems. https://core.ac.uk/download/190382646.pdf (accessed on 21st October 2025).

23. Ferreira, L., Marquez, L., & Smith, N. (2002). E-business impacts for urban freight: Results from an Australian study. https://core.ac.uk/download/10874375.pdf (accessed on 21st October 2025).

24. Gray, G. E., Kelley, N., & Larwin, T. F. (2006). Bus Rapid Transit: A Handbook for Partners, MTI Report 06-02. https://core.ac.uk/download/70409909.pdf (accessed on 21st October 2025).

25. Haloi, M., & Jayagopi, D. B. (2014). Characterizing driving behavior using automatic visual analysis. http://arxiv.org/abs/1503.04036 (accessed on 21st October 2025).

26. Henrickson, K., Pereira, F. C., & Rodrigues, F. (2019). Multi-output Gaussian processes for crowdsourced traffic data imputation. http://arxiv.org/abs/1812.08739 (accessed on 21st October 2025).

27. Hossen, M. S., Kappes, C., Morris, B., Park, S., et al. (2019). Design and preliminary testing of demand-responsive transverse rumble strips. https://core.ac.uk/download/250588061.pdf (accessed on 21st October 2025).

28. Kamble, P. P. (2024). Intelligent transportation systems: Fusing computer vision and sensor networks for traffic management. https://core.ac.uk/download/603899121.pdf (accessed on 21st October 2025).

29. Kopardekar, P. H., & Sridhar, B. (2025). Towards autonomous aviation operations: What can we learn from other areas of automation? https://core.ac.uk/download/pdf/286393131.pdf (accessed on 21st October 2025).

30. Kotian, A. L., Abhishek, B., Allapur, A. R., Gowda, A., & Gowda, A. (2025). A comprehensive review of different frameworks for ensuring data privacy and security for IoT networks in smart city. 2025 3rd International Conference on Intelligent Data Communication Technologies and Internet of Things (IDCIoT). https://www.semanticscholar.org/paper/d650a91313e6fb22e15f8145a0bf2e26c2c4f41b (accessed on 21st October 2025).

31. Lauwers, D., & Papa, E. (2015). Smart mobility: Opportunity or threat to innovate places and cities. https://core.ac.uk/download/55823526.pdf (accessed on 21st October 2025).

32. Liu, R., & Tate, J. E. (2004). Network effects of intelligent speed adaptation systems. https://core.ac.uk/download/53168.pdf (accessed on 21st October 2025).

33. Luo, K., Wang, Y., Liu, Y., & Zhu, K. (2025). Collaborative integration of vehicle and roadside infrastructure sensor for temporal dependency-aware task offloading in the Internet of Vehicles. International Journal of Intelligent Systems, 2025. https://www.semanticscholar.org/paper/4ef9e4e648b7897e7c11202ab15cab499ed9a237 (accessed on 21st October 2025).

34. Mahbub, A. M. I., & Malikopoulos, A. A. (2020). Concurrent optimization of vehicle dynamics and powertrain operation using connectivity and automation. http://arxiv.org/abs/1911.03475 (accessed on 21st October 2025).

35. Mangina, E., & Vlachos, I. (2005). EDI and intelligent agents' integration to manage food chains. https://core.ac.uk/download/17298965.pdf (accessed on 21st October 2025).

36. Milito, R., Nemirovsky, M., Roca, D., Valero Cortés, J., et al. (2020). Advances in the Hierarchical Emergent Behaviors (HEB) approach to autonomous vehicles. https://core.ac.uk/download/187265623.pdf (accessed on 21st October 2025).

37. N/A. (2009). Moving Cooler: An Analysis of Transportation Strategies for Reducing Greenhouse Gas Emissions. https://core.ac.uk/download/71353321.pdf (accessed on 21st October 2025).

38. N/A. (2013). Community as Canvas: The Power of Culture in the Emergence of Intelligent Communities. https://core.ac.uk/download/71364783.pdf (accessed on 21st October 2025).

39. Okorie, E. E., & Anyaragbu, H. (2024). Leveraging IoT and big data for sustainable transportation in smart cities. https://core.ac.uk/download/636457450.pdf (accessed on 21st October 2025).

40. Qureshi, K., Jeon, G., Hassan, M. M., Hassan, M. R., & Kaur, K. (2023). Blockchain-based privacy-preserving authentication model intelligent transportation systems. IEEE Transactions on Intelligent Transportation Systems, 24, 7435–7443. https://www.semanticscholar.org/paper/dd67e9bf12fe6d770033df096368ec267141b539 (accessed on 21st October 2025).

41. Sercaianu, M., & Luca, O. (2025). Mobility management in European projects. Lessons learned for Romania. https://core.ac.uk/download/pdf/6234329.pdf (accessed on 21st October 2025).

42. Shepherd, S. P. (1992). A review of traffic signal control. https://core.ac.uk/download/54287.pdf (accessed on 21st October 2025).

43. Yigitcanlar, T. (2009). Managing ubiquitous eco cities: The role of urban telecommunication infrastructure networks and convergence technologies. https://core.ac.uk/download/10895362.pdf (accessed on 21st October 2025).

CHAPTER 7

DATA-DRIVEN PUBLIC TRANSPORTATION PLANNING

Contents

INTRODUCTION

The increasing complexity of city life demands strategic public transportation planning that is dominated by data-based strategies. As cities grow, the demand for efficient and reliable transport systems increases. Public transport is not only a means of mobility but also a critical enabler of economic vitality, social engagement, and environmental resilience. By applying data analytics, city planners are able to see patterns of commuting, identify areas of deficiency in service, and streamline routes and schedules to better accommodate populations. Forward-looking in scope, it emphasizes the importance of consolidating various sources of data to create adaptive and resilient transportation infrastructure that accommodates multiple communities (Bierbaum et al., 2011). The notion of evidence-based public transit planning is becoming more widely accepted as a key to upgrading living standards in urban areas. Not only does data inform the transportation system but also relate to larger societal objectives, such as education and fiscal sustainability. For this purpose, the creation of a common vision and metrics serves to ensure that investments optimizing prosperity and access to transportation are better aligned.

Figure 7.1. Data-driven public transportation planning.

Source: https://socialdhara.com/wp-content/uploads/2021/02/urban-congetion-1024x576.jpg.

Effective planning can lead to increased connectivity, reduced traffic congestion, and an enhanced sustainable city. Advanced analytics enable cities to prioritize projects strategically that provide the maximum good to the maximum number of people and ensure that transport systems contribute value to the social fabric (Bierbaum et al., 2011). Another new trend is the incorporation of energy efficiency in public transport systems, which aligns with data-driven planning. The necessity for systems to transition into renewable power sources has an important role to play in creating sustainable urban environments. This also requires not only a general comprehension of geographical and morphological constraints but the enforcement of programs such as the Covenant of Mayors (CoM). In this context, municipalities need to develop Sustainable Energy Action Plans (SEAP) outlining actionable strategies for achieving challenging sustainability goals by 2020 and beyond. Addressing the challenging transport-energy resources nexus offers a double task, allowing urban planners to gain access to economic opportunities while establishing regulatory compliance (Figure 7.1).

Definition of Data-Driven Public Transportation Planning

Data-driven public transit planning has a process that relies on quantitative and qualitative data for decision-making purposes. The process allows transport planners to develop insights from various datasets such as ridership numbers, traffic patterns, and population development. By using the data, planners are able to identify inefficiencies, establish demand forecast, and optimize resource allocation. One of the most important factors is the infusion of live data and historical trends, which enables adaptive and responsive transit operations to be developed. For example, public transit organizations may analyze peak travel times to re-balance service frequency or routes, hence enhancing customer satisfaction and operational efficiency. Thus, the essence of data-driven urban planning of public transport lies in the utilization of comprehensive data analysis to develop effective transportation networks that cater to the evolving needs of urban residents. The use of technology in public transport planning has also significantly transformed data-driven planning. As cities grow ever increasingly, so does the demand for efficient transit plans, hence the imperative of utilizing state-of-the-art tools to collect data and analyze it. Technologies such as geographic information systems (GIS), mobile applications, and automated transit system observation allow planners to more clearly see transportation movement and identify trends. The technologies facilitate the collection of data on user behavior, safety accidents, and service quality, which are required in sound transportation policy-making. One of the most important things data-driven planning does is assist in making transportation systems accessible and equitable to multiple community needs. Planners are using these new technologies to drive the development of sustainable transit choices that enhance mobility and overall quality of life in metropolitan areas. Lastly, public transport planning must consider the issue raised by the insufficient data, especially in rural areas where data collection may not be as rigorous. For instance, rural transportation authorities may lack the data they require to demonstrate safety requirements for purposes of securing funds, such as traffic safety upgrades using past crash data and traffic volume. The development of specific applications, such as the UAF Traffic application, demonstrates a proactive approach to this problem through the ability of localized data retrieval. It further suggests the importance of qualifying qualitative data in such fields, thereby increasing the scope of data employed in planning. The recognition of these challenges indicates that a proper understanding of data-driven public transportation planning is also a matter of filling gaps and drawing on innovative solutions in order to offer broad-ranging planning initiatives that can enjoy long-term benefits for communities irrespective of their geographical location (Bennett et al., 2019).

Importance of Public Transportation in Urban Development

Public transportation plays a crucial role in urban growth by providing a consistent means for citizens to reach necessary services, employment, and social contacts. Its importance is greater than mobility; well-designed public transit networks can have a profound effect on the livability of the city. In fact, effective public transportation relieves road congestion and reduces greenhouse gas emissions, making cities healthier. Also, access to efficient means of transportation fosters economic growth through the attraction of business and investment to cities. If areas can be efficiently linked by public transit, they are more likely to thrive as vibrant, economically sound communities. This connection between transit provision and urban vitality highlights the need to link public transportation planning with broader city development plans in a way that economic revitalization and quality of life for residents are incorporated into one another. The incorporation of evidence-based practice into public transportation planning also enhances its role in urban development. With data analytics, it is possible to view patterns of transit usage, most traveled routes and hours of the day, and make educated decisions that optimize service efficiency. Not only does it optimize current public transit systems but also helps with predictive modeling of future demand and hence proactive planning of cities. Evidence-based findings improve the development of targeted investments in transportation, ultimately resulting in the planning of transportation services according to community need. For example, as pointed out, establishing a common vision and indicators can enhance the coordination and communication among different actors, leading to more coordinated and efficient transportation plans integrated with economic growth and community health (Bierbaum et al., 2011). Additionally, innovative transport models such as Delivery as a Service (DaaS) indicate how evolving business patterns have the potential to optimize public transportation networks and cities. DaaS emphasizes efficiency, customer-oriented delivery of services, and enhanced operations that are attuned to the on-demand economy, illustrating how public transportation can innovate to stay abreast of changing urban dynamics. By adding new technologies such as the Internet of Things and transport optimization methods, city planners can enhance inter-operability between public transportation systems and logistics. Not only does an integrated system benefit customers, but it also promotes environmentally friendly urban logistic solutions with reduced traffic and carbon footprint. As cities start emphasizing sustainability, understanding and complying with these new business models as well as existing transport systems will be essential to the achievement of general urban development goals (Huschebeck et al., 2020) (Figure 7.2).

Figure 7.2. Importance of public transportation in urban development.

Source: https://www.addressofchoice.com/aoc_assets/blog/1584607541_Latest_Transport_System_that_can_transform_Indian_Cities.jpg.

Overview of the Role of Data in Enhancing Transportation Systems

Data integration into transportation systems can facilitate a transformational realignment towards more efficient public transit planning. At the heart of the transformation is the collection and analysis of various sources of data, from traffic patterns to demographic information and user behavior. Well-functioning cities that integrate these data sources can anticipate transit needs more accurately, facilitating the precise design of services to meet the precise demand of their communities. For instance, the Chicago Metropolitan Agency for Planning (CMAP) takes a data-based approach towards making recommendations on strategies as in the case of its GO TO 2040 plan, emphasizing the significance of data use integration and regional economic growth (Kosarko et al., 2011). This planning has the potential to significantly increase operational efficiency, ensure effective resource utilization, and enhance overall user experience on public transport networks, making the local environment more user-friendly. Moreover, the use of data extends beyond issues of operational effectiveness; it is critical in encouraging equitable access to transportation. Disproportionate access to public transport can perpetuate socioeconomic gaps, driving mobility between neighborhoods and local development. Through data analysis, cities can identify under-served areas, comparing the provision of transit with the resident labor force's demographic makeup. The findings from these analyses can guide investments that expand access to safe transportation options, thereby promoting economic and community strength. The merit of connecting transportation investment to school and economic growth can be discerned in instances where measurement has guided targeted improvement and intervention tactics (Bierbaum et al., 2011). Appropriate utilization of data can thus lead to an inclusive public transport policy, social equity, and overall quality of life for all citizens. Furthermore, the utilization of real-time data significantly enhances the capacity of the transport systems to respond to anticipated and unanticipated problems. With an ever more dynamic urban landscape, access to real-time data regarding traffic, transit boardings, and system events makes it possible for transport agencies to be responsive in real time and adjust accordingly. Not only does this keep delays at bay, but it also increases safety and customer satisfaction. Systems such as real-time passenger information systems show the way data can reimagine customer engagement with transit and foster a culture of reliability and trust. As cities look to waves of new arrivals—like the projected increase of over two million in the Chicago metropolitan region—it more and more becomes important to employ data in support of keeping quality-of-life levels high and economic development in the face of growing strains on transit systems (Kosarko et al., 2011). Continued development and implementation of data-informed insight will thus be essential to public transit planning efforts to come.

7.1. HISTORICAL CONTEXT

Planning for public transport has been transformed by many different historical contexts that inform urban landscapes and work commutes. During the mid-20th century, for instance, the rise of automobile culture transformed transport planning considerably. With cities expanding outward, urban planners spent more time building roads and parking lots than on transit systems. This transformation also left investment in rail transport and buses as the forgotten aspects of urban transportation. Excessive dependency on personal vehicles not only changed the mode of travel but also produced issues such as traffic congestion, pollution, and urban sprawl. All these changes in urban transport infrastructure need data-driven planning that incorporates knowledge from history regarding the existing systems so that planners can set routes more accurately to the needs of today's populations. Historic traffic patterns and trends provide critical information for effective public transit planning. Various studies have highlighted the extent to which large events—like sporting events—impact traffic behavior, and these can be applied as case studies for use in broader city planning. An example is that one of the major studies on Nebraska Cornhuskers football games over five years employed INRIX data to examine the cross-section of event timing and highway congestion. With methods such as Multi-EigenSpot to identify hotspots, scientists detected patterns that indicate how some events contribute to aggravating underlying congestion issues. The results indicate the need to condense historical information into traffic management in order to provide informed interventions that can better enhance commuter experience and maximize public transportation use in peak hours (Ahsani et al., 2019). Further, the blending of arts, culture, and public transport provides more layers of complexity to context in planning history. Cultural practice has been seen to have the power to convert urban neighborhoods into a more unified and more balanced pattern of development. Historical social dynamics can be employed for planning public transport and open spaces so planners can learn the cultural context of the areas served. Local artist projects that connect to local narratives can become vital in the creation of transportation systems that are not only effective but also culturally sensitive. If the culture's expressions are known to be rich in history, then transportation projects can be created that project community identity while addressing broader social equity issues (Liu et al., 2017). Hence, historical context knowledge allows planners to craft evidence-led solutions that are attuned to the specific characteristics and trends of urban living.

7.1.1. Evolution of Public Transportation Planning

The evolution of public transport planning has helped significantly to foster urban development and mobility in modern society. Historically, transport planning was done with minimal regard to information or the public's demand, with outcomes varying from unsuitable routes to unused services. However, as cities grew and populations rose, the necessity for more effective public transport arose. The shift to evidence-based decision-making began to gain traction in the later half of the 20th century, as planners came to understand that relying on data would lead to improved delivery of service, increasing ridership, and maximized allocation of resources. This shift involved combining technologies, such as Geographic Information Systems (GIS) and ridership information, to enable planners to maximize route and schedule optimization using actual user behavior and patterns. In the past few years, intergovernmental collaboration among public transport agencies and community organizations has been more evident than ever. Evidence-based collaborations permit more detailed insight into transportation demand in various groups. ExpandED Schools is an example of the way collaborative efforts strategically can enhance educational opportunities along with transit use because community groups understand local needs best. These models demonstrate how computer programming can be data-driven not only in transportation but in any area, supporting overall coordination of community services.

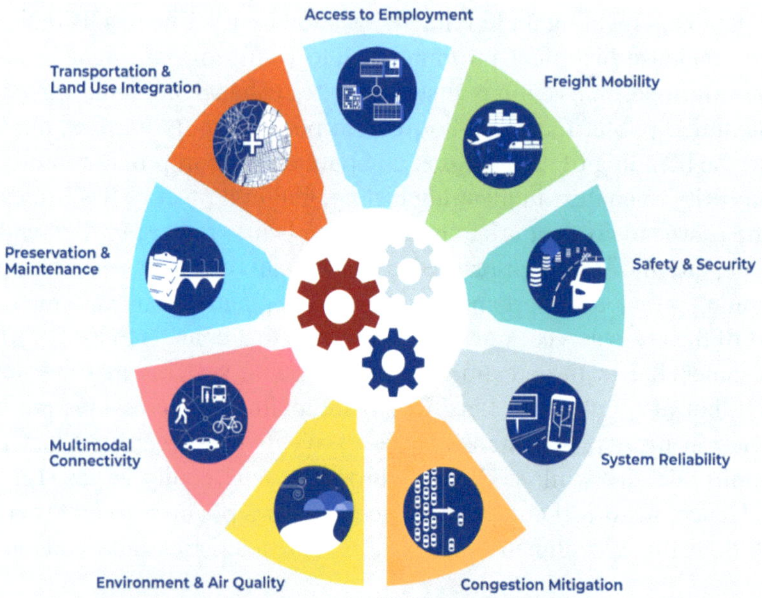

Figure 7.3. Evolution of public transportation planning.

Source: https://planrva.org/wp-content/uploads/2019/04/Diagram8revised411.png.

Their potential is not without its limitations, though, such as obtaining secure funding and continuing to collect data in order to sharpen planning efforts. This subtlety points to the necessity of dynamic frameworks that can adapt to rising urban demands. Moreover, data analytics in public transport planning extends far beyond the short-term logistics issues. Research like the Georgetown Climate Center's report shows that successful adaptation is frequently contingent upon having the ability to understand and to communicate sophisticated impacts of transportation systems. This approach facilitates resilience through measuring constraints and inducing efforts to bypass them. Data-based intelligence not only allows for increased operational efficiency but also an enhanced ability to respond to shifting environmental conditions. Emphasizing the cycle of transport planning, in which data inform policy development and public engagement, points to a new paradigm for more sustainable and adaptive urban infrastructure. As planners continue to tackle these complexities, the public transport of the future will be guided by an unshakeable commitment to practices built on evidence and guided by inclusivity and responsiveness to the evolving demands of urban residents (Figure 7.3).

7.1.2. Early Methods of Data Collection and Analysis

In early data collection and analysis for public transit planning, researchers relied primarily on qualitative methods such as surveys, interviews, and observation studies. These early methods provided rudimentary details regarding patterns of ridership, behavior of commuters, and socio-economic structures. For instance, surveys of users may identify preferences and problems, while direct observations may measure traffic volumes and peak periods. The limited technology of the day required that the information were often collected manually and thus created related issues of accuracy and timeliness. Those early methods were nevertheless important to the development of a crude basis for understanding urban mobility, which is necessary to inform subsequent transportation planning decisions and policy development intended to improve system efficiency and customer satisfaction.

As planners' need for more powerful planning methods grew, so did the intricacy of data collection tools and methods. Computational methods in the earlier times incorporated statistical analysis that minimized the interpretability of large databases. This marked the start of the transition towards the utilization of quantitative data, where planners could contrast ridership fluctuations against several factors such as time, weather patterns, and demographical shifts. Besides, Geographic Information Systems (GIS) creation allowed spatial analysis so that planners could visualize transit networks and growth or potential for improvement opportunities. All these advances in methods laid the groundwork for a data-driven approach to public transportation planning, with which it became simpler to transition to inclusive models that influenced policy and investment strategies. The combination of early data gathering methods with new technologies paved the way for more inclusive public transportation planning methods. The era also witnessed the development of planning support systems that not only facilitated centralization of data but also facilitated stakeholders' participation. Participants in transport projects could utilize shared datasets to influence their perspectives and contribute feedback to decision-making, which catered to the interests of the population diversity. The shift to visualization technologies, as studied in contemporary planning paradigms, enhanced stakeholder participation by accurately illustrating transport occurrences and accompanying project effects. Notably, as outlined in Batty et al. (2007), the transition towards a coordinated, data-driven methodology continues to revolutionize the way planners forecast urban issues, enabling public transport infrastructure to match changing demands of the populace it caters to.

7.1.3. Impact of Technology on Transportation Planning

The use of technology in transport planning has transformed the way in which cities react to mobility, sustainability, and efficiency challenges. Advanced data analytics assist city planners in scheduling and routing transit according to actual demand rather than past patterns. Such a realignment enhances the flexibility of public transportation systems in responding to changing passenger demands and urban rhythms. By means of technologies such as mobile apps and GPS, transport organizations can provide precise information and personalized travel options to users.

This, in turn, stimulates public engagement and satisfaction, which are crucial in expanding sustainable transport modes. Moreover, the technologies facilitate the vision of Smart Cities since technology powers better urban life and addresses infrastructural demands effectively (Monte et al., 2017). In developing modern transportation systems, information is of central concern. With data-driven strategies, public transit planners can craft higher levels of accessibility and fairness for different populations, aligning public transit to address a broad demographic. Collaboration is critical in ensuring coordination between investments and community demands, thereby driving economic growth. Coordination usually accompanies generating shared visions and measurement metrics that recognize the intersection of transportation, education, and community development.

With the application of smart technologies, agencies can identify underprivileged areas and adapt services to optimize access by low-income citizens, thereby furthering the greater aim of social equity. Transport technology thus makes possible an integrative approach to planning, enabling organizations to expand the reach of public transportation modes and construct evaluations that benefit all citizens (Bierbaum et al., 2011). Also, the introduction of Automated Vehicles (AVs) will further alter the landscape of transportation planning. AVs' introduction has the potential to revolutionize mobility by providing excellent integration with existing public systems and heal traffic congestion and safety concerns. However, for these technologies to be successfully integrated, urban planners must develop effective regulatory systems that account for the unique challenges of AVs. This includes assessing potential impacts on land use, public safety, and investment in infrastructure. Hopefully, innovation emanating from robotics technology can create new solutions that solve urban challenges so that transportation systems are efficient and passenger-centric. Hence, as cities open arms to technological revolution, the collective engagement of stakeholders will be crucial when developing a transportation system that will serve the current and future mobility demands (Monte et al., 2017) (Figure 7.4).

Figure 7.4. Impact of technology on transportation planning.

Source: https://www.cartrack.ng/wp-content/uploads/2024/05/65e80aeaaad625e74b513d93_Main-Image-1200x480.png.

7.2. TYPES OF DATA UTILIZED

When planning public transportation, there are various types of data that must be examined in order to make decisions that enhance efficiency and access. Geographic Information Systems (GIS) is a fundamental tool in the process because it allows planners to use the ability to study patterns and spatial associations. This way, planners can study in-depth distance factors that affect modes of transportation, especially in urban areas. Understanding the impact of land use changes on transportation dynamics is beneficial in service gap identification as well as in support of infrastructure planning. For example, as reported in studies, increasing regionalization of education has resulted in lower walkability for students, indicating the necessity to include land use data in the planning of public transport. With the use of GIS data, planners are able to create solutions to address issues of uncontrolled city expansion and the consequences for public transportation accessibility (Arthur et al., 2014). Quantitative information, such as traffic counts and transit passenger data, also increase the database used for public transportation planning. These data provide tangible statistics that reflect usage patterns, which can inform service adjustments, route scheduling, and transit policies. For example, traffic counting software has been invaluable in gathering information about transport types other than car traffic, including pedestrians and bicycles. This multi-faceted approach makes it possible to have a full appreciation of traffic flow so that transport systems are of value to all. In addition, through the use of historic crash records, planners are able to identify high-risk spots and prioritize priorities for safety improvements, therefore making informed funding requests for funding necessary improvements to the infrastructure. Last but not least, the integration of several quantitative sources of data helps better appreciate public transportation needs (Bennett et al., 2019). Qualitative data also plays a very important role in the public transportation planning process, particularly if there is little or no quantitative data. In rural areas, where traffic records may not exist, planners generally have to rely on qualitative analysis through Road Safety Audits and community member comment. These may uncover specific issues and local attitude towards transport provision so that planners can respond to needs directly from the community. For instance, public meetings and questionnaires can indicate social attitudes towards public transport and economic factors on ridership. With the incorporation of qualitative information into planning frameworks, agencies can design initiatives that reflect the preferences of the community and increase public satisfaction levels with transportation systems. This unified process, incorporating the aspects of qualitative information and quantitative analysis, strengthens the pillars of sound public transportation planning, eventually yielding more fair and efficient services for all (Bennett et al., 2019).

7.2.1. Demographic Data and Its Significance

Demographic information is necessary in the field of public transportation planning because it offers key information regarding the population's make-up, mobility requirements, and tastes. The information includes several factors such as age, income, race, and employment status that drive how residents use transportation systems. Through the analysis of these demographics, planners are able to determine certain groups that might need more resources or specialized service. For instance, higher concentrations of low-income households in a neighborhood could indicate the need for affordable transportation to access employment opportunities. Subsequently, the utilization of demographic data not only enhances fairness of transportation access but also optimizes overall system efficiency by aligning service with community demand. Additionally, demographic data play a crucial role in addressing public health concerns regarding transportation. As highlighted by various studies, the interconnectivity between transportation systems and health effects is vital, particularly with regards to air quality and respiratory disease. Research has shown that populations living in areas of high transit availability, coupled with auto dependency, experience higher health vulnerabilities due to exposure to pollution, most prominently in densely urbanized areas like Dallas-Fort Worth and Los Angeles. Using demographic data, planners can enhance their understanding of the spatial connections between transit access, socioeconomic status, and health outcomes, as in Arabi et al. (2019). Strategic interventions to reduce pollution and its impact on health can be guided by such data, thereby enabling a healthier environment to live in for urban residents. Demographic data is also essential in enhancing public transit access to individuals with varying transport modes. For example, evidence indicates that while walking remains the most prevalent method of travel to transit stops, mode choice is largely influenced by neighborhood traits and family income. Planners must consider these forces as highlighted in Mallon et al. (2013) in order to develop policies that encourage sustainable modes of transport such as biking or carpooling for the car owners. Recognizing the interplay of demographic factors and transport choice allows the development of targeted policies that optimize first and last-mile connectivity. As a result, there can be efficient public transportation systems that not only accommodate the median consumer but also several population segments, optimizing total ridership and user satisfaction (Figure 7.5).

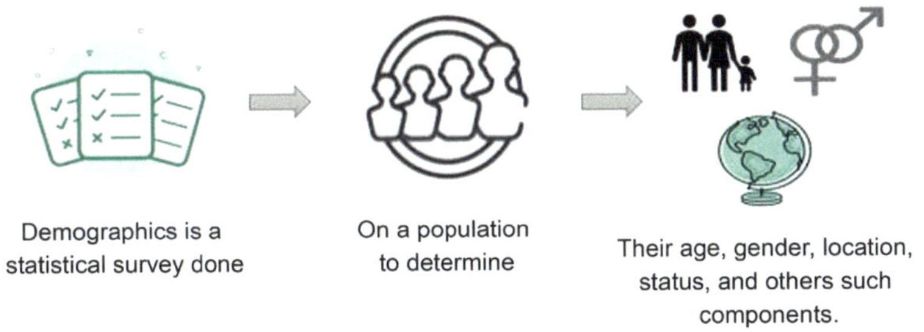

Figure 7.5. Demographic data and its significance.

Source: https://wallstreetmojo-files.s3.ap-south-1.amazonaws.com/2022/10/Demographics.png.

7.2.2. Traffic Patterns and Ridership Statistics

Traffic movement knowledge is crucial in planning public transportation. With traffic congestion and environmental issues in the city, knowing how cars move and the associated ridership becomes increasingly important. Knowing these patterns assists transportation agencies in identifying peak travel periods, busy corridors, and the drivers influencing passengers' choices. This fact-based approach not only optimizes for operations but also optimizes the delivery of service, ultimately resulting in increased ridership. As stated, investing today in a more diversified transit system — that provides tolerable alternatives to walk and bicycle, and use public transport in addition to drive — will yield much more effective use of transportation resources (N/A, 2008). These findings are crucial when creating interventions towards higher use of public transport and lower usage of single-occupancy vehicles. Ridership information also offers useful insights into the effectiveness of public transit systems and ways in which they can be improved. These data are utilized by agencies to quantify not only daily ridership but also demographics, traveler patterns, and satisfaction rates. These measurements help to identify underserved areas and potential barriers to access, such as connectivity to primary employment centers or social hubs. Tracking ridership over time allows transit agencies to adjust schedules and service frequencies to better align with the evolving needs of the public. From research on the transit system in New Jersey, the determination of public transport-specific weaknesses and strengths can enhance overall performance and crisis response planning (Bartin et al., 2017).

Figure 7.6. Traffic patterns and ridership statistics.

Source: https://content.govdelivery.com/attachments/fancy_images/ILCMAP/2022/12/6912025/traffic-and-transit-ridership-graphic_original.png.

By means of utilization of this data, public transport planning can also be responsive to current needs and eventual emergencies. Ultimately, the integration of traffic patterns and ridership data is a step towards public transportation planning that is more inclusive and responsive. This data-centered approach allows for robust examination of the influence of a variety of factors on public transport effectiveness, from urban infrastructure to aggregate socioeconomic conditions. By remaining data-centered, transit planners can make informed decisions that prioritize sustainability and responsiveness. Not only does this model position cities for today's challenges, but it positions them on a trajectory towards resilient transit systems that can meet future pressures. Efficient integration of active transport options further reinforces this system, establishing a multi-faceted transit approach for urban transport that benefits all residents (N/A, 2008). Ongoing application of these data-based methods solidifies their effect on the development of a sustainable urban transit culture (Figure 7.6).

7.2.3. Geographic Information Systems (GIS) in Transportation Planning

In recent years, Geographic Information Systems (GIS) have transformed transport planning, providing planners with powerful platforms to model and measure spatial data. The ability to track routes, assess traffic flows, and assess the accessibility of public transportation underscores the application of GIS in urban mobility issues. Through the integration of varied datasets—demographics, land use, and transport facilities—planners have been able to develop comprehensive models that forecast transport needs. As is especially well demonstrated, GIS use enables stakeholder participation through enabling members of the community to view maps of proposed plans and how they impact them, making for a more democratic planning process. Both capable of generating realistic models as well as eliciting feedback from the population in terms of transportation development, GIS is a priceless tool in today's urban planning initiatives (Batty et al., 2007). Further, GIS implications even extend beyond current transportation policy to influence future land-use policy making and planning. As urban population sizes increase, the linkage between land use and transportation systems increases its significance. Studies on school districting and transportation needs explained how geographic factors influence children's accessibility (Arthur et al., 2014). By using GIS, planners are able to make the best optimal school location decisions, ensuring equitable access while minimizing reliance on car travel. This integration of transportation and school infrastructure is part of a broad trend in city planning, as GIS is embedded into decision-making. Through spatial relation analysis, planners can resolve in advance issues pertaining to mobility, equity, and sustainability in the development of public transit. Finally, web GIS technologies have democratized access to transport data to enable real-time analysis and stakeholder co-planning. With planners and decision-makers operating from cloud-based platforms, sharing, and disseminating information easily revolutionizes the game of transport planning. For instance, web-based collaborators founded on GIS can facilitate urban development and transport policy discourses between stakeholders as diverse as local government executives and citizens. This participatory process not only enhances transparency but also enables a more responsive planning process that can respond to the needs of the community. Thus, GIS is not just an analysis tool but a dynamic framework that aids data-driven public transportation planning and engages constituents substantively (Batty et al., 2007) (Figure 7.7).

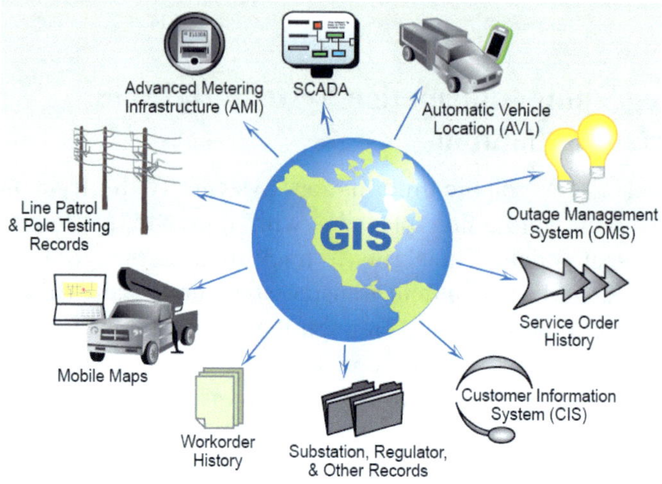

Figure 7.7. Geographic information systems (GIS) in transportation planning.

Source: https://www.transindo-training.com/wp-content/uploads/2018/03/GIS.jpg.

7.3. DATA COLLECTION METHODS

Data gathering techniques in public transportation planning form an integral component of decision-making and system improvement. Various techniques from qualitative surveys to intricate quantitative analysis are necessary in capturing urban transit complexity. For instance, qualitative data may be complemented with interviews from stakeholders and surveys from the community in order to obtain a sense of public opinion and unmet demands. Conversely, quantitative data, such as transit ridership figures and traffic volume, permits good analysis of service effectiveness and demand. Employing a blend of these methods, planners can create a comprehensive image that supports quality policy and funding decisions, enabling transportation infrastructure to be responsive to the needs of urban dwellers. This data approach not only serves planning but promotes community involvement and accountability in decision-making as well. Application of technology in data collection has revolutionized public transport planning, offering better and faster access to information. Technologies such as mobile traffic counting software help in the collection of precious data that would otherwise be difficult to obtain, particularly in rural parts where there are limited sources of information. The UAF Traffic app, for example, can record traffic flows easily as well as other transport modes. This skill is crucial in raising funding since giving evidence of a harsh need through empirical evidence is crucial in making valid proposals to funding bodies. Qualitative findings and road safety audits have been shown to be also be accepted by the funders, though quantitative evidence makes them more effective (Bennett et al., 2019). Thus, incorporating these technological innovations is crucial in improving public transport systems. Further, looking at the interaction of transportation with health outcomes illustrates the need for some data collection methods in public transportation planning. The impact that transportation systems have on respiratory health can be explained by data measures such as air quality indicators and socioeconomic conditions. Research that examines urban form in relation to respiratory disease risk emphasizes the application of models like Ordinary Least Squares and Geographically Weighted Regression that detect spatial connections guiding public health policy (Arabi et al., 2019). Through integrating health-based indicators into transportation planning, planners can develop countermeasures that mitigate risks and enhance community health. This shows that robust data gathering and analysis are not simply logistical necessities but are also central to creating healthier city spaces via smart transportation investments (Figure 7.8).

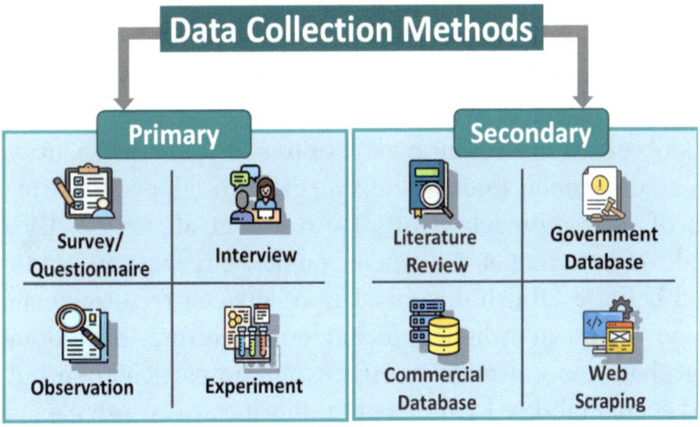

Figure 7.8. Data collection methods.

Source: https://cdn.educba.com/academy/wp-content/uploads/2023/09/Data-Collection-Methods.png.

7.3.1. Surveys and Public Feedback Mechanisms

Surveys and mechanisms for public feedback are an integral component of data-driven public transportation planning, providing valuable information on user behavior, needs, and preferences. Collecting feedback in real-time via surveys allows transportation agencies to gauge public sentiment and identify areas of concern. By using carefully designed surveys that focus on specialized groups, agencies have a better understanding of how various segments of the population use transportation services. These results enable planners to tailor service, enhance route efficiency, and optimize resource use. In addition, using surveys not only generates greater public involvement but also guarantees that transportation systems are designed in accordance with the wants of the citizens it carries. Moreover, using focus groups in addition to surveys may facilitate the process of feedback by offering qualitative as well as quantitative data. Focus groups leave space for more in-depth discussions, allowing participants to express views on transportation issues and hence expose hidden issues that could be hidden in surveys. For instance, discussions on particular themes, such as access and security, can offer detailed narratives that planners can utilize to inform them on what exactly users value. In codriving behavior studies, focus groups were employed alongside surveys to gauge consumer response to transport plans and demonstrated significant behavior change among individuals who were subjected to knowledge of driving behavior (Finson et al., 2012). Therefore, blending multiple feedback sources may increase data validity and usefulness for more effective decision making. Furthermore, the effectiveness of public comment processes can be measured through longitudinal studies that track trends in public opinion over time. These are useful for determining whether transport projects work and where the areas of improvement are. However, as observed in a recent review of transit agencies' security awareness campaigns, some agencies consistently neglect to adopt robust means to test the success of their outreach efforts (Edwards et al., 2010). By developing standardized measurements to gauge public engagement and awareness, agencies can better establish the correlation between feedback campaigns and actual change in transit riding or behavior. Therefore, developing effective feedback mechanisms not only reenforces public participation but also leads to increased accountability by transport authorities to ensure plans are not just data-driven but react to the needs of the people.

7.3.2. Use of Sensors and Smart Technology

The infusion of smart technology and sensors in public transportation networks is an innovative means of employing data to structure urban spaces. Through the placement of sensors of various types along corridors of transportation—from GPS-equipped tags on buses to environmental sensors along transportation corridors—cities can track transit activity and rider patterns in real time. This continuous flow of data is the basis for smart transportation systems that maximize speed and responsiveness. For instance, cities can utilize this information to optimize their routes, allocate resources according to actual demand, and eventually reduce waiting times for travelers. The depth of this information promotes a proactive way that promises to bring significant improvements in the quality of services provided, together with sustainable practices within urban areas, within the broader goal of constructing smart cities that utilize data effectively to solve public problems (Stahl et al., 2017). Furthermore, smart technology is not only a means of passive data gathering but also actively enables interaction between the transport infrastructure and users.

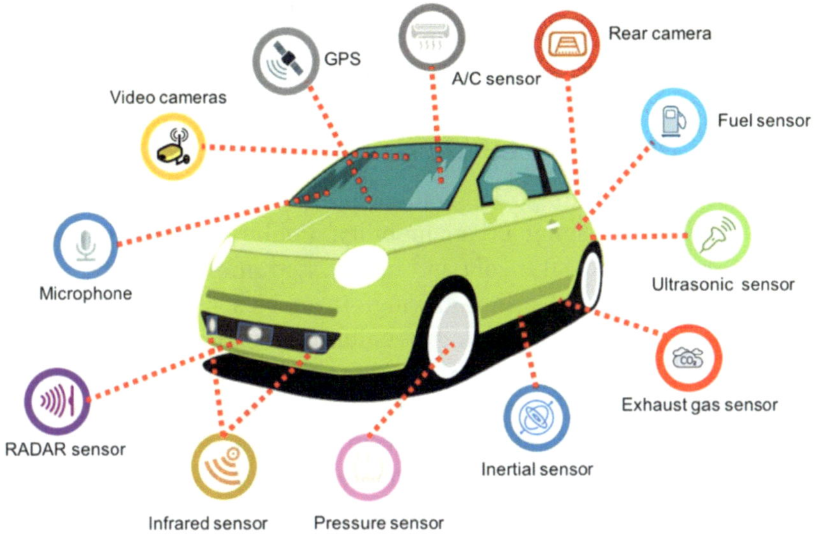

Figure 7.9. Use of sensors and smart technology transportation systems.

Source: https://www.mdpi.com/sensors/sensors-18-01212/article_deploy/html/images/sensors-18-01212-g001.png.

With mobile apps that analyze sensor data, a two-way communication becomes possible, where passengers are informed in real time of service alterations, delays, and the best routes of travel. This real-time supply of information empowers users to make wise decisions about their travel schedule, thereby leading to user satisfaction and promotion of public transport systems usage. In addition, through the implementation of smart ticketing systems with contactless payment systems, boarding can be optimized, access point crowding reduced, and analysis provided to better understand ridership. These technological developments are required to meet the evolving needs of urban residents and demonstrate the virtues of data-driven decision-making in public transportation (Monte et al., 2017). Smart technology and sensors also introduce possibilities for new urban policies and practices. For instance, data collected from intelligent systems is able to detect inequalities in mobility access, and metropolitan areas can then reorganize services to disadvantaged sections to offer equitable mobility. Additionally, results from data analysis can help promote sustainable forms of transport, for instance, by promoting public transport instead of the use of private vehicles. Lastly, the incorporation of intelligent technology and sensors not only optimizes existing infrastructure but also informs future city planning and transport policy to create stronger and more adaptable transport systems. The cyclical loop of data collection, analysis, and use is a good example of the trend of creating wiser urban systems that revolve around serving the needs of all citizens under the data-orientated public transport planning paradigm (Stahl et al., 2017) (Figure 7.9).

7.3.3. Integration of Social Media and Mobile Applications

Social media and mobile app integration into public transport planning is a paradigm shift in managing urban mobility. Public transport systems now incorporate real-time social media and mobile app information to understand passenger behavior and decision-making. It enables cities to engage with commuters proactively, establishing a culture of participation in decision-making. By leveraging crowdsourced data, not only are traffic forecasts improved, but relieving congestion and delay becomes more manageable as well. Mobile technology penetration allows transport authorities to draw on users' intelligence firsthand, providing a more comprehensive view of mobility patterns and commuter needs, ultimately leading to improved offerings and improved operations. Further, mobile applications are key enablers of intermodal connectivity. By incorporating features such as real-time information, route planning, and payment options, mobile apps make the experience for users simpler and promote multimodal travel. Bike or ride sharing services such as Uber and Zipcar use social media connectivity to enhance exposure and usage. Such applications not just facilitate easy travel but also stimulate users to choose more sustainable means of mobility and hence contribute to overall urban mobility policy. Integration of various modes of transport into a single mobile platform urges customers to experience and experiment with alternative modes, reducing dependency on solo vehicles and de-congesting the city. Lastly, the function of geographic information systems (GIS) in combination with mobile technology and social media cannot be overemphasized in data-driven planning for public transport. Miller (2013) opines that development of VGI allows planners to collect user-generated information that accurately captures usage and consumer patterns and hence is a powerful tool for making decisions. As Smart Cities grow, in which data drives infrastructure development, GIS enables one to visualize and analyze complex data sets. Incorporating crowdsourced data with classic data collection techniques can yield more sustainable mobility results because planners can see trends unfolding before they become fully apparent and react accordingly to changes in commuter behavior. Therefore, the partnership of social media and mobile apps is not merely a technological innovation, but a paradigm shift towards more responsive and human-centered urban transportation systems, consolidating the role of data in planning future public transport (Figure 7.10).

Figure 7.10. Integration of social media and mobile applications.

Source: https://img.freepik.com/premium-photo/social-media-integration_ 1029 473-208108.jpg.

7.4. DATA ANALYSIS TECHNIQUES

The significance of data analysis techniques in public transport planning is increasingly becoming so as planners and policymakers strive to enhance service efficiency and accessibility. Data analysis tools encompass techniques such as spatial analysis, predictive modeling, and simulation techniques, which assist planners in efficiently interpreting intricate sets of data. Spatial analysis uses geographic information systems (GIS) to map and analyze the spatial extent of transport services so that planners are able to identify the disadvantaged areas and rationalize routes. Predictive modeling helps in forecasting future demand based on parameters such as population growth and demographic changes, giving planners valuable feedback regarding potential variations in service delivery. With the implementation of these approaches, public transportation agencies can make their systems responsive to shifting requirements of urban residents.

Incorporation of data-driven approaches involves that public transportation planners must contend with both technological innovation and societal phenomena. As cities change in adapting to population increase and climatic instability, it is significant to understand land use-transportation interaction. The application of land use-transportation models (LUTMs) is one illustration of this interconnectedness, which enables planners to model outcomes for a range of scenarios and test the consequences of alternative policy choices for transportation networks.

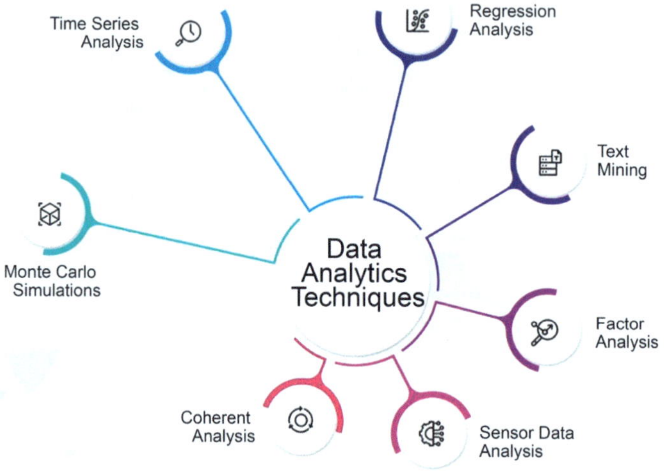

Figure 7.11. Data analysis techniques.

Source: https://d2ms8rpfqc4h24.cloudfront.net/types_of_data_analytics_0911 7b 6d85.jpg.

Scenario analysis and what-if modeling are some of the methods that offer planners vital tools for investigating possible future states of the transportation network under alternative conditions. These analysis planning frameworks facilitate evidence-based decision-making through enabling stakeholders to think through the impact of their decisions, making planning data-driven and people-centered, and thus optimizing the general performance of urban transport systems. Furthermore, emphasis on visualization in planning has enhanced stakeholder engagement and participatory decision-making. Natural interfaces that translate extensive datasets into comprehensible visualizations facilitate citizens to contribute more easily to transport planning initiatives. Not only does this democratization of information encourage transparency, but also public involvement, and thus enhances planning mechanisms to better capture the needs of people. For instance, web-based interfaces enable spatial data to be made available in real time, which supports collaborative approaches to identifying urban transport issues. As highlighted in the latest breakthroughs, such collaborative efforts mark a shift towards pluralist integration of perspectives in the planning for public transport systems in order to guarantee their sustainability and robustness in the face of changing urban environments (Ashcraft et al., 2018; Batty et al., 2007). By the utilization of data analysis techniques, public transport planning can be a flexible system sensitive to the aspirations and realities of diverse communities (Figure 7.11).

7.4.1. Statistical Methods for Interpreting Transportation Data

Sound understanding of transport data is more and more reliant on powerful statistical techniques that enhance our appreciation of subtle dynamics in urban mobility. Techniques such as Principal Component Analysis (PCA) and Ordinary Least Squares (OLS) regression are now stock-in-trade techniques for researchers and planners interested in analyzing determinants of transport systems. For instance, when examining how urban planning influences air quality and public health, PCA can help to identify principal variables responsible for respiratory disease risk. Reducing multifaceted data to main components, planners are then in a better position to relate socioeconomic characteristics and transit accessibility to health outcomes. This method emphasizes the need to integrate statistical analysis into transportation planning, with the goal of guiding decisions responsive to public health and city sustainability, a tangible concern in the case of urbanized populations facing environmental challenges (Arabi et al., 2019). Geographically Weighted Regression (GWR) is a statistical technique that becomes increasingly important in the analysis of transportation data with a goal to solving local patterns of variation across geographic space. As opposed to traditional regression procedures, GWR facilitates an examination of spatial heterogeneity, therefore suggesting the manner in which variable relationships might differ geographically. This is particularly significant in studies linking transportation infrastructure and motor vehicle access to public health outcomes. As proven in studies conducted in the Dallas-Fort Worth and Los Angeles MSAs, GWR captures the nuances of the effect of accessibility to transit on the risk of respiratory illnesses across different socioeconomic populations. Hence, the application of GWR enables transport planners to design interventions to enhance equal access to public transportation while limiting related health risks from air pollution (Arabi et al., 2019). Incorporation of survey data in statistical analysis further enhances the information about public attitudes and behavior towards transport policies. National surveys, for example, those indicating public favorability towards transportation funding through tax adjustment, use varying statistical procedures to analyze attitudinal data by demographic segments. With reactions analysis to different types of taxation illustrating how public support can emerge strongly in respect of perceived impacts of proposed taxes, ranging from the ways that roads are enhanced to the ways that the transport network is modernized, statistical analysis of the data informs policymakers about the willingness of society to fund transport improvements and thus plan more targeted and effective funding strategies. As can be observed, the general support for specific tax alternatives differs profoundly, thereby creating the need for evidence-based approaches that balance fiscal policies with urban and public goals for development.

7.4.2. Predictive Modeling for Future Transportation Needs

Predictive modeling is the solution to addressing the evolving demands of urban transport networks. As cities continue to expand and experience demographic changes, there has never been a more pressing need for transport solutions that can adapt. By utilizing historic data and current trends, predictive models enable transport planners to anticipate future passenger demand and plan infrastructure ahead. This approach enables more effective service delivery, better resource utilization, and increased user satisfaction. It also helps solve environmental problems through the use of sustainable modes of transport. Therefore, the application of predictive modeling in public transport planning is not only a futuristic concept but a commitment toward building more sustainable urban environments which are able to support evolving patterns of mobility and usage. Technology advancements, particularly in data analysis and collection methods, augment the ability of predictive modeling. Planners can utilize vast amounts of real-time data on traffic patterns, commuter behavior, and environmental factors through the use of big data analysis. There is sufficient data to design more sophisticated models that can examine various scenarios and forecast their potential outcomes. The predictive modeling process also involves iterative steps that ensure the predictions become increasingly refined through time, which guarantees constant adaptation to new patterns. As postulated in current arguments over planning support systems, switching to visualization and the transferability of data across platforms significantly enhances the participatory potential of the engaged parties. Public Transport Support (PSS) systems can utilize this technology to revamp planning processes into more reflective and knowledge-based processes (Batty et al., 2007). Predictive modeling must take into account the compound nature of transport needs, as opposed to merely numerical data. Understanding of socio-economic impacts of transport planning is critical to efficient and equitable delivery of services. Models that include accessibility and equity dimensions ensure public transportation systems are attuned to the needs of all community members but particularly marginalized communities that may experience disproportionate challenges. Active inclusion of these factors helps to foster community support and acceptance for public transportation initiatives. Furthermore, as Cloyd et al. (2007) suggests, the combination of predictive modeling with ecosystem methods of management is pivotal in effectively managing coastal and urban environments. Such a holistic perspective facilitates setting transportation plans that are environmentally conscious as well as economically viable, paving the way for an integrated solution to urban planning in the context of imminent challenges (Figure 7.12).

Figure 7.12. Predictive modeling for future transportation needs.

Source: https://mygodigi.com/wp-content/uploads/2024/09/1704221493501.png.

7.4.3. Geographic Analysis for Route Optimization

Geographic analysis plays an important role in route optimization for data-driven public transit planning because it enables agencies to have a visual comprehension of spatial data and base transit system decisions on that. With Geographic Information Systems (GIS) as a platform, transportation planners are able to overlay various datasets, including demographic, environmental, and infrastructural, and identify patterns that might not be easily determined with traditional analysis. Not only does this geo-spatial perspective reflect areas of high demand for public transportation, but it also supports the ability to see potential restrictions such as traffic or physical barriers. This ability to visualize based on population density, land use, and existing transit infrastructure makes route planning more accurately and more efficiently, leading to enhanced overall delivery of service with less operational inefficiencies. Further, network analysis, a core subset of GIS, provides other levels of comprehension that are essential in an effort to refine routes. The method allows one to study transportation networks through a consideration of aspects such as travel time, distance, and accessibility with the ultimate aim of creating efficient and trustworthy routes. Various algorithms are applied with this purpose in mind, specifically solving specific issues such as the Traveling Salesman Problem and time-constrained vehicle routing, which further optimizes public transportation systems' performance. Within a state like New Jersey, for instance, where complex city patterns and diverse commuter demand exist, implementing advanced network analysis methods can usher significant progress in transit performance. Effective application of these tools enables transit systems to become more attuned to the needs of people while enabling effective allocation of resources, a key component in sustainable public transport systems (Fischer, 2025). Moreover, the incorporation of geographic analysis in public transport planning offers advantages in recovery and resiliency, particularly during disaster scenarios. An effective analysis framework can evaluate transit system and infrastructure performance and enable proactive action to improve their disruption resiliency. For instance, real-time integration via structural health monitoring and remote sensing enables timely checks of bridge condition and transit network performance after an event. This purpose is served by geographic analysis in charting weaknesses in the transit system and informing recovery efforts that prioritize the most critical routes and modes of transport. It is from the case of New Jerseys transit system analysis that one can chart the intersection of efficiency and resiliency, underlining the function of spatial data in creating adaptive transport options that are capable of absorbing and recovering from unexpected shocks (Bartin et al., 2017).

7.5. CASE STUDIES

In data-informed public transit planning practice, case studies become central to explaining the complicated dynamics between urban design, transportation behavior, and community outcomes. As an example, a careful examination of various districts is employed to illustrate how previous land use development patterns, particularly the development of car-oriented philosophies, have reshaped school accessibility. The report highlights that increased distance to schools, driven by regionalization and low-density development, significantly lowers the ability of children to walk to school. This is not just a reflection of wider society trends toward car dependency but also emphasizes the importance of forward-looking planning giving precedence to walkability and equitable access to education. Such case study learnings emphasize the pivotal importance of integrative planning models operating counter to the mechanisms of self-reinforcing land use patterns and, in the long term, developing healthier communities. Furthermore, case studies provide useful lessons on transportation systems' health impacts, particularly in urban communities. An investigation of the relationship between transportation accessibility and respiratory disease in urban settings like Los Angeles and Dallas-Fort Worth reveals robust links between socioeconomic status and health. Employing advanced statistical techniques such as Geographically Weighted Regression and Principal Component Analysis, researchers were able to identify distinctive patterns that illustrate the function of transportation intensity in raising the risk of air pollution conditions. This connection reflects an imperative to public transit planning that not only facilitates access to fundamental services but also takes into account environmental sustainability and public health. These comparisons compel urban planners to consider the broader effects of transport policy on community health and promote designs that preclude contact with poisonous pollutants. Ultimately, case studies capture the actual application of theoretical paradigms to evidence-based public transportation planning, illustrating the necessity of an integrated and context-oriented approach. They are a bridge between empirical research and policy, demonstrating the utility of data obtained within specific contexts for informing best practices in transportation planning. Integrating social, economic, and environmental data ensures that interventions are community-specific in response to specific community challenges. Moreover, they add to quantitative analysis of data in order to forecast the overall impacts of proposed planning projects. Thus, these case studies not only educate us more effectively regarding existing systems but also serve blueprints for future projects in order to create lasting, equitable, and sustainable transport systems in diverse urban environments.

7.5.1. Successful Data-Driven Transportation Projects in Major Cities

The application of data-driven public transport planning has introduced revolutionary advances in big cities, which have led to improving urban mobility and sustainability. Data analysis has enabled cities to optimize transport services, both in terms of efficiency and in user experience. For instance, the California High-Speed Rail (HSR) project shows the application of data to aid urban development around transit stations. The project encourages the creation of transit-supportive density nodes by analyzing travel behavior and population statistics to see where the services would be most useful. This approach, as described in research on HSR, maximizes complementary activities of first-tier and second-tier cities and offers a solution that cuts down congestion and greenhouse gases, thus being more sustainable in the long run (Cuff et al., 2012). Besides, data-driven solutions extend beyond traditional transportation systems, evolving into new models like Delivery as a Service (DaaS). This model aims at enhancing the efficiency of urban logistics by coordinating delivery operations in line with customers' needs. By leveraging advanced technologies such as the Internet of Things, DaaS optimizes transportation routes and consolidates urban logistics, thus alleviating problems of the on-demand economy. Cities that have integrated these concepts note evident improvements in supply chain effectiveness and overall city planning. Integrating the data collection measures not only offers better travel choices but also promotes the deployment of clean vehicles, thus promoting environmental sustainability (Huschebeck et al., 2020). Successful data-informed transport programs greatly depend on coordination between technology, policy, and public participation. Solid communication channels between parties, including governments and tech companies, help data insights yield actionable improvements. Cities, for example, that use real-time data to optimize transit frequency, service area, and vehicle type experience a sharp boost in public confidence and ridership. By involving citizens in the planning process in a more direct way and utilizing feedback loops, planners can implement transportation solutions that actually fit the needs of the community. This sort of integrated strategy not only enhances greater efficiency in public transit but also improves social equity and access, further making urban transport systems inclusive and responsive to diverse populations (Figure 7.13).

Figure 7.13. Successful data-driven transportation projects in major cities.

Source: https://media.licdn.com/dms/image/v2/D5612AQHqqShCyb0sRw/article-cover_image-shrink_720_1280/article-cover_image-shrink_720_1280/0/1730901783249?e=2147483647&v=beta&t=EcJRBxnvyqIB-aNnmqpPsMyAhyxiic__qW8H00ikCV4.

7.5.2. Lessons Learned From Failed Transportation Initiatives

The failures of past transportation projects provide useful lessons that can render future evidence-based planning for public transport more efficient. Shared across most cases is the lack of stakeholder involvement, particularly for marginalized groups impacted by transit planning. In the majority of cases, such projects are centered on the technical aspects of planning transportation and compromise the social infrastructure upon which community acceptance and usage rely. A successful initiative will thus have to recognize and address the place-based intricacies of community dynamics, as N/A (2013) explains. The lesson is clear: residents must be worked with—not as passive recipients of services, but as real participants in the decision-making process—in order to establish trust and ensure that the initiative addresses the actual needs of its users. In addition, lessons gleaned from previous failure highlight the necessity to adopt an overall, data-informed approach within public transportation endeavors. For instance, endeavors that failed to incorporate significant data were likely to dismiss important trends of ridership and demographic patterns, leading to endeavors that underperformed among the target marketplace. Effective planning requires meaningful analysis of available data in an effort to determine where true gaps in transportation are located. This analysis tool can guide efforts to enhance route planning, service frequency, and access, hence creating an inclusive model for all. As seen in the Making Connections program featured in N/A (2013), employing a multi-faceted approach, including good data-based strategies, will be effective in community-focused projects. Furthermore, the New Orleans case shows us yet another crucial lesson regarding systemic fairness in public transportation projects. Transportation inequities are often in response failure, resulting in marginalization of some groups, particularly those in lower incomes. Initiatives that fail to address such differences risk exacerbating existing inequalities. As in, reform initiatives need not be centered only on increased efficiency but also on fostering greater equity of access to transportation services. Learning from past failures is necessary to develop transit systems that are equitable and inclusive so that all community members can benefit from improved transit opportunities. Focusing on equity and using strong data, planners are able to design transportation systems that serve more than one population while enhancing overall transit networks.

7.5.3. Comparative Analysis of Different Cities' Approaches

Cities all over the globe are unique in public transport planning by their own approach tailored to their respective problems, infrastructure, and technological context. For example, Amsterdam focuses heavily on bike infrastructure, greatly reducing motor vehicle reliance. Not only does this kind of policy enhance congestion, but also promotes sustainability goals. While other major cities like Los Angeles have unique issues with their car culture and sprawl, more integrated use of transit systems needs to happen. The varying investment cultures by city are seen in city policy in terms of funding and project priorities. The varying methodologies highlight the comparison of success of various approaches to accessibility, efficiency, and environmental sustainability of public transport. By means of data-driven solutions, cities can enhance their public transport efficiency, as shown in the comparative analysis of international experience. For instance, while some cities apply cutting-edge predictive analytics to forecast passenger demand and optimize route timetables, others use geographical information systems (GIS) to better understand the spatial dynamics of urban transport needs. The integration of these technologies not only contributes to planning decisions but also addresses existing inefficiencies and, therefore, improves service delivery as a whole. Benchmarking against cities that have successfully leveraged data analytics reflects the key success factors that can be replicated elsewhere. Moreover, innovative solutions such as the application of delivery drones and air taxis, which a recent literature review illustrates, provide city planners with the chance to envision urban mobility solutions beyond conventional boundaries (Biehle et al., 2020). The importance also cannot be overemphasized in placing the transport emissions inventories in the best representation and integration into climate planning. Many cities have problems of methodology that compromise the integrity of their emissions data, leading to skewed assessments of their public transport capacity to respond to environmental goals (Bishins et al., 2010). A proper and thorough inventory can inform policy changes that are needed and investment strategies so that cities will be adequately prepared to respond to climate challenges. As more and more cities depend on integrated analysis of data for transit planning, the need to add environmental considerations to the equation will be of top importance. Thus, through ongoing learning from diverse approaches, cities do not only improve public transportation but also contribute meaningfully to worldwide sustainability.

7.6. CHALLENGES AND LIMITATIONS

Evidence-based public transit planning implementation involves various challenges and limitations that have the potential to undermine successful results. One major challenge lies in the quality and availability of data itself. Public transportation authorities in most areas rely on outdated or incomplete datasets, which can result in erroneous analysis and ineffective policy decisions. Additionally, blending various data sources such as traffic volumes, socioeconomic profile, and ridership pattern requires sophisticated infrastructure that is not available to most agencies. For this reason, without reliable data and integration capacity, transportation planners cannot formulate plans that actually cater to the needs of concern to the community. Such ineffectiveness not only limits the potential to upgrade transit facilities but can also disillusion public confidence in the planning process. Another. critical concern is the financial constraints that public transit agencies are likely to face. Budget constraints constrain the ability of agencies to invest in advanced technologies required to conduct data collection and analysis. The majority of agencies are compelled to compete for available funds against a plethora of public priorities, and hence it is challenging to acquire funds required to undertake extensive data-informed programs. Moreover, as the synthesis report to The Kresge Foundation highlights, financial shortages can impede transformation in transport agencies because they are unable to make innovative suggestions that could enhance service provision and efficiency (Grannis et al., 2014). This ongoing lack of money exacerbates problems at hand, leading to stagnation in enhancing services and decreased public satisfaction. Socio politics also introduce layers of complexity to data-driven planning for public transportation. The stakeholders for whom the planning is done, including community members and local governments, may have competing interests that make it difficult to achieve consensus and make decisions. While data can enlighten ridership demand and rider preference, politics of the case might determine what data becomes priority or is interpreted in a particular manner. The combined educational frameworks report highlights the necessity of collaborative strategies to dispel hindrances, just as transport agencies themselves will need to cooperate with other stakeholders in the community to come up with viable planning solutions (Russell et al., 2016). In the end, unless these socio-political forces are tackled, data-driven public transit plans are bound to face resistance, minimizing their impact and usefulness for society.

7.6.1. Data Privacy Concerns and Ethical Considerations

In the current era, the incorporation of data-driven approaches in public transport planning raises genuine data privacy concerns and ethical challenges. With advancing transport models that increasingly rely on mobile and personal data collection and usage, particularly from intelligent technologies and mobile devices, privacy violation risks increase. The public remains in the dark about their usage of location and the way they are moving, and therefore, there is loss of public confidence in the ruling powers. As it has been posed in recent analyses, while the benefits of greater transportation efficiency may be tempting, a careful analysis must review the societal impacts associated with such data-innovations. These issues highlight the need for specific frameworks to address privacy protection requirements without restricting innovation in transportation planning (Biehle et al., 2020). Moreover, the ethical aspects of making use of personal data extend beyond the information collection itself; they also involve the scrutiny of how this data is shared and processed. Ethical frameworks should involve consent and openness, ensuring that individuals are agents of data use rather than passive targets of surveillance. Most of the discussion emphasizes that privacy-sensitive applications can enable crucial information on mobility patterns, revealing a possibility of ethical use of data, respecting individual rights, while staying useful to society at large. This strategy falls in line with the belief that compliance with the law is not enough; what is required is a transition towards best practice where the emphasis is on balancing privacy and the benefits derived from data analysis for public transport (Boman et al., 2013). Additionally, involving the public in dialogue regarding the use of data not only makes information more transparent but also instills a sense of ownership from the citizens for their information. Involving community stakeholders in the discussion of the same can ease fears and foster an environment of collaboration between authorities and citizens. Through active participation of citizens, transport endeavors based on data will not only be efficient in operational practices but will also imbibe social values and ethics. Lastly, the effectiveness of evidence-based public transport planning relies on establishing trust, and this can be achieved by infusing data ethics into the very core of transport policies and practices and, consequently, fostering responsible governance and efficient decision-making processes (Biehle et al., 2020; Boman et al., 2013).

7.6.2. Limitations of Data Accuracy and Reliability

In public transport planning, reliability and precision of information are column elements to facilitate informed policy decisions. However, underlying limitations consistently denigrate these attributes and therefore lead to potentially erroneous policy-making. A key issue stems from the often outdated or deficient data sets utilized to analyze transport trends. For example, public transit agencies can be basing planning on earlier ridership data that does not take into account present fluctuations in social habits or city design. Such lapses in data integrity can result in planning that fails to satisfy users as they are today, eventually influencing efficiency and usability of services. Thus, reliance on missing or out-of-date data is a significant shortcoming in rendering data-driven public transportation planning effective. In addition, the type of data collection methods can create other reliability problems. Most transportation agencies use automated data collection approaches, which, as good as they are, sometimes oversimplify complex human activity and traffic movement too much. For instance, approaches that consider only GPS traces may not account for nuances such as road closures or special events that have a huge effect on travel time. As noted in the example of traffic management for Nebraska Cornhuskers football games, temporary congestion patterns are typically not uncovered through standard data-gathering processes, leading to less-than-optimal response for peak events. In the absence of integrated data that captures real-time dynamics, transit planners will struggle to optimize schedules and routes in an effective manner, which goes to increase the effects of these data shortcomings in terms of reliability.

Relevance of data is equally critical in transportation planning. With evolving city mobility patterns, with factors like micro-mobility and ride-sharing becoming the standard, conventional data parameters may be unable to capture significant trends. Studies have indicated that the dynamics of user behavior in large events like sporting events or festivals require tailored data analysis to respond adequately to congestion. Examination of patterns of change in traffic flows around specific events can offer vital intelligence to inform resource allocation. However, the question remains as to how one can make such data timely and relevant to broader transport planning. There needs to be an understanding of the shortfalls of data reliability and precision; there needs to be a need for responsive data collection and analytical methods that keep pace with dynamic metropolitan mobility trends, such as applied in methods like the Multi-EigenSpot and DBN applied in related studies (Ahsani et al., 2019) (Figure 7.14).

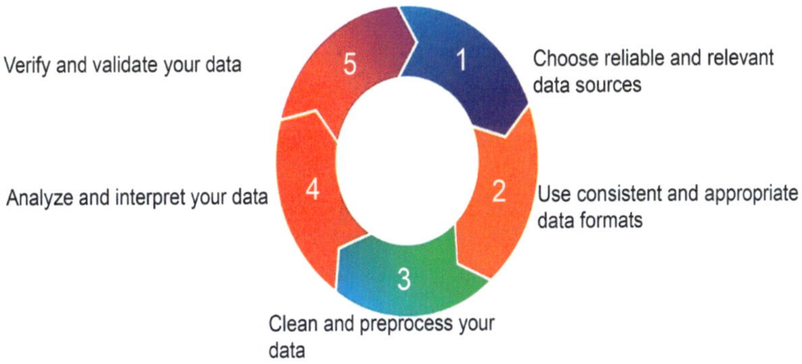

Figure 7.14. Limitations of data accuracy and reliability.

Source: https://fastercapital.co/i/Cost-Model-Updating--How-to-Incorporate-New-Data-and-Information-into-Your-Cost-Model-Over-Time--How-to-ensure-the-accuracy-and-reliability-of-your-data.webp.

7.6.3. Financial Constraints in Implementing Data-Driven Solutions

Implementation of data-driven public transportation planning solutions is usually hindered by the large financial constraints that cities face. These constraints normally arise from budget constraints as well as prioritizing immediate operating needs over spending money on technology in the future. Investment in data collection, processing hardware, and analytic tools in most cities is limited. As such, agencies generally end up bogged down from taking innovative data-driven projects that could be utilized to enhance service performance and customer satisfaction. The Covenant of Mayors (CoM) is also an initial attempt at addressing such economic challenges by promoting the use of sustainable energy and transport means, yet lacking a structured finance mechanism, its effectiveness in spearheading necessary improvements in public transportation network data usage can be compromised (Berghi et al., 2017). In addition, the difference between the theoretical benefits of coordinated transport planning and its use in practice also reflects more universal budget constraints on city governments. Even though research proves that need-based models founded on real-world evidence are superior, the necessary expenditure on resources such as technology and staff training is often too expensive for most municipalities. This separation hinders marrying planning for transport to genuine needs of communities, causing over-reliance on traditional approaches. As has been demonstrated through studies in most urban areas, such as Berlins research into e-mobility, there may be limited funds that can prevent genuinely innovative solutions that cater to genuine mobility needs of citizens. Hence, with limited funding, the majority of cities cannot transition from traditional models to adaptively more responsive, data-informed models (Hoor et al., 2019). Notwithstanding these funding issues, policymakers and transport planners need to prioritize resource allocation into investments in data-centric technologies. A funding approach can assist municipalities in overcoming the limitations and leveraging the full capacity of integral transport planning. Contracting partnerships with private institutions, accepting grants, and engaging in collaborative work may provide sources of funding necessary for the implementation of advanced data-centric solutions. This would not only enable a more data-centric strategy but also improved-quality public transportation services provided to citizens. Finally, financial considerations must be addressed in order to unlock the potential that data-driven public transport planning has to provide, resulting in more sustainable and efficient urban mobility systems.

SUMMARY

In a conclusion of data-driven planning of public transportation, it becomes evident that employing robust data analytics significantly enhances the efficiency of transport systems. As cities expand and populations densify, it is essential that effective, reliable public transport be present. Data analytics not only assists in identifying transport trends and user behavior but helps make knowledge-based choices when it comes to investing in infrastructure. By identifying common vision and measures, stakeholders can harmonize their goals so that transportation projects can result in economic and community prosperity and ultimately contribute to a prosperous urban landscape (Bierbaum et al., 2011). Also, the ability to predict and control congestion becomes highly vital in optimizing transportation networks. Studies have demonstrated that historical analysis of traffic data can yield valuable insights into patterns and outliers, including those caused by significant events. The particular challenges of major sports events, for instance, demonstrate the utility of adaptive traffic management strategies. By applying algorithms like Multi-EigenSpot to identify traffic congestion hotspots, as witnessed by traffic flow in Nebraska Cornhuskers football games, planners can create timely interventions that bypass clogging and improve overall public transit efficiency, with forecasting accuracy achieved using methods like Dynamic Bayesian Networks (Ahsani et al., 2019). Ultimately, data-driven public transportation planning implications extend far beyond short-term logistical concerns. It is a force behind broader community development and resilience. Incorporating strong data analysis in planning not only improves transit reliability but also strengthens civic engagement by working to address the needs of communities. Placing transportation equity at the top enables planners to gain transportation benefits for all citizens through urban transit systems. As cities confront increasing challenges, an evidence-based approach will be essential in creating responsive, sustainable transportation systems that address the diverse interests of the people and enable the sustainable urban way of life (Bierbaum et al., 2011).

Information is important in transportation planning to create effective public transit systems that are sensitive to community needs. Good-quality information enables planners to discover more about travel habits and demand, and therefore plan routes and schedules that will make it easier and more efficient to get around. For example, extensive freight analysis by the Oregon Department of Transportation illustrates how making multiple-sources of information come together improves long-range planning. With traffic flow analysis, demographic information, and land use patterns, planners are able to discern best-course options that anticipate growth while curbing congestion. Decision-making in conjunction with data not only maintains transportation projects in step with economic realities but also invites stakeholders to engage in smart, policy-relevant discussions, thereby facilitating open and sustainable transit solutions (Knudson et al., 2022). Moreover, the interface of public health with data adds an extra value to smart transportation planning. Transportation in urban areas has been empirically linked with several environmental issues, such as compromised air quality and subsequent health consequences. Research indicates that transportation infrastructure is capable of influencing respiratory disease outcomes through factors like transit coverage as well as socioeconomic status. Within Metropolitan Statistical Areas like Los Angeles and Dallas-Fort Worth, geospatial statistical techniques like Geographically Weighted Regression reveal underlying spatial correlations that are employed for purposes of estimating populations at risk. Through their examination, planners are able to design transport infrastructure that maximizes efficiency as well as public health. This holistic approach confirms the pivotal role of data in addressing both transport efficiency and the general well-being consequences of city mobility on public health (Arabi et al., 2019). Moreover, effective transport planning relies on making trade-offs and preventing unintended consequences, which is best facilitated by good-quality data analysis. Testing for possible public policy allows decision-makers to reconcile conflicting objectives and judge the capacity of intended strategy to resist forces beyond the powers of the agency. Data serve as a source on which these judgments can be based, ensuring transport planning to be responsive and adaptive to changing cityscapes. Planners with good data tools can simulate various circumstances and foresee future demand, hence optimizing resource allocation and mitigating adverse impacts. This evidence-based approach fosters responsibility and responsiveness cultures in public transport systems that, in turn, translate into improved mobility solutions responding to the evolving needs of city dwellers (Knudson et al., 2022).

REFERENCES

1. Agrawal, A. W., & Nixon, H. (2011). What do Americans think about federal transportation tax options? Results from year 2 of a national survey, Research Report 10-12. https://core.ac.uk/download/70409923.pdf (accessed on 21st October 2025).
2. Ahsani, V., & Sharma, A. (2019). Assessing the impact of game day schedule and opponents on travel patterns and route choice using big data analytics. https://core.ac.uk/download/286729950.pdf (accessed on 21st October 2025).
3. Arabi, M., Famili, S., Hyun, K. (K.), Mattingly, S., et al. (2019). 17-09 Assessing the impact of air pollution on public health along transit routes. https://core.ac.uk/download/287016823.pdf (accessed on 21st October 2025).
4. Arthur, A. C. E., Eulberg, N. M., & O'Malley, K. C. (2014). Wheels on the bus: School transportation as a reflection of sprawl. https://cupola.gettysburg.edu/cgi/viewcontent.cgi?article=1303&context=student_scholarship (accessed on 21st October 2025).
5. Ashcraft, C. M., Brewer, J. F., Burakowski, E., Coffin, A., et al. (2018). Preparing for a Northwest Passage: A workshop on the role of New England in navigating the new Arctic. https://core.ac.uk/download/215537367.pdf (accessed on 21st October 2025).
6. Attard, M., Capineri, C., & Haklay, M. (2016). The potential of volunteered geographic information (VGI) in future transport systems. https://core.ac.uk/download/83020172.pdf (accessed on 21st October 2025).
7. Bartin, B., Deka, D., El-Awar, A., Lou, Y., et al. (2017). Performance measures to assess resiliency and efficiency of transit systems. https://core.ac.uk/download/129532928.pdf (accessed on 21st October 2025).
8. Batty, M. (2007). Planning support systems: Progress, predictions, and speculations on the shape of things to come. https://core.ac.uk/download/1683968.pdf (accessed on 21st October 2025).
9. Bennett, F. L., Metzgar, J. B., & Perkins, R. A. (2019). Enabling data-driven transportation safety improvements in rural Alaska. https://core.ac.uk/download/289066736.pdf (accessed on 21st October 2025).
10. Berghi, S. (2017). Energy use in urban transport sector within the Sustainable Energy Action Plans (SEAPs) of three Italian big cities. https://core.ac.uk/download/127587543.pdf (accessed on 21st October 2025).
11. Biehle, T., Fischer, L., & Kellermann, R. (2020). Drones for parcel and passenger transportation: A literature review. https://core.ac.uk/download/323212148.pdf (accessed on 21st October 2025).

12. Bierbaum, A. H., McKoy, D. L., & Vincent, J. M. (2011). Opportunity-rich schools and sustainable communities: Seven steps to align high-quality education with innovations in city and metropolitan planning and development. https://core.ac.uk/download/71358710.pdf (accessed on 21st October 2025).
13. Bishins, A., & Duduta, N. (2010). Citywide transportation greenhouse gas emissions inventories: A review of selected methodologies. https://core.ac.uk/download/71350130.pdf (accessed on 21st October 2025).
14. Bolea Monte, Y., Casanovas Garcia, J., Grau Saldes, A., Puig-Pey Clavería, J., et al. (2017). Public entities driven robotic innovation in urban areas. https://core.ac.uk/download/224975022.pdf (accessed on 21st October 2025).
15. Boman, M., Bylund, M., Hirsch, B., Sanches, P., et al. (2013). Knowing your population: Privacy-sensitive mining of massive data. http://arxiv.org/abs/1412.2247 (accessed on 21st October 2025).
16. Cloyd, E. T., Leonardi, A. P., Scheurer, D. L., Turner, R. E., et al. (2007). Establishing National Ocean Service priorities for estuarine, coastal, and ocean modeling: Capabilities, gaps, and preliminary prioritization factors. https://core.ac.uk/download/11017864.pdf (accessed on 21st October 2025).
17. Cohen, A., Davis, R., Dowd, M. K., Shaheen, S. A., et al. (2019). A framework for integrating transportation into smart cities. https://core.ac.uk/download/289097213.pdf (accessed on 21st October 2025).
18. Cuff, D., Higgins, H., & Loukaitou-Sideris, A. (2012). Planning for complementarity: An examination of the role and opportunities of first-tier and second-tier cities along the high-speed rail network in California, Research Report 11-17. https://core.ac.uk/download/70410032.pdf (accessed on 21st October 2025).
19. Edwards, F., Haas, P. J., & Rohlich, N. (2010). Exploring the effectiveness of transit security awareness campaigns in the San Francisco Bay Area, Research Report 09-19. https://core.ac.uk/download/70413457.pdf (accessed on 21st October 2025).
20. Finson, R. S., Martin, E. W., & Shaheen, S. (2012). Eco driving and carbon foot printing: Understanding how public education can reduce greenhouse gas emissions and fuel use. https://core.ac.uk/download/70413482.pdf (accessed on 21st October 2025).
21. Fischer, M. M. (2025). GIS and Network Analysis. https://core.ac.uk/download/pdf/7040225.pdf (accessed on 21st October 2025).

22. Grannis, J., Stumberg, R., Hoverter, S., & Arroyo, V. (2014). Preparing for climate impacts: Lessons from the front lines. https://core.ac.uk/download/75778399.pdf (accessed on 21st October 2025).
23. Hoor, M., & Schwedes, O. (2019). Integrated transport planning: From supply- to demand-oriented planning. Considering the benefits. https://core.ac.uk/download/286083113.pdf (accessed on 21st October 2025).
24. Huschebeck, M., Leonardi, J., & Leonardi, A., et al. (2020). Approaching delivery as a service. https://core.ac.uk/download/305119210.pdf (accessed on 21st October 2025).
25. Kingsland, N. (2015). The hidden connection in city-wide reform: How charter schools can bring equity to public education. https://core.ac.uk/download/75784195.pdf (accessed on 21st October 2025).
26. Knudson, B. (2022). Freight moves the Oregon economy. https://core.ac.uk/download/542625672.pdf (accessed on 21st October 2025).
27. Kosarko, G., & Weissbourd, R. (2011). Economic impacts of GO TO 2040. https://core.ac.uk/download/71345969.pdf (accessed on 21st October 2025).
28. Liu, J., Rose, K., & Daniel, M. H. (2017). Creating change through arts, culture, and equitable development: A policy and practice primer. https://core.ac.uk/download/80510640.pdf (accessed on 21st October 2025).
29. Mallon, Y., Thakuriah, P., & Tilahun, N. (2013). Factors determining transit access by car owners: Implications for intermodal passenger transportation planning. https://core.ac.uk/download/9652213.pdf (accessed on 21st October 2025).
30. Masvidal Andreu, C. (2017). Inventory drivers in a pharmaceutical supply chain. https://core.ac.uk/download/157811883.pdf (accessed on 21st October 2025).
31. N/A. (2008). Active Transportation for America. https://core.ac.uk/download/71343026.pdf (accessed on 21st October 2025).
32. N/A. (2013). Community Change: Lessons from Making Connections. https://core.ac.uk/download/84911967.pdf (accessed on 21st October 2025).
33. Russell, C. A., Hildreth, J. L., & Stevens, P. (2016). ExpandED Schools National Demonstration: Lessons for scale and sustainability. https://core.ac.uk/download/75785743.pdf (accessed on 21st October 2025).
34. Stahl, E., & Dupuis, N. (2017). Trends in smart city development. https://core.ac.uk/download/83111937.pdf (accessed on 21st October 2025).

CHAPTER 8

SAFETY AND RISK ANALYSIS USING DATA

Contents

INTRODUCTION

Today's society has had its safety and risk analysis heightened significantly by the rapid pace of development of data collection and analysis techniques. Data-driven techniques are very important in understanding multi-dimensional threats across various industries, including healthcare, engineering, and environmental management. Not only do these techniques assist in easy identification of expected dangers but also refine decision-making processes through predicting future outcomes based on historical trends and existing information. As companies are increasingly putting more priority on evidence-based practices, introducing sophisticated analytical tools becomes a must. With the application of advanced modeling and simulation software, safety professionals can better quantify uncertainties as well as assess risk scenarios that were difficult to quantify previously. This development reflects a trend toward an entire picture of safety that interacts with genuine variability and uncertainty. The utilization of Dynamic Probabilistic Risk Assessment (DPRA) methods is a milestone in safety analysis. DPRA methods involve dynamic simulation and system performance modeling to account for real-time operation complexities. With the use of system simulator codes such as RELAP5, combined with monitoring and control simulations, analysts are readily capable of incorporating both deterministic and stochastic factors influencing system behavior. In a standard DPRA analysis, there is a structured process used, beginning with parameter sampling of the uncertainty space.

Figure 8.1. Safety and risk analysis using Data-driven approaches.

Source: https://www.ksolves.com/wp-content/uploads/2024/07/Security-Risks-Posed-by-Generative-AI.jpg.

This is complemented by system behavior simulation, output assessment, and correlating the visualization of parameter values and results. These processes form a comprehensive framework for the analysis of possible safety hazards and total system resilience to uncertainties (Yilmaz et al., 2013). Also, the dual relationship between risk perception and community involvement foregrounds the critical role of participatory approaches in safety analysis. Engagement of marginalized individuals, such as the youth, in discussions around safety and risk improves the understanding of their day-to-day lives and motivates them to participate in decision-making at the community level. With creative approaches such as visual storytelling and multimedia platforms, individuals can voice their concerns and opinions regarding safety issues in their environment. Such approaches have been proven to empower the participants by enabling them to interact and counter the social processes underlying their vulnerability. The integration of arts-based activities within public safety discourse has the ability to stimulate serious discussion, which yields learning that can be leveraged to support risk management practice and community resilience planning (Cutts et al., 2011). Hence, integrating people's voices through data-informed risk analysis not only makes the findings pertinent but also initiates social change (Figure 8.1).

Definition of Safety and Risk Analysis

The essence of safety and risk analysis is the scientific assessment, estimation, and prioritization of risks with the ultimate goal of minimizing potential harm to humans, the environment, and assets. Safety analysis is tasked with ensuring that the processes of operation are designed to minimize hazards, while risk analysis comparatively quantifies the chances and impact of undesirable events. The interface between these analyses is crucial in facilitating informed decision-making, especially in industries like water recycling and waste management, where new technologies and methods are being invented. Public health and ecosystem safety have seen a paradigm shift when it comes to perceiving and managing risk, with a need for systematic evaluation that not only considers immediate risk but also long-term environmental impact. Thus, the definition of safety and risk analysis in this context involves the identification of the ever-changing industry norm and regulatory landscape. Further, as industries are gravitating more towards complex processes and technologies, the procedures of performing safety and risk analysis must also shift to support such efficacy. Evident recent developments in the waste sector highlight the need for a more nuanced awareness of risk, particularly in light of increased public and state-level surveillance already outlined. In this regard, managing risk is essential in ensuring that newly adopted technologies such as biomass gasification and thermal treatment technologies are employed safely and sustainably. The integration of risk assessment into such technology allows operators to prevent potential dangers in advance, enhancing the confidence of the public in waste management systems. The essential element of this analysis is establishing robust planning and permitting frameworks that meet regulatory needs as well as social expectations, thereby ensuring effective implementation and uptake of innovative solutions in the sector (Eduljee et al., 2006). In the water management sector, the Water Safety Plan (WSP) is a prominent example for the understanding of risk analysis application. The WSP methodology supports the application of an integrated approach to the management of the risks associated with water reuse by stressing the incorporation of diverse contexts and objectives. With growing demand for good water reuse systems, the need for risk management procedures that are sensitive to uncertainty and prioritizing risk has never been higher. Reconciling the WSP into specialized forms, such as the Water Reuse Safety Plan (WRSP), offers a remedy to address such demands appropriately. Through emphasizing communication, decision support, and stakeholder involvement, the WRSP aims to enhance risk management related to water reuse in a way that overall public health and safety objectives are met while causing innovative solutions in the water management system (Goodwin et al., 2015).

Importance of Data in Risk Assessment

The effectiveness of risk assessment across different industries mainly depends on the analysis of data to recognize, measure, and reduce risk. Risk evaluation with a systematic approach starts with data aggregation, which increases the capability to perform thorough analyses. For instance, Job Safety Analysis (JSA) centers on the determination of job-related hazards, whereby data on previous accidents and existing working conditions inform risk classification into high, medium, and low. Despite thorough analyses conducted at PT. Indo Acidatama Tbk. and other organizations, the incidence of work accidents underscores the need for additional data exploration in order to enhance workers' awareness of safety procedures. Finally, applying empirical data together with intensive training has the capability to generate a safety culture, which generally reduces the incidence of workplace injuries significantly, thereby enhancing overall operation safety. Data plays a major role in enhancing safety measures, particularly with respect to systematic safety risk assessment of complex systems. The essence of system safety analyses, as stated by NASA, is the utilization of fact-based data to analyze potential hazards inherent in placing humans at risk and environmental sustainability. Using many methods of safety analysis, including hazard analyses and probabilistic risk assessment, allows companies to develop and review control measures. This formal process helps not only to streamline familiar threats but also safeguard against unfamiliar vulnerabilities. It is by aggregate risk analysis that organizations are encouraged towards an active risk management culture. By the integration of such intense evaluations into decision-making, organizations are able to assure maximum safety with ongoing improvement. Furthermore, reducing uncertainties in risk analyses reaffirms intensive understanding of data to aid decision processes. Effective risk assessment needs to review several results since not all risks can be quantified. Qualitative methods and bounding analyses have a tendency to augment data analysis, with robust measures of safety even under uncertainties. NASAs guidance facilitates a more collaborative safety risk assessment method beyond single hazard analysis, paving the way for comprehension of total risks. This multi-angled process not only enables compliance with designated safety objectives but also enhances the organizational structure for risk driver identification. Ultimately, a data-driven risk assessment process is required in order to support safety claims, ultimately bolstering the basis for operational integrity and mission success in any safety-critical environment (Figure 8.2).

Figure 8.2. Importance of data in risk assessment.

Source: https://www.lepide.com/blog/wp-content/uploads/2024/11/steps-to-perform-data-risk-assessment.jpg.

8.1. HISTORICAL CONTEXT OF SAFETY AND RISK ANALYSIS

Safety and risk analysis heritage is firmly embedded in the evolution of engineering practice, particularly because demands from the general public for safer buildings have increased with the passage of time. Early risk assessment practices were rudimentary and even predominantly reactionary, developed through the experience of specific failures rather than as a method of pre-emptive prevention. As industries developed, particularly with the event of extensive catastrophes, there developed a realization of the need to carry out systematic analysis in order to ascertain areas of risk and implement safety measures according. Formal methodologies, such as the bowtie approach and fault tree analysis, illustrate this development, which is an evolution towards quantitative methods that utilize data in informing decisions. This paradigm shift emphasizes the importance of embedding empirical evidence in making failure predictions and hence creating a foundation for contemporary practice of data-driven safety and risk assessment with data flowing from operating systems and past mishaps. Moreover, there are some methods that enhance the credibility of risk assessment through effective application of data. For example, empirical Bayes usage in safety analysis is a new approach to system reliability estimation in developmental stages. Apart from utilizing the heritage experience in other comparable systems, the approach also makes provisions for adjusting early reliability estimates as per new experience. Such adaptability is essential, particularly for industries as they strive to minimize risk in dynamic operating conditions. By reviewing early test results and integrating them into the models, safety analysts can have some idea about potential weaknesses and have them addressed ahead of time. The evolution over time towards data-driven methods thus reflects a tendency in general to merge empirical data with analytical tools to refine risk prediction and enhance safety procedures in various industries. Moreover, the shift from deterministic to probabilistic methods in risk analysis has transformed how vulnerabilities in historic buildings, in particular, are assessed. Traditional approaches of depending on comprehensive inspections became impossible with historic centers offering unique challenges in regard to size and complexity. The Bayesian approach to risk and safety analysis has been an appealing choice, as analysts are able to base significant data—e.g., inter-storey heights—upon measurable parameters rather than on expensive field surveys. By iteratively refining initial hypotheses as data become available, this probabilistic methodology allows a better understanding of risk, in this case seismic vulnerabilities in listed buildings. This not only adheres methodologically to the value of efficiency in data collection but also to the value of the continuous improvement of risk estimates, thereby instilling a culture of proactiveness for safety assurance in accordance with the data culture then existing in safety and risk analysis.

8.1.1. Evolution of Safety Standards

The creation of safety standards has revolutionized the practice of risk analysis in most sectors, particularly fire safety regulation and air traffic management. Fire safety has traditionally been governed by prescriptive codes that identified the action to be taken, inclined towards strict compliance but failing to address the complexity of modern buildings. The use of performance-based design, to which the Italian National Fire Rescue and Services' new Fire Prevention Code of 2015 is a testament, represents a shift towards flexibility. This method allows safety professionals to develop tailored actions which take into account personal risk profiles of different buildings, promoting better fire safety outcomes. These progressive standards demonstrate a shift from generic policy towards advanced practices that leverage data to optimize security in more complex environments (Caciolai et al., 2018). Even in air traffic management, safety standards have undergone substantial development due to improvements in technology and rising air traffic volumes. The commitment to reducing accident risk has inspired the development of innovative approaches and tools for risk assessment that are philosophically distinct from other forms of transport. Initiatives like the USAs Next Generation Air Transportation System (NextGen) and Europe's Single European Sky ATM Research Program (SESAR) are a new paradigm in air traffic safety. These initiatives put extra emphasis on system resilience and human factors, necessitating a thorough evaluation of future risks involved with new technologies and modes of operations. With increasing air traffic, these shifting standards have to keep pace, never at the expense of safety while dealing with the intricacies of contemporary air transport (Brooker et al., 2007). The continually evolving process of safety standards addresses the very interdependent nature of regulatory models and risk analysis methodologies. Contemporary standards must combine data-driven intelligence to assess and control risks more efficiently, moving beyond mere compliance towards a proactive safety management. By analyzing data, organizations can identify hazards and weaknesses better and adopt a more reactive approach in response to emerging risks. This is accomplished both in fire safety and aviation, where the utilization of data impacts design for safety and operational practice. Safety standard development thus not only reflects a chronology of development in the past but also the ongoing commitment to refining safety performance through the application of risk analysis innovation and knowledge in decision-making (Caciolai et al., 2018; Brooker et al., 2007).

8.1.2. Key Milestones in Risk Analysis Development

Risk analysis has evolved based on a chain of milestones that have closely impacted its applications and methodologies. One of these milestones is the increasing interest in incorporating environmental and social governance (ESG) considerations into risk assessment methodologies. This is due to a significant transformation, driven by impact investors, who are requiring corporations to care not only about the bottom line but about the social and environmental impacts of business. In 2012, the Total Portfolio Activation study introduced the concept of a framework under which investors might pursue measurable impacts on various asset classes, thus making risk assessment in relation to ESG considerations more robust. The momentum generated by this study has helped in improved comprehension regarding how participation and advocacy can generate significant social effects, proving the merit of interlinked and dynamic risk analysis in investment scenarios (Johnson et al., 2014). Another critical point in constructing risk analysis is captured in healthcare models, particularly in the formation of accountable care organizations (ACOs). As can be seen from the Brookings-Dartmouth ACO Pilot Program, fusion of performance measurement and reporting into healthcare risk analysis has been revolutionary. These organizations apply risk analysis systems not only for fiscal management, but to optimize patient care outcomes by forging strategic alliances with payer organizations.

Figure 8.3. Key milestones in risk analysis development.

Source: https://cdn.shopify.com/s/files/1/0403/8760/8730/files/key-rate-indicator-development-process_1024x1024.png?v=1718984355.

This integration allows for a more methodical evaluation of health-related risks, which allows organizations to streamline resource allocation while guaranteeing patient safety is given priority. By placing greater focus on the application of performance measures, health systems have the ability to improve risk analysis functions in order to deliver a spectrum of health outcomes and thereby become accountability and effectiveness models of risk management (Citters et al., 2012). As the face of risk analysis continues to evolve, the application of robust data analytics is a prime development. Increased access to big data has equipped analysts with the ability to discover patterns and correlations that were not previously accessible. Such a situation is of unprecedented significance in fields such as finance and public health, where data-driven decision-making can lead to better safety outcomes. For instance, public equity investment has been one of the prime areas where data analytics is utilized to assess the social and environmental impacts actively. Despite the challenge in quantifying the impact of widely held equity, emerging initiatives like the Impact of Equity Engagement (IE2) are striving towards frameworks that can quantify engagement impact on investment, thereby expanding the risk analysis horizons and paving the way for deeper societal impacts (Johnson et al., 2014) (Figure 8.3).

8.1.3. Impact of Historical Events on Current Practices

The evolution of safety and risk analysis procedures is deeply rooted within the record of events that have shaped present systems of regulation. A case in point is the early 21st-century banking crisis, which brought to light egregious shortcomings in risk management within financial institutions. The New Basel Accord was then crafted with the aim of harmonizing regulatory capital with underlying risk as well as facilitating systematic risk management procedures. This is a turning point as a historical moment when the importance of solid methodologies in risk analytics, such as model validation and portfolio aggregation, is reaffirmed. By examining such regulatory measures, it is evident that previous crises have a direct impact on current practice in risk analysis in banking, ultimately seeking increased stability and resilience within financial systems, as evidenced in the objectives set by the Basel Committee on Banking Supervision (Saidenberg et al., 2025). In health care, nursing education experience in the past has a direct impact on current practices in patient safety and outcomes. Research indicating that nurses' educational levels are associated with outcomes in patients supports this effect. In important ways, the implementation of evidence-based practices has been spurred by past controversy over adequacy of nursing education. It is demonstrated through research that hospitals with larger proportions of baccalaureate-prepared registered nurses report fewer patient falls and mortalities. Historical determinants of such results inform current practice by emphasizing the necessity of higher education and regular professional training to guarantee well-educated medical staff, thereby informing patient safety initiatives in terms of well-trained medical staff. Such attitudes reinforce that improved enhancement in the quality of healthcare results is not only a product of technology but is also a result of learning from previous inadequate nursing practice and training (Coto et al., 2016). The response of authorities to past failures is a pointer to the wider trend across many fields, where past exerts influence future policies and practices. For instance, failures in risk management observed in banking as well as medical industries resulted in reforms calling for greater accountability and transparency. These reforms speak of an emerging consensus on the necessity for evidence-based approaches to safety and risk assessment. Through historical lessons, stronger risk analyses and use of data are now encouraged by regulatory systems so that past errors are not repeated. By doing this, current practices guarantee learning from the past through analysis methods developed to ensure that the risks, both operational and systemic, are sufficiently covered, and hence contributions are made toward enhanced safety outcomes by sectors. The evolution of these traditions illustrates the continuous process of learning, accommodation, and improvement driven by precedents.

8.2. TYPES OF DATA USED IN SAFETY AND RISK ANALYSIS

In risk and safety analysis, various types of data are essential for decision-making and avoiding risks. Quantitative data, which includes numerical measures, is usually utilized to quantify risks and yields systematically. Statistical analysis, frequency of occurrence, and probabilistic models are popular types of this category of data that allow analysts to provide predictive conclusions on potential hazards. Equally significant is qualitative data, as it provides insight through narrative and the opinions of professionals. These can be crucial when quantifying human factors or intricacies in the operations which quantitative data would leave out. The use of both quantitative and qualitative data guarantees safety issues are comprehensively observed, thereby depicting a more accurate model for risk analysis. Furthermore, the use of advanced technologies has transformed the safety and risk analysis environment to include the use of real-time data. Incident reporting and tracking systems produce enormous amounts of temporal data that can enhance the knowledge of the time-evolving behavior of risks. As noted in recent research, machine learning (ML) usage has evolved as a primary tool for understanding complex data sets and improving safety systems across various industries, including transportation. Specifically, techniques like decision trees allow analysts to predict and classify safety results efficiently. Through the utilization of real-time data combined with historical trends, risk decision-makers can create proactive measures tailored to evolving risks, thus optimizing safety measures and enhancing overall effectiveness. Moreover, specialized types of data such as environmental data and engineering evaluation data are critical to complete safety analyses. For example, for engineered nanomaterials, as discussed in recent literature, the ability to make predictive estimates of potential environmental impacts is dependent upon the availability and quality of existing hazard and exposure data. Such predictions are used by decision-makers in risk assessment within specific settings and frameworks, thereby closing the gaps in data quality that otherwise can lead to ineffective risk management. Moreover, integrating data from various sources, like incident reports and environmental surveys, provides the multi-dimensional view of risks, allowing organizations to make sound decisions that are in harmony with safety standards and public health. Lastly, various types of data not only enhance the analysis but also contribute to the construction of sound frameworks for effective risk and safety management (Figure 8.4).

Figure 8.4. Types of data used in safety and risk analysis.

Source: https://assets-global.website-files.com/5efc3ccdb72aaa7480ec8179/606ffdb3cb9a00161a6c5f1a_types%20of%20digital%20risks.jpg.

8.2.1. Quantitative Data Sources

Quantitative data sources play an important role in safety and risk analysis, with the information being used as a foundation for informed decision-making and risk factor analysis. Numerical data on incidents, hazards, and safety performance metrics are typical data sources of this type. Analyzing such data consistently enables organizations to identify trends and potential risks which might not be apparent based on qualitative considerations alone. For instance, in the case of vaccine risk communication, empirical data revealing adverse events can guide the methods of effective communication of safety information to the population (Bostrom et al., 1997). Based on quantitative evidence not only enhances the validity of risk communication but also raises public trust, and hence, it is a critical element of safety programs. In addition, the systematic approach to safety assessment gives emphasis to quantitative data being utilized as a measure of system safety. Organizations like NASA promote systematic and disciplined evaluation that involves utilization of a combination of safety assessment tools including quantitative measures to measure hazards and associated controls. This methodological strategy allows for a holistic evaluation of the overall safety hazards, going beyond the focus on individual incidents to examine the overall effects of the potential accumulated risks within systems (Benjamin et al., 2025). With the use of quantitative information, organizations can derive reasonable results that assist in balancing risks, with the measure of safety being such that it not only guards against known hazards but also provides an intervening cover against unexpected risks. The value of quantitative data sources lies beyond mere conformity with safety precautions; they enable the optimization of risk management practices by evidence-based interventions. Quantitative data sources allow safety analysts to model probable impacts of various circumstances and thus offer forethought-based answers rather than merely reacting to situations. Organizations do this by enhancing their abilities to anticipate and counter threats arising from sophisticated systems. The intensity of analysis that is associated with these quantitative measurements confers advantages such as earlier identification of risk drivers and improved resource allocation for safety programs. The application of quantitative sources of information is, therefore, an important characteristic of building safety practices that seek to improve risk management in an attempt to achieve levels of optimum safety in different settings of operations (Benjamin et al., 2025) (Figure 8.5).

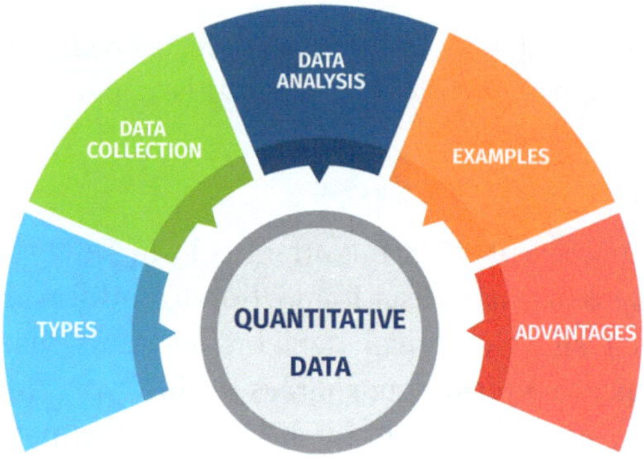

Figure 8.5. Quantitative data sources.

Source: https://www.questionpro.com/blog/wp-content/uploads/2018/05/Quantitative-data_2_FI.jpg.

8.2.2. Qualitative Data Sources

In the field of risk and safety analysis, qualitative sources of information are critical in the sense that they offer vibrant descriptions that are not in quantitative data. Qualitative data are likely to originate from formal interviews, focus groups, and field notes, which offer researchers the possibility of exploring the perceptions and experiences of the persons who are involved in healthcare practice. For instance, in examining high-risk prescribing in primary care, researchers utilized semi-structured interviews and field notes to better understand clinicians' interactions with prescribed medications and how practices can be maximized. Qualitative approaches such as these can illuminate the complexities of provision of healthcare, and of the determinants underpinning it that may impinge on patient safety. In addition, through the comprehension of healthcare professionals' perspectives, interventions can be tailored uniquely to overcome the specific barriers and challenges they face, ultimately resulting in enhanced quality of care and better patient outcomes. To add, qualitative sources of data are of substantial importance in establishing the effectiveness of safety interventions and assessing their effect on clinical practice. For the data-driven quality improvement in primary care (DQIP) intervention, qualitative clinician feedback provided rich context about implementation process and perceived value of high-risk prescribing interventions. The qualitative analysis revealed that clinicians valued being reminded to review potentially risky prescriptions, which impacted their clinical decision-making. These kinds of findings reveal that qualitative data can enhance safety intervention design and sustainability by identifying motivational factors and points of resistance among healthcare workers. Hence, integrating these qualitative evaluations into safety and risk analysis models means interventions that are not only evidence-based but also culturally and contextually sensitized in the healthcare settings where they are being put in place, thereby triggering long-term safety process improvement. Besides that, including qualitative data sources allows for an enhanced understanding of patients' attitudes and preferences regarding safety interventions. Patients' attitudes towards healthcare may be the determining factor in influencing compliance with their treatments and outcomes, in general. For instance, research has established that a well-educated patient will more likely be active with their care plan, thus minimizing potential risks associated with medication errors. By examining patients' experiences with health providers and opinions about the quality of care, qualitative information can provide insight into areas of patient-provider communication that need to be improved. This is congruent with the findings of research focused on the adequacy of available information on the quality of providers, and the type of information that patients would like to utilize when choosing health providers (Harris et al., 2008). Hence, qualitative data not just increases the understanding of safety and risk analysis but also enables the stakeholders to use precautions that ensure patient safety while creating improvement culture in healthcare settings (Figure 8.6).

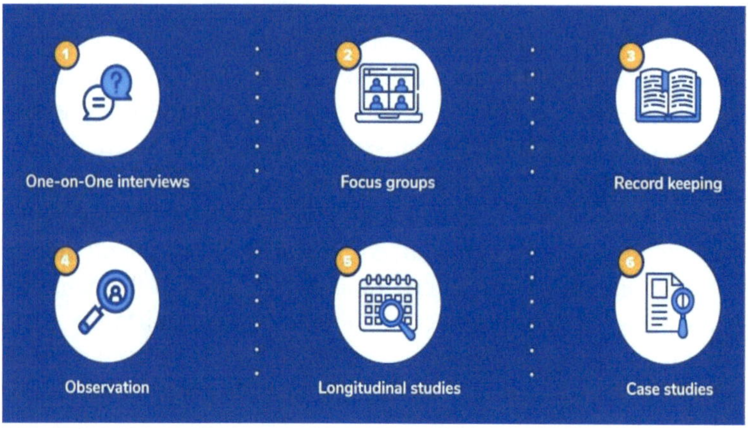

Figure 8.6. Qualitative data sources.

Source: https://kerjayuk.com/wp-content/uploads/2022/01/Qualitative-Data-Collection.jpg.

8.2.3. Emerging Data Technologies

Emergence of new data technologies is rapidly transforming the safety and risk analysis dynamics, particularly in those sectors where accuracy and real-time sensitivity are critical. Artificial intelligence, machine learning, and big data analytics offer unprecedented tools to enhance decision-making processes. Using predictive modeling and advanced simulation, players can anticipate contingent risks and implement forward-looking prevention strategies. This capability is essential in such industries as health care, where data-driven insights allow for the identification of vulnerabilities, so interventions are made early enough. In addition, convergence among new technologies provides a platform that facilitates concerted effort among various actors, including academia, regulatory bodies, and companies, thereby enhancing the system under which safety and risk assessment take place. Though full of potential, data technologies also pose issues for consideration, most notably governance and ethical concerns. The hybrid governance model currently espoused by regulators, particularly where nanotechnology comes into play, is used to describe the challenges of addressing such innovations effectively. As suggested, the model consists of adaptive methods complemented by traditional governance premised on different regulatory practices that tend to be inconsistent when addressing new threats. One of the central challenges is that this case-by-case system of regulation is vulnerable to abrupt shifts in policy and hence becomes less effective at pre-empting harm before it actually happens. All stakeholders need to work through such complexities while simultaneously making sure that there is space for innovation but not a compromise in public safety or morals, as reflected in recent calls for finance and co-regulatory interventions to address these issues (Davies, 2008). Besides, enabling a uniform mode of safety and risk assessment through data technologies requires integrating anticipatory criteria at an earlier point in the development process. Current approaches depend more on large amounts of data collection, which can delay response to prompt intervention. Through integrating upstream factors such as human rights and social justice, systems of regulation can be fortified to be able to handle mounting problems better. The European Union's past experience with regulation in the form of evolutions toward self-regulation and prescriptive regulation manifests the necessity of employing a balanced governance approach that is both adaptive and robust (Ruggiu et al., 2015). In conclusion, successful integration of new data technologies into risk and safety analysis depends on an intersocietal process that is concerned with moral priorities and risk management anticipatory in nature that defends the environment and public health (Figure 8.7).

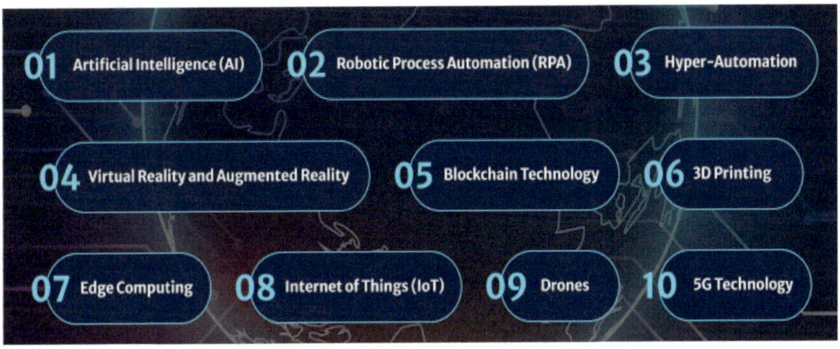

Figure 8.7. Emerging data technologies.

Source: https://www.founderjar.com/wp-content/uploads/2023/01/Top-10-New-Technology-Trends-for-this-Year.png.

8.3. METHODOLOGIES FOR ANALYZING SAFETY AND RISK

A systematic review of safety and risk analysis methodologies reveals that it is essential to be systems-focused, particularly in dealing with complex environments such as health care. The Safer Clinical Systems program highlights that enhancing patient safety involves looking to clinical processes and system-level interactions with the potential to cause failures. Reliability measurement of clinical workflows and defect detection enable practitioners to identify the sources of safety violations. This review demonstrates that root causes of incidents more often lie not in individual faults but in the broader organizational context. Since different components contribute to safety outcomes, taking a systems approach helps build evidence-based interventions for mitigating the risk of harm through continuous improvement of healthcare processes to eventually develop a safer healthcare system (Cooke et al., 2010). In high-hazard sectors, such as nuclear power, the methods applied to safety analysis are further developed to include investigation of incidents post-event for optimal risk control measures. It is vital in formulating safety precautions that lessons from experience are incorporated. The approaches followed by nuclear sites utilizing a 'toolbox' of event investigation techniques are a classic case of the need for qualitative consideration of the technical and process factors of safety. By comparing many event investigation approaches derived from various high-risk industries, lessons can be derived regarding the implementation of best practices. Comparative analysis not only identifies the strengths of various approaches but also puts their weaknesses into perspective, offering the grounds for informed recommendations for specific operational environments (Marc et al., 2011). Finally, the establishment of systematic methods for safety and risk assessment relies on continued advances in data-rich practices in healthcare as well as high-hazard industries. The emphasis on system awareness, drawn out of initiatives like the Safer Clinical Systems program, portends the entry into full-fledged safety appraisal that incorporates multiple sources of data and organizational learning. With methodologies continuously changing in organizations, the integration of lessons learned from past experiences is an indicator of a proper risk management strategy. Focusing methodologies capable of resilience and responsiveness, stakeholders are enabled to manage the complexity in safety-critical environments, hence fortifying the overall safety culture and reducing the occurrence of undesirable events in different industries (Cooke et al., 2010; Marc et al., 2011).

8.3.1. Statistical Analysis Techniques

Statistical analysis techniques in safety and risk assessment applications serve as pillars for data interpretation and decision-making. Statistical analysis techniques allow organizations to determine risks and enhance safety levels by transforming vast amounts of data into meaningful information. Descriptive statistics, for instance, provide a means of summarizing the most important characteristics of data sets and identifying trends and patterns that might reflect potential safety hazards. Inferential statistics facilitate analysts to draw conclusions about the implications of the sample data for the broader population, and thereby impact preventive action and operational policy. With organizations having to deal with the complexity of big data, the task is not only in gathering vast amounts of information but also in applying the same to drive risk management decision-making (Koronis et al., 2018). In addition, advanced statistical techniques such as predictive modeling and data mining have gained prominence in risk analysis so that companies can predict probable risks more efficiently. Predictive modeling uses historical data to predict future events and hence enables companies to proactively respond to risks before they develop into full-fledged issues.

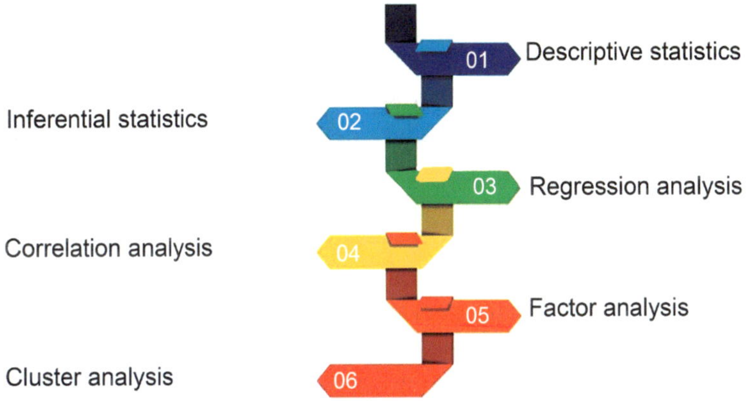

Figure 8.8. Statistical analysis techniques.

Source: https://fastercapital.co/i/Data-analysis--How-to-Use-Data-to-Make-Informed-Decisions--Statistical-Analysis-Techniques.webp.

This is quite relevant in sectors such as utilities, where continuous real-time monitoring of asset health is crucial. The convergence of big data analysis and machine learning represents a paradigm shift, empowering analysts with advanced tools for the detection of subtle trends in large sets of data that may be less visible to more traditional methods. The application of these approaches is critical to developing an active safety culture that focuses on risk avoidance while promoting operational effectiveness (Márquez et al., 2016). Further, statistical analysis techniques not only help to identify risks but also form a critical part of developing detailed safety management systems. By incorporating statistical techniques into tools like Prognostics and Health Management (PHM), organizational decision-making can be improved significantly. The new science provides a more effective interpretation of system reliability and useful life left, thereby affecting maintenance policies under Condition-Based Maintenance (CBM) policies. The integration of predictive analytics into Reliability Centered Maintenance (RCM) programs benefits by enhancing the accuracy and effectiveness of maintenance work. While continued upgrades are made to safety and risk analysis processes, embracing good statistical practices will be a major step towards maximizing data's capability, ultimately resulting in safer destinations for all concerned (Koronis et al., 2018) (Figure 8.8).

8.3.2. Predictive Modeling Approaches

Predictive modeling techniques have emerged as essential methods in risk and safety analysis, particularly in dynamic environments where traditional methods may fall short. Predictive techniques rely on past facts to forecast the future so that decision-makers can anticipate risks prior to their occurrence. Applying machine learning techniques and statistical processes, organizations are able to create models that identify patterns and anomalies within data sets, which in turn enhances their ability to predict safety problems. Merging of sources of data increases the strength of these predictive models, providing an integrated view that can guide organizational decision-making. Therefore, predictive modeling not just aids in the identification of future risks but also lays a ground for intervention techniques that minimize recognized risks in a timely manner. One of the most hopeful advancements in predictive modeling is the addition of uncertainty analysis, which helps to reconcile the inherent frailties of standard models. By taking into account the uncertainty and variability of different input data, predictive models may render their outputs more precise to simulate actual circumstances. The Uncertainty-aware Attention mechanism, for instance, enhances deep learning models using input-dependent uncertainty to better make risk predictable based on electronic health records. This approach recognizes that a few cases may have a higher degree of uncertainty, allowing for more efficient distribution of attention during the training of models. By doing this, deploying UA not only improves the accuracy of prediction but also provides more in-depth insights which correlate with the experts' interpretations in the field, thus improving confidence in the output of the models (Huh et al., 2019). Further, with sectors such as aviation continuously integrating their processes using new technologies, predictive modeling in safety assurance is increasingly gaining significance. The shift towards proactive risk management ability involves recognizing and resolving issues for safety prior to the onset of these becoming serious. Using model-based and data-driven methods, organizations can monitor precursors and trends that signal emerging safety concerns. For example, anticipatory testing can provide insights on how automation will lead to non-deterministic behaviors that were not anticipated within the initial design stage. Such forecasting is what drives the process of developing timely corrective measures, either in the form of revised designs or automated operational corrective measures, towards ultimately securing the environment in advanced operational spaces. Priority given to predictive modeling not only aids in ensuring safety but also lays the groundwork for prospective research and development of safety practices (Ancel et al., 2025).

8.3.3. Qualitative Risk Assessment Methods

Qualitative risk assessment methods are of crucial value within the domain of safety and risk analysis for identification, estimation, and regulation of potential hazards. The methods in question are subjective, opinion based, and stakeholder data to identify risks that are non-numeric in nature. Professionals are capable of uncovering insights into the complex interactions of safety-critical systems by applying instruments such as structured interviews, focus groups, and expert panels. This approach is particularly valuable within industries where empirical data may be hard to find or access. The data obtained from qualitative assessments gives organizations the required information for effective decision-making regarding risk management policies and thus enhances their ability to manage uncertainties more effectively. Here, qualitative approaches are a core supporting instrument to quantitative assessments, since they create a thorough comprehension of risk in particular contextual terms for safety-critical activities. One of the most important qualitative risk assessment techniques is scenario analysis, which allows teams to paint a picture of possible future states that may influence safety performance. The process allows for group dialogue among stakeholders on the possible risks and consequences, with greater understanding about how the various elements interact with each other in the real world. By building various scenarios, businesses can prepare and prepare for a set of possibilities and enhance their ability to resist unexpected events. Importantly, scenario analysis is more likely to reveal observations that statistical analysis cannot detect, such as organizational culture and human factors that can influence safety practice. With increasingly more industries adopting agile principles, qualitative risk assessment methods like scenario analysis will become the norm in order to tackle the new challenges that arise in high-intensity settings, particularly in industries that handle safety-critical components, according to recent studies regarding agile development practices (Calikli et al., 2018). Qualitative risk assessment methods also intersect with educational courses, particularly internet safety for children. Research indicates there exists a great divide between what children know about the internet online and doing things with that knowledge (Scott et al., 2015). That divide can be filled through qualitative assessments that gather opinions of children's online experiences and perceptions of online risks to inform tailored safety training. By raising questions among pupils about their experiences on the internet, educators can construct improved pedagogies of learning for proper use of the internet and secure online culture. These qualitative approaches not only make learning material more relevant and individualized but also enable kids to convey themselves better concerning matters and experiences, thereby leading to stronger and more interactive safety education frameworks. By doing so, qualitative methods of risk assessment form part of making safety controls work according to the realities of life for the people it seeks to secure, hence enhancing the quality of risk management outcomes (Figure 8.9).

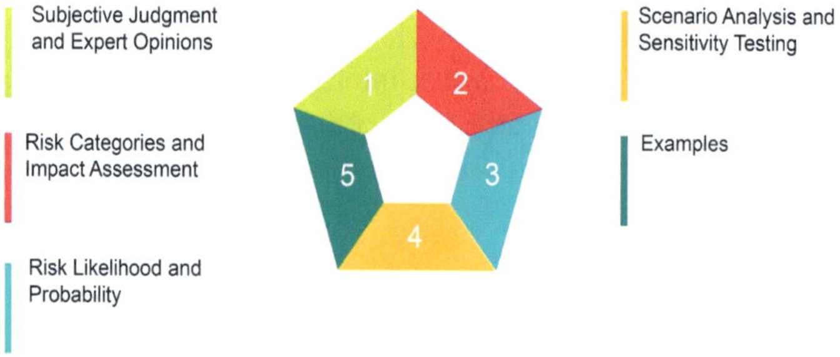

Figure 8.9. Qualitative risk assessment methods.

Source: https://fastercapital.co/i/Financial-Risk--Financial-Risk-Management-Frameworks-and-Tools--Qualitative-Risk-Assessment.webp.

8.4. APPLICATIONS OF SAFETY AND RISK ANALYSIS

The application of risk and safety assessment transcends multiple disciplines, with excellent utility in agriculture, experimentation with genetically modified organisms (GMOs). As experimentation with these technologies continues, there is a necessity to ensure thorough risk assessment to balance scientific innovation and public acceptance. Since their inception, GMOs have been largely focused on boosting agricultural production; however, consumer skepticism, particularly in Europe, has deterred their acceptance. Studies indicate that despite the perceived benefits of enhanced nutrition and environmental sustainability, the consumer remains unconvinced about the hazards of GMOs. They want transparency and in general are not satisfied with labeling only, which clearly illustrates the need for intensive data review and effective communication mechanisms to bridge the science-consumer acceptance gap (Marin et al., 2025). With this, there is a need to understand consumer attitudes by means of safety and risk analysis so as to create a favorable environment for technology uptake within agriculture. Elsewhere, the virtual reality (VR) technologies' environment also has much to benefit from adopting systematic safety and risk analysis. With technology advancements in immersive VR learning environments such as vSocial, so increase the potential vulnerabilities within their adoption. These VRLEs come up with new solutions for training and learning people with learning barriers but come with certain risks relating to security, privacy, and safety (SPS). A systematic approach through the use of attack trees is required to quantify and counter these threats. By quantifying threats more systematically, instructors can not only enhance the user experience but also ensure safe distribution of learning content. Risk score analysis, which is based on the duration and intensity of threats, highlights the importance of risk control measures taken in advance so institutions are able to deal with vulnerabilities in a better way while building educational use benefits of VR technology (Calyam et al., 2018). Agricultural and technological applications also highlight the manner in which data-driven safety and risk evaluation can enhance decision-making. By way of statistical modelling and simulation approaches, stakeholders can predict probable risks and implement particular solutions directed at particular public fears. In agriculture, for example, this could involve re-examining GMO research according to consumer values or improving labelling processes that provide product clarity and assurance. Similarly, in digital learning environments, the use of data to track behavior and risk exposures can lead to enhanced security measures that eventually safeguard users while optimizing their learning processes. With the increasing realization of the significance of security and risk analysis in diverse industries, its application, powered by effective data, will lead next-generation innovation that not only addresses technical needs but also resonates with values and social norms.

8.4.1. Industrial Safety Management

Industrial safety management is an imperative risk analysis area, particularly in hazardous operations such as chemical process industries. The history of accidents in chemical process industries has made it a necessity to have systematic processes for identifying, analyzing, and controlling risks. Data analysis is a cornerstone in all these by enabling safety managers to understand patterns and interrelations among variables associated with workplace hazards. By using modern data techniques, organizations can identify important risk factors, predict potential accidents, and formulate safety improvement plans. This pre-emptive approach not only guards' workers but also saves economic expenses incurred due to accidents. As identified in the work of various authors, the increase in industrial accidents indicates the need for the implementation of total safety management systems that are in a position to address the particular challenges that exist in such changing environments (Gurjar et al., 2002). Moreover, technology application in industrial safety management is revolutionizing the area of risk assessment. Employment of real-time data monitoring devices, artificial intelligence, and machine learning algorithms allows for continuous evaluation of safety protocols and practices. Sophisticated analysis makes aware the dynamic hazards and thus allows prompt modification of safety protocols as circumstances change. Data-driven insights allow organizations to transition away from reactive to proactive approaches and significantly improve workplace safety. For instance, predictive analytics may identify potential points of failure before turning into catastrophic accidents. This approach is indispensable in the provision of a secure working environment and compliance with regulations. Improved choices regarding the security of employees and the sustainability of the organization can be achieved through the enhanced understanding of the operation risks (Procházka et al., 2018). Finally, effective training and education are a part of industrial safety management along with data analysis in promoting a safety culture. The employees must be able to recognize hazards and respond with the proper reactions to emergency conditions. Professional development through data-driven intelligence on a continual basis allows the employees to know the best practices as well as regulatory obligations. Safety management needs to integrate findings from data analysis into training programs that are focused on real-life scenarios, thus integrating safety protocols into actuality. This mode of training also fosters employee engagement, providing employees with a sense of agency over there and others' safety. Lastly, the merging of data analysis with education programs is a holistic method for creating a more secure industrial environment with lower risk and higher preparedness for unforeseen crises (Gurjar et al., 2002) (Figure 8.10).

Figure 8.10. Industrial safety management.

Source: https://d1aeya7jd2fyco.cloudfront.net/blog/industrial-safety.webp.

8.4.2. Public Health and Safety

The intersection of public health and safety becomes increasingly data-analytically driven through the implementation of predictive risk intelligence. This method applies the use of Smart Information Systems (SIS) in order to forecast future health hazards and safety risks within communities. Based on large databases collected from such sources as medical records and social media, researchers and public health officials are able to identify emerging trends and patterns. For instance, predictive analytics can warn of the outbreak of infectious diseases to allow for anticipatory intervention to reduce transmission rates. Such practices, however, regarding data are unethical, particularly concerning data privacy and consent. According to recent studies, data transparency and accountability in predictive practices are essential in keeping the populace well-informed and guaranteeing ethical compliance in public health interventions (Jiya et al., 2019). Furthermore, evidence-based strategies also need to tackle the environmental determinants of public health and well-being. The staggering statistics involved with food wastage are an acute example of such a relationship; America throws away $218 billion annually on food, which has substantial economic as well as environmental impacts (N/A, 2016). By leveraging data analytics, the agencies can enhance their ability to determine the factors leading to food waste and its implications on public health, thereby proving the need for targeted interventions. Effective communication of data findings has the ability to bring about sustainable actions, educating consumers and businesses to reduce waste and ensure healthier food systems. This approach not only enriches public health but also assists in making the environment clean by mitigating the harmful effects of waste on natural resources. Finally, the use of data analysis in public health and safety calls for an interdisciplinary strategy in merging technology, ethics, and community involvement. With collaborative relationships among stakeholders like policymakers, health professionals, and data scientists, it is possible to have a well-integrated approach to address complex public health matters. By fostering responsible innovation, as laid out by the Responsible Data Science framework, ethics are incorporated into designing health data systems. Together, these functions are likely to yield more system transparencies that eventually serve health outcomes and safety practices in communities while balancing the intricacies of data application (Jiya et al., 2019).

8.4.3. Environmental Risk Assessment

Environmental risk assessment (ERA) is a major part of the general appraisal of risks generated by various environmental factors. It is particularly vital in engineered nanomaterials (ENMs), in which conventional hazard and exposure information can be insufficient. Effective risk assessment tools are required for decision-makers who are supposed to evaluate such implied risks. There exists a critical need to create frameworks to allow for quick assessments, thus enabling decision-makers to make informed choices relying on the distinct contexts under which various uses of ENMs exist. Current frameworks are mostly inadequate to cater to the diverse needs of decision-makers, justifying the gaps that need to be filled in order to move towards sustainable nanotechnology (Bilal et al., 2017). Thus, development of ERA techniques is of utmost importance for safeguarding both environmental health and public well-being from impacts of emerging technologies. Incorporation of health and safety assessments into environmental risk appraisal is another vital domain of this research area. For instance, a study dealing with safety and health risks at swimming pools applied Hazard Identification (HAZID) and As Low As Reasonably Practicable (ALARP) techniques to classify and evaluate various risks.

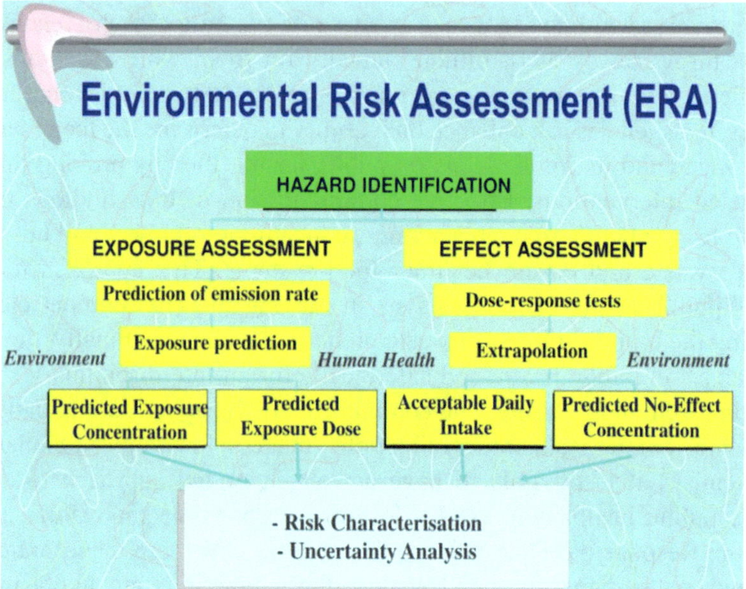

Figure 8.11. Environmental risk assessment (ERA).

Source: https://image2.slideserve.com/4067802/environmental-risk-assessment-era-l.jpg.

This approach revealed a significant amount of safety hazards and health risks associated with environmental factors and equipment safety in recreational environments. The result reflected the importance of systematic data gathering and analysis in that they led to the identification of 69 safety risk categories and classifying unacceptable risks which were subsequently managed by intervention. Such studies are a common example of the applications in the field of ERA protocols for identification of trends and devising corrective measures where public safety may be at stake (Etesam et al., 2020). Moreover, the long-term goal of environmental risk assessment is not merely to establish and classify risks but also to provide an environment within which effective mitigation and management of risks occur. By proper involvement of stakeholders like operators and pool managers, there can be the establishment of a safety and health culture. Training programs that attempt to enhance the level of awareness of health and safety protocols can contribute to significant advancements in the management of risks. These training programs empower individuals who are given the responsibility of maintaining safety standards while driving a shared sense of responsibility towards stewardship of the environment. Therefore, ERA is more than mere formation of risk identification, evolving into an active strategy that places emphasis on constant improvement in maintaining public health and security in varied settings, hence promoting overall welfare in society (Figure 8.11).

8.5. CHALLENGES IN SAFETY AND RISK ANALYSIS

The safety and risk analysis discipline is faced with numerous challenges that complicate effective decision-making and policy execution. One such major challenge is the inherent uncertainty in data collection and analysis processes. Data quality is vulnerable to a number of factors, such as sampling errors and reporting deficiencies, which are prone to conceal the real risks associated with particular activities or environments. Furthermore, using more than one data source brings into play the conflicting data, and it is difficult to make a comprehensive notion of safety risk. With growing efforts from companies to become more data-driven in safety and risk assessment, these data integrity issues are essential to address. Without reliable data, any risk assessment can lead to misplaced priorities and, in the end, jeopardize safety. Another critical challenge in safety and risk assessment is the need for continuing training and education of personnel in this specialty. Safety regulations and risk assessment methods continually evolve, and practitioners must keep abreast of developments. The complexities of using advanced data analysis tools reinforce the requirement for continuing professional development. Most of them will be bogged down by the pace of change and the increasing need for sophisticated data analysis instruments, which might suppress their reaction towards managing safety hazards. Thus, organizations must consider training programs at the highest level to transfer knowledge and skills in their personnel on how to utilize data to plan safety effectively, implement a risk-conscious culture, and exercise proactive decision-making. In addition, the integration of diverse stakeholder opinions presents yet another great challenge to safety and risk analysis. Government institutions, industry, and affected communities who are involved in safety policies may have diverse opinions on risk tolerance as well as safety priorities. The diversity may complicate the development of consensus-making processes and the integration of one safety strategy. Furthermore, in the absence of collaborative approach employed, some of the most important risk factors can be left out, thereby leading to inefficient safety enhancement. For instance, accountable care organization development, as presented within the framework of healthcare delivery models, illustrates the intricacy of balancing differing stakeholder requirements within safety initiatives (Citters et al., 2012). Multiple stakeholders' participation in public debate is required in order to address such challenges and to facilitate the development of effective risk analysis frameworks that, ultimately, contribute towards community security (Ayala-Rivera et al., 2017).

8.5.1. Data Quality and Reliability Issues

Quality and reliability of data are the top concerns in the field of safety and risk analysis that have a direct influence on results integrity and subsequent decision-making process. The challenge is not only obtaining high-quality data but ensuring the systems that generate such data operate with high reliability levels as well. When examining healthcare systems, for instance, evidence shows clinical processes more frequently lead to safety breakdowns than errors in individuals. This necessitates healthy systems that can effectively measure data quality and detect defects in processes. By implementing frameworks intended for the measurement of clinical system reliability, stakeholders can understand why safety issues arise, paving the way for interventions to be applied to mitigate the risk of care-related complications (Cooke et al., 2010). Finally, by offering quality improvement in data with reliable systems, the frequency of adverse events in healthcare organizations can be reduced. Furthermore, the existence of barriers to successful incident reporting also complicates the issues of reliability and data quality. Learning organization relies heavily on reporting incidents without fear of retribution, but many healthcare professionals remain reluctant to employ formal reporting mechanisms. It is crucial to understand the reasoning behind this reluctance in order to develop a safety culture within healthcare organizations. By exploring other methodologies, such as using Plan-Do-Study-Act (PDSA) cycles, organizations are able to establish processes of continuous improvement that not only improve data quality but also encourage staff engagement. Qualitative feedback obtained from such exercises often results in a positive shift in safety culture and supports the notion that identification of process failure can lead to enhanced reliability in data collection and analysis (Sujan et al., 2012). If individuals feel that their work is valued, overall safety data integrity is enhanced. Further, enhancing systems and procedures underlying data generation directly relates to minimizing errors and increasing the reliability of ensuing analysis. The complexities of healthcare systems complicate total understanding as a total strategy of understanding how multiple components relate to each other to impact safety outcomes. Efforts like the Safer Clinical Systems initiative describe the potential of systematic improvement, whereby an emphasis on measuring process reliability not only identifies defects but encourages a greater understanding of integrated clinical activity. By thorough examination of these related systems, health care organizations can break down silos that suppress the sharing of critical information and intelligence. Thus, with significant improvement in safety comes a concerted effort to enhance the quality of information through a rigorous framework to ultimately realize improved patient outcomes and a culture of continuous improvement (Cooke et al., 2010).

8.5.2. Ethical Considerations in Data Use

In the quest to utilize data for risk and safety analysis, ethical implications come to the fore. The possibility of using data to drive decision-making raises the need to cautiously analyze how it is being utilized, especially in relation to privacy and consent. In educational settings, institutions tend to compile vast amounts of datasets that may be applied for other uses, such as managing libraries and improving user experience. But the ethical considerations of employing this information cannot be overlooked. Collection, analysis, and utilization of the data must be brought out into the open, since stakeholders must be confident that their data is handled ethically. As pointed out in (Showers et al., 2015), the success of scholarly library initiatives is contingent upon the identification of such ethical considerations, so that libraries not only fulfill their institutional roles, but also uphold superior ethical standards in data stewardship. Second, the challenge of sharing and accessing data imposes additional ethical considerations, specifically regarding the integrity and protection of sensitive data. When institutions invest in full-service data analytics offerings, the potential for erroneous data breaches or abuses is higher. The architecting of data infrastructures needs to be informed by ethics so that privacy is encoded from the start. Librarians and data administrators also need training to navigate technical and cultural dynamics of utilizing data because its impacts extend beyond institutional policy to broader societal consequences. Through the understanding of the sophistication of such challenges, organizations can promote a culture of ethical data use that incorporates an aspect of accountability and regard for people's rights, as exemplified by Showers et al. (2015). Moreover, ethical thought has to become paramount in the area of possible data abuse, particularly in sensitive contexts such as parenting evaluation where actions have an impact on children's welfare. deals with the issues surrounding the assessment of risk in relation to inappropriate internet behavior, stating that there is a need for serious reflection about how data should be interpreted and applied in such cases. Misinterpretation could lead to severe damage, not only to individuals being judged but also to their families. Hence, it is an urgent requirement for professionals working with data to adopt ethical guidelines that emphasize the welfare of all concerned parties. By taking into consideration the effects of the use of data and codified stringent ethical standards, professionals are able to ensure their analyses contribute to the overall well-being of society, thus enhancing the quality of safety and risk analysis programs (Figure 8.12).

Figure 8.12. Ethical considerations in data.

Source: https://daxg39y63pxwu.cloudfront.net/images/blog/ethical-considerations-in-data-science/image_21268895971706097383016.png.

8.5.3. Limitations of Current Analytical Models

The limitations of current analytical models used for safety and risk analysis are profound, based on an underlying misunderstanding of their capacities by policymakers and analysts. Many decision-makers misinterpret these models as definite tools for predicting outcomes, reading them as answer machines that give non-negotiable facts rather than probabilistic estimates. This misperception has the potential to lead to far-reaching judgment errors, as choices are made by simplifying complex systems. With policymakers unaware of the inherent uncertainties in such models, they perpetuate a vicious circle that discourages model improvement and sensitivity within regulative frameworks. Its implications are multidirectional, but above all, they establish environments where stakeholders can utilize misconceptions to strategic advantage, ultimately undermining proper safety and risk analysis techniques (Fisher et al., 2010). Other than misconstruction, the methods employed in current analytical frameworks are not nimble enough to adequately analyze new or state-of-the-art systems. Traditional probabilistic risk assessment methods are too cumbersome to be applied in dynamic environments such as conceptual system design. Though the analyses attempt to calculate risks based on historical failure rates, they do not manage to factor out important human factors responsible for safety failures. Present models may emphasize component reliability without much consideration for the complexities of human interaction that create operational failures. This divergence points toward the necessity of new approaches that combine statistical modeling with qualitative assessments of human performance, as these are the foundation of a rich appreciation of risk within complex systems (Nieberding et al., 2025). Secondly, the convergence and application of advanced analytical models face challenges in the form of standardization and acceptance within the safety and risk analysis community. Since various industries approach designing specialized practices tailored to their unique needs, it is difficult to lay down a standardized framework. Variations in collection, analysis, and application of data may result in disproportionate measurements across the field and make it hard to make conclusions regarding risk as a whole. Moreover, as more advanced methodologies are generated to bridge gaps created by traditional models, there is limited consensus that may deter shared use, thus diminishing their impacts on safety outcomes. This is defeated by having researchers, practitioners, and policymakers collaborate to rationalize methodologies and have a more coordinated approach for risk analysis so that enhanced safety practices are standardized across fields (Fisher et al., 2010; Nieberding et al., 2025).

8.6. FUTURE TRENDS IN SAFETY AND RISK ANALYSIS

Looking to the future of what is in store for safety and risk analysis, one of the most outstanding trends is integration of advanced data analytics. In an era of explosive expansion in the quantity and type of data available, organizations across the globe are making greater use of big data technologies to develop more robust risk assessment capabilities. Through the analysis of huge datasets, companies can identify likely dangers, quantify risk in real-time, and actively take precautions against accidents before they happen. In the illustration of the financial trading sector, near-miss event analysis has highlighted large operator performance and system responsible for error detection and avoidance. These results prove that a data-driven approach not only enhances operational safety but also creates generalized organizational resilience. The ability to analyze data patterns and trends allows organizations to move from a reactive to a preventive approach, making safety and risk analysis relevant in a dynamic environment (Griffiths et al., 2018). The second prominent trend is machine learning and artificial intelligence (AI) application in safety and risk analysis.

These technologies are revolutionary because they allow predictive analysis, which can forecast future risk based on past data. As companies adopt AI algorithms, they are in a position to identify patterns that are not easily noticeable to human analysts, thereby more accurate predictions of risk. The implications extend beyond mere discovery; companies are in a position to simulate various scenarios, evaluate the likely impact of various risks, and prioritize resources accordingly. This technological shift is particularly relevant in sectors such as environmental management, where climate change is posing more catastrophic risks. As recent summaries have pointed out, the embedding of climate projections in risk analysis models necessitates the use of new statistical methods, which is a shortfall that has to be supplemented by organizations to ensure sufficient safety measures in an increasingly uncertain world (Kjeldsen et al., 2013). Finally, the future of risk and safety analysis will likely involve increased emphasis on building a safety culture that entails stakeholder engagement and collaboration. Since recognizing systemic failures often stem from poor communication and coordination, organizations are more placing teamwork at the forefront of risk management. By creating forums where employees, management, and external partners can discuss issues, organizations can gain a better understanding of safety issues. The change in strategy is supported by studies that indicate teamwork and situation awareness reduce errors by greatly in high-stakes environments such as financial trading. With this in mind, the development of incident-monitoring systems that translate near-miss information not only helps to enhance the skills of the individual but also improves collective safety measures. Focusing on cooperative measures, organizations are best placed to be in a position to address future risks in risk analysis and safety, ensuring sustainable operations (Griffiths et al., 2018).

8.6.1. Integration of Artificial Intelligence

The integration of artificial intelligence (AI) in risk and safety analysis is a paradigm shift in organizational determination of potential risk and management of operational risk. Traditional risk policy is generally developed using history-based information and subject matter expertise, which is often limited by human bias and knowledge constraints. AI, on the other hand, can process vast amounts of information, identifying patterns and relationships that can elude even veteran analysts. Utilizing machine learning algorithms, organizations can enhance their predictive ability, allowing for proactive mitigation to mitigate risks before they become catastrophic. Furthermore, the adaptability of AI systems allows for continuous learning, thus refining the accuracy of risk identification with the eradication of time as new information is added. This reflects a growing awareness within various sectors, especially in risk-exposed industries, that the integration of AI can significantly enhance safety measures and risk management systems' resilience (Burov et al., 2020). Apart from technical uses, the application of AI technologies for safety analysis extends beyond technical means and considers the implicit understanding of human behavior in risky scenarios. With organizations becoming more active in the realms of complex, data-centric worlds, understanding the interaction between technological elements and human aspects becomes imperative. For example, AI can be used to enable cognitive vaccination against social engineering attacks, enabling employees to detect and refuse manipulative efforts by harmful actors. This method emphasizes the need to educate staff members regarding weaknesses within the virtual world and improve decision-making mechanisms.

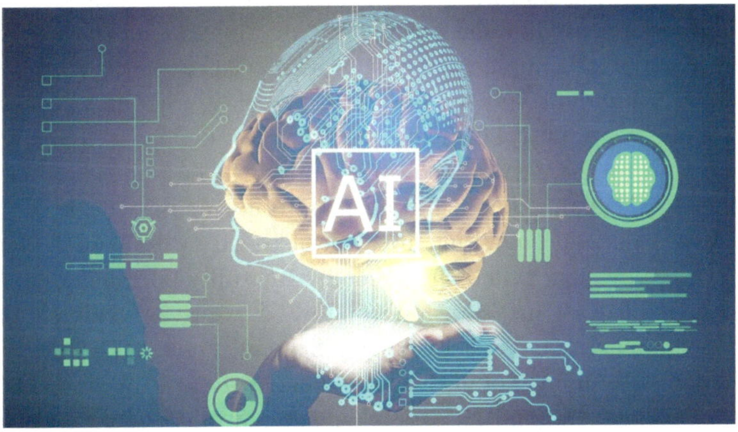

Figure 8.13. Integration of artificial intelligence.

Source: https://mechdb.com/wp-content/uploads/2023/12/Integration-of-Artificial-Intelligence-for-Autonomous-Data-Management-in-Mechanical-Systems.webp.

As pointed out by recent studies, young people, being the most susceptible component in cybersecurity safety and risk systems, are most susceptible to cognitive dangers. Addressing these shortcomings through AI-enabled training programs can render an organization's safety and risk culture more resilient (Burov et al., 2020). The broader implications of integrating AI into safety and risk analysis go into geopolitical forces as well. As various nations increase their AI development, the technology is presented as a deciding factor in warfare and non-warfare. The competitive landscape is shifting, with nations race to establish dominance in AI capabilities that enhance national power and economic power. Experts say that advanced AI technologies are like catalytic technologies like electricity, where they power not only war development but also influence world affairs and the balance of power. The rapid diffusion of AI in the civilian domain renders these advancements complex, creating opportunities and challenges for military use. Last but not least, integration of AI into safety systems must consider these global dynamics because technology growth consequences will have far-reaching implications on strategic decisions and risk management worldwide in the future (Figure 8.13).

8.6.2. Real-Time Data Monitoring Systems

Real-time monitoring data systems have central roles to play in securing and mitigating risks across various industries like aerospace and the transportation of hazardous materials. They facilitate immediate data capture and processing, enabling the stakeholders to respond rapidly by making logical decisions. In high-risk environments like the ones under investigation within the National Transonic Facility, the implementation of advanced monitoring technologies becomes of paramount importance. The aeroelastic facility test focused on the requirement for robust data systems, leveraging in-flight dynamic instrumentation to harvest valuable information. In line with the test safety analysis and risk evaluation, real-time monitoring of data does not merely safeguard experimental model integrity but also that of the human personnel. Implementation of such technologies thereby rightfully falls under preemptive risk management systems, illustrating how data-driven technologies can enhance operational standards of safety. Secondly, the application of data monitoring in real-time is not only within the aerospace sector but also within transport of dangerous goods. The implementation of on-board units (OBUs) with sensor networks, particularly Controller Area Network (CAN) bus-based technology, is a complete solution towards monitoring. The systems make sure that they transmit information regarding the condition of hazardous materials that are being transported on an ongoing basis to enable instant responses in the event of any abnormalities or extreme situations. A GSM/GPRS networking subsystem can efficiently send this critical information to monitoring centers in a centralized manner, giving relevant stakeholders timely information. The proposed secure communication interfaces that meet IEC 61508 standards reflect the growing concern of the importance of transmission of safety-related data. These technologies support sophisticated risk analysis and modeling techniques that are essential in ensuring compliance with regulations and general safety optimization in potentially risky environments. Incorporation of such real-time monitoring systems into the safety plans is not just a technological advance but also a change in risk management culture. For example, the inclusion of automatic and manual interlock systems enhances the safety of operation by creating fail-safes that prevent catastrophic outcomes. With the ability to evaluate and address threats on an ongoing basis, organizations are increasingly able to foresee and counterbalance threats before they escalate. By utilizing data analytics in conjunction with live monitoring, all stakeholders are able to better define the nitty-gritty of would-be weaknesses, thus allowing for effective assessment of risks associated with specific operations. As operation matrices evolve, so too will the adoption of dynamic data monitoring systems become a central aspect of effective safety and risk management strategies, immensely contributing to the development of safer working environments and safeguarding of assets as well as human lives (Figure 8.14).

Figure 8.14. Real-time data monitoring systems.

Source: https://contact-centres.com/wp-content/uploads/2017/04/ultra.comms_.real_.time_.image_.april_.2017.jpg.

8.6.3. Collaborative Risk Management Frameworks

Increased organization complexity and interconnectedness of risk have spurred the evolution of collaborative risk management models. These models encourage mutual strategies, where stakeholders consult and establish collective insight to recognize, analyze, and manage risks that undermine safety and operational integrity. Through the fusion of diverse views and abilities, collaborative models are more effective in recognizing vulnerabilities that would otherwise be overlooked under traditional risk management. Therefore, they are able to establish a learning environment for collective learning and continuous improvement, which allows organizations to adapt actively to risks. The combination of public participation and participatory evaluation methods produces a more holistic conception of risk, ultimately reducing organizational resilience and long-term safety (Bachelor et al., 2006). Utilizing data in decision-making is a crucial aspect of building the effectiveness of collaborative risk management systems. A reliance on data analytics enables organizations to examine past trends, evaluate present conditions, and forecast upcoming risks, thereby influencing their collaborative activity. These systems can be able to tap real-time information to update stakeholders on a continuous basis and make corresponding adjustments in strategies, thereby enhancing responsiveness to emerging risks. Through systematic processes in data gathering and analysis, organizations are well positioned to collaborate efficiently, promoting transparency and accountability. Such a reliance on measurable data not only solidifies the cooperative process but also provokes stakeholders to engage intensively in risk discussions, which in turn facilitates safer and more sustainable operating strategies (Calyam et al., 2018). Future collaborative risk management models will have greater integration with technological developments. The integration of state-of-the-art platforms, such as Social Virtual Reality based Learning Environments (VRLEs), is a good illustration of this phenomenon. By applying immersive experiences, stakeholders can engage with simulations that highlight probable threats to allow for interactive assessment and immediate feedback. New methods such as these can assist in instilling deeper understanding of threats, especially where threats reside in scenarios involving complex, multi-dimensional exposures. As models continue to advance with the addition of emerging technologies, the probability that stakeholders can better understand one another will grow, resulting in eventually improved risk management methods. They emphasize the imperative of cooperation in being able to effectively address safety and risk analysis through data, opening the door to an organizational culture more at ease with risk (Calyam et al., 2018).

SUMMARY

In evaluating the profound implications of safety and risk analysis using data, it is clear that such techniques become invaluable for the creation of well-informed decision-making across various industries. Medical record retrospective analysis, for instance, has become a valuable tool for identifying organizational weaknesses and strengths. As demonstrated in evidence, controlled data analysis can provide considerable improvement in clinical documentation quality and, consequently, enhancing patient care outcomes (Bontempi et al., 2017). Beyond the significance of compliance itself, these findings reflect a forward-thinking risk management initiative, synchronizing a culture of ongoing improvement in health care. The ability to adequately analyze information indicates a commitment to safety that any company seeking to solve contemporary challenges must make. Secondly, the interplay between advantages and disadvantages remains the core of safety and risk analysis. The study of the prophylactic use of aspirin for primary prevention of cardiovascular illness and cancer also reinforces the need for ensuring proper risk assessments. Although aspirin has been commended for lowering mortality and major cardiovascular events, it is equally essential to examine risks associated with it, such as increased bleeding rates (Clarke et al., 2013). This conservative approach serves to underscore the merit of careful data analysis so that stakeholders can weigh likely benefits against unfavorable outcomes in a systematic fashion. Ensuring that such analyses are grounded in methodologically robust research further enhances their utility and relevance to informing clinical practice and public health policy. Overall, the integration of data-driven safety and risk analysis gives organizations valuable tools for simplifying complicated decision-making processes. Through the creation of techniques that assess the benefits and possible threats of various interventions—such as those witnessed in healthcare and pharmaceutical environments—organizations will be better able to advocate good risk management practices. In the future, continuous examination of data will be paramount in steering future projects aimed at improving standards of safety. Ongoing study endeavors, particularly in lines remaining open to specification, such as the long-term effects of prophylactic treatments, record the constant nature of risk determination and affirm the need for continued attention to empirical data in initiating policy updating and enhancing protection measures (Bontempi et al., 2017; Clarke et al., 2013).

For risk and safety analysis, data-driven methods form the core of the identification of patient safety incidents (PSIs) that remain undetected as well as overall healthcare quality enhancement. One such example is the Trigger Review Method (TRM) that has been adopted by Scotland as a systematic way of screening clinical notes. TRM not only facilitates the recognition of PSIs but also facilitates healthcare professionals, and most significantly general practitioners (GPs), to be engaged in continuous learning and improvement. The method supports the necessity to develop a structured training intervention to ensure that the clinical workforce is adequately prepared to identify and act on safety issues in an effective manner. This approach emphasizes the need for adaptive education systems that can grow in order to meet the challenges of an increasingly complex healthcare system and establish a culture of safety in practice (Academy of Colleges et al., 2013). Child safety is also another critical area of risk analysis, particularly the use of roads. The systematic review of Cattan et al. (2008) cites some promising factors responsible for pedestrian injury rates among children and young adults and emphasizes their vulnerability as road users. Road safety behavior among children is highly influenced by their parents' practices, with parents playing a pivotal role in developing the consciousness and skills among their children concerning how to navigate traffic. The study has established that direct contact, in the form of hand-in-hand road crossing or providing directions, can significantly contribute to a child's road safety knowledge. This emphasizes the importance of educating parents and guardians on how to engage children in safe use of roads and preventing the incidence of accidents (Ayrton et al., 2008). In conclusion, the assessment of safety and risk based on evidence is critical in planning effective interventions in most areas of applications like healthcare and children's road safety. The TRM evidence displays how systematic methodology can shape the enhancement of patient safety issues, while findings based on parental influence on children's road behavior reveal how safety education is multidimensional. Both disciplines indicate that organized training and education can significantly enhance safety outcomes, thus necessitating the holistic intervention of various stakeholders. With data-led approaches increasingly developing, their capacity to offer insight into best practice and initiatives for the minimization of risks will continue to be critical to enhancing environments for the most at-risk populations (Academy of Colleges et al., 2013; Ayrton et al., 2008).

REFERENCES

1. Academy of Royal Medical Colleges, Royal College of Surgeons of Edinburgh, NHS Institute for Innovation and Improvement, De Wet, C., Zwart, D. L. M., Griffin, F. A., et al. (2013). Applying the trigger review method after a brief educational intervention: Potential for teaching and improving safety in GP specialty training? https://core.ac.uk/download/17049883.pdf (accessed on 21st October 2025).
2. Alawad, H., An, M., & Kaewunruen, S. (2019). Learning from accidents: Machine learning for safety at railway stations. https://core.ac.uk/download/286269815.pdf (accessed on 21st October 2025).
3. Ancel, E., Corbetta, M., Darafsheh, K., Dill, E., et al. (2025). Architecture and information requirements to assess and predict flight safety risks during highly autonomous urban flight operations. https://core.ac.uk/download/pdf/288485506.pdf (accessed on 21st October 2025).
4. Ayala-Rivera, M., Barry, F., Bryg, R., Cheng, E., et al. (2017). Randomized controlled trial of a coordinated care intervention to improve risk factor control after stroke or transient ischemic attack in the safety net: Secondary stroke prevention by Uniting Community and Chronic care model teams Early to End Disparities (SUCCEED). https://core.ac.uk/download/323078454.pdf (accessed on 21st October 2025).
5. Ayrton, R., Cattan, M., Cross, R., Green, J., et al. (2008). Child–parent interaction in relation to road safety education: Part 2–main report. https://core.ac.uk/download/4147367.pdf (accessed on 21st October 2025).
6. Bachelor, L., & Grieve, M. (2006). Community development evaluation story map and legend. https://core.ac.uk/download/71340775.pdf (accessed on 21st October 2025).
7. Bedford, T. J., Quigley, J. L., & Walls, L. A. (2009). Empirical Bayes estimates of development reliability for one shot devices. https://core.ac.uk/download/9024579.pdf (accessed on 21st October 2025).
8. Benjamin, A., Dezfuli, H., Everett, C., Smith, R., et al. (2025). NASA System Safety Handbook. https://core.ac.uk/download/pdf/10566584.pdf (accessed on 21st October 2025).
9. Bilal, M., Cohen, Y., Godwin, H. A., Romero-Franco, M., et al. (2017). Needs and challenges for assessing the environmental impacts of engineered nanomaterials (ENMs). https://core.ac.uk/download/323076183.pdf (accessed on 21st October 2025).
10. Boer, D. P., Connell, M., Merdian, H. L., Witt, P. H., et al. (2010). Assessing parental risk in parenting plan (child custody) evaluation cases

involving internet sexual behavior. https://core.ac.uk/download/1645782.pdf (accessed on 21st October 2025).

11. Bontempi, C., Cacchio, D., Catanzaro, S., Chiarini, A., et al. (2017). Quality assessment of medical record as a tool for clinical risk management: A three-year experience of a teaching hospital Policlinico Umberto I, Rome. https://core.ac.uk/download/188828083.pdf (accessed on 21st October 2025).
12. Bostrom, A. (1997). Vaccine risk communication: Lessons from risk perception, decision making and environmental risk communication research. https://core.ac.uk/download/72056389.pdf (accessed on 21st October 2025).
13. Brooker, P. (2007). Air traffic safety: Continued evolution or a new paradigm. https://core.ac.uk/download/138278.pdf (accessed on 21st October 2025).
14. Burov, O., Krylova-Grek, Y., Lavrov, E., Lytvynova, S., et al. (2020). *Cybersecurity in Educational Networks* [Кибербезопасность в образовательных сетях]. https://core.ac.uk/download/322373029.pdf (accessed on 21st October 2025).
15. Butler, T. W., Hergert, D. W., Herring, F. M., Young, J. R., et al. (2025). Buffet test in the National Transonic Facility. https://core.ac.uk/download/pdf/42811963.pdf (accessed on 21st October 2025).
16. Caciolai, M., Cancelliere, P., Gai, G., Gissi, E., et al. (2018). Italian hybrid fire prevention code. https://core.ac.uk/download/127587254.pdf (accessed on 21st October 2025).
17. Calikli, G., Kanagwa, B., Kasauli, R., Knauss, E., et al. (2018). Safety-critical systems and agile development: A mapping study. http://arxiv.org/abs/1807.07800 (accessed on 21st October 2025).
18. Calyam, P., Gulhane, A., Hoefer, G., Hoque, M. A., et al. (2018). Security, privacy and safety risk assessment for virtual reality learning environment applications. http://arxiv.org/abs/1811.12476 (accessed on 21st October 2025).
19. Campostrini, G. P., Da Porto, F., Marson, C., Modena, C., et al. (2016). The application of a Bayesian approach to assess the seismic vulnerability of historical centers. https://core.ac.uk/download/80155669.pdf (accessed on 21st October 2025).
20. Clarke, A., Connock, M., Freeman, K., Grove, A., et al. (2013). Aspirin for prophylactic use in the primary prevention of cardiovascular disease and cancer: A systematic review and overview of reviews. https://core.ac.uk/download/17180813.pdf (accessed on 21st October 2025).

21. Cooke, M. (2010). Safer Clinical Systems: Interim Report, August 2010. https://core.ac.uk/download/46327.pdf (accessed on 21st October 2025).
22. Coto, J. A., & Wilder, C. (2016). An exploratory study of patient falls. https://core.ac.uk/download/144553583.pdf (accessed on 21st October 2025).
23. Crespo Márquez, A., Guillén López, A. J., Gómez López, F. J., Macchi, M., et al. (2016). On the role of prognostics and health management in advanced maintenance systems. https://core.ac.uk/download/190375225.pdf (accessed on 21st October 2025).
24. Cutts, W., Fenge, L. A., & Hodges, C. E. M. (2011). Seen but seldom heard: Creative participatory methods in a study of youth and risk. https://core.ac.uk/download/4899589.pdf (accessed on 21st October 2025).
25. Darnoto, S., Tarwaka, T., Kusumasari, K., et al. (2014). Penilaian risiko pekerjaan dengan Job Safety Analysis (JSA) terhadap angka kecelakaan kerja pada karyawan PT. Indo Acidatama Tbk. Kemiri, Kebakkramat, Karanganyar. https://core.ac.uk/download/148604891.pdf (accessed on 21st October 2025).
26. Davies, J. C. (2008). Nanotechnology Oversight: An Agenda for the New Administration. https://core.ac.uk/download/71353341.pdf (accessed on 21st October 2025).
27. Dreischulte, T., Grant, A. M., & Guthrie, B. (2014). Developing a complex intervention to improve prescribing safety in primary care: Mixed methods feasibility and optimization pilot study. https://core.ac.uk/download/30660277.pdf (accessed on 21st October 2025).
28. Eduljee, G. H., Hall, D. H., Longhurst, P. J., Pollard, S. J. T., et al. (2006). Recent developments in the application of risk analysis to waste technologies. https://core.ac.uk/download/140608.pdf (accessed on 21st October 2025).
29. Etesam, A., Mohammadzadeh, M., Hesami Arani, M., Jaafarzadeh, N., Tazik, M., et al. (2020). Health and safety hazards identification and risk assessment in the swimming pools using combined HAZID and ALARP. https://core.ac.uk/download/338417214.pdf (accessed on 21st October 2025).
30. Fisher, E., Pascual, P., & Wagner, W. E. (2010). Misunderstanding models in environmental and public health regulation. https://core.ac.uk/download/211330199.pdf (accessed on 21st October 2025).
31. Franeková, M., Vaculík, M., Vestenický, M., Vestenický, M., et al. (2008). On-board unit and its possibilities of communications on safety and security principles. https://core.ac.uk/download/8986873.pdf (accessed on 21st October 2025).

32. Goodwin, D., Jeffrey, P., Raffin, M., Smith, H., et al. (2015). Applying the water safety plan to water reuse: Towards a conceptual risk management framework. https://core.ac.uk/download/42144305.pdf (accessed on 21st October 2025).
33. Griffiths, A., Leaver, M., & Reader, T. W. (2018). Near misses in financial trading: Skills for capturing and averting error. https://core.ac.uk/download/157800644.pdf (accessed on 21st October 2025).
34. Gurjar, B. R., & Mohan, M. (2002). Environmental risk analysis: Problems and perspectives in different countries. https://core.ac.uk/download/72056551.pdf (accessed on 21st October 2025).
35. Harris, K. M., & Beeuwkes Buntin, M. (2008). Choosing a health care provider: The role of quality information. https://core.ac.uk/download/71346325.pdf (accessed on 21st October 2025).
36. Horowitz, M. C. (2018). Artificial intelligence, international competition, and the balance of power (May 2018). https://core.ac.uk/download/211333744.pdf (accessed on 21st October 2025).
37. Huh, J. (2019). Uncertainty-aware attention for reliable interpretation and prediction. https://core.ac.uk/download/213600587.pdf (accessed on 21st October 2025).
38. Jiya, T. (2019). Ethical implications of predictive risk intelligence. https://core.ac.uk/download/228181605.pdf (accessed on 21st October 2025).
39. Johnson, B., Electris, C., Humphreys, J., Lang, K., & Korn, S. (2014). The impact of equity engagement: Evaluating the impact of shareholder engagement in public equity investing. https://core.ac.uk/download/80510499.pdf (accessed on 21st October 2025).
40. Kjeldsen, T. R., Lang, M., Lawrence, D., Madsen, H., et al. (2013). A review of applied methods in Europe for flood-frequency analysis in a changing environment. https://core.ac.uk/download/16747798.pdf (accessed on 21st October 2025).
41. Koronis, A., Zhou, C., Tamilselvan, P., Khuntia, S. R., Khuntia, S. R., Khatri, V., & Peng, X. (2018). Smart asset management for electric utilities: Big data and future. http://arxiv.org/abs/1706.09711 (accessed on 21st October 2025).
42. Marin, F., & Notaro, S. (2025). Consumer attitudes toward GM food with hypothetical functional characteristics. https://core.ac.uk/download/pdf/7055784.pdf (accessed on 21st October 2025).
43. N/A. (2016). A Roadmap to Reduce U.S. Food Waste by 20 Percent. https://core.ac.uk/download/75784647.pdf (accessed on 21st October 2025).

44. Nieberding, J. J., Ross, L. J., & Williams, C. H. (2025). Prognostic launch vehicle probability of failure assessment methodology for conceptual systems predicated on human causal factors. https://core.ac.uk/download/pdf/286816701.pdf (accessed on 21st October 2025).

45. Noel, M., & Ziedelis, S. (2011). Comparative analysis of nuclear event investigation methods, tools and techniques. https://core.ac.uk/download/38621375.pdf (accessed on 21st October 2025).

46. Procházka, J., Procházková, D., & Rusko, M. (2018). Principles for management of risks of critical infrastructure. https://core.ac.uk/download/287889332.pdf (accessed on 21st October 2025).

47. Ruggiu, D. (2015). The consolidation process of the EU regulatory framework on nanotechnologies: Within and beyond the EU case-by-case approach. https://core.ac.uk/download/53525097.pdf (accessed on 21st October 2025).

48. Saidenberg, M., May, C., & Schuermann, T. (2025). The New Basel Capital Accord and questions for research. https://core.ac.uk/download/pdf/6649673.pdf (accessed on 21st October 2025).

49. Scott, J. (2015). Children and the internet: An exploration of Year 5 pupils' online experiences and perceptions of risk. https://core.ac.uk/download/30734254.pdf (accessed on 21st October 2025).

50. Showers, B., & Stone, G. (2015). Safety in numbers: Developing a shared analytics services for academic libraries. https://core.ac.uk/download/30729680.pdf (accessed on 21st October 2025).

51. Sujan, M. A. (2012). A novel tool for organizational learning and its impact on safety culture in a hospital dispensary. https://core.ac.uk/download/16664107.pdf (accessed on 21st October 2025).

52. Van Citters, A. D., Larson, B. K., Fisher, E. S., Nelson, E. C., Gbemudu, J. N., Kreindler, S. A., & Shortell, S. M. (2012). Norton Healthcare: A strong payer-provider partnership for the journey to accountable care. https://core.ac.uk/download/71358697.pdf (accessed on 21st October 2025).

53. Yilmaz, Y., Alfonsi, A., Wang, B., Rabiti, C., Smith, C., Maljovec, D., et al. (2013). Dynamic PRA: An overview of new algorithms to generate, analyze and visualize data. https://core.ac.uk/download/54491663.pdf (accessed on 21st October 2025).

CHAPTER 9

SUSTAINABLE AND SMART MOBILITY SOLUTIONS

Contents

INTRODUCTION

The growing intricacies of urban life have dictated that urban mobility solutions need to be redefined within the urban landscape, placing emphasis on intelligent and sustainable mobility practices. With increasing urbanization, traditional transport infrastructure grapples with accommodating population growth while simultaneously dealing with environmental concerns. It is, then, the integration of technology in transport planning that has emerged as a popular trend, thus making intelligent mobility a focus area for city planners and policymakers. However, the term smart mobility has tended to be applied in a narrow way, emphasizing technology solutions without necessarily grappling with broader issues. As was brought out in the debate on this term, there is an evident disconnect between technological vision capability and reality of technology application in urban mobility systems (Lauwers et al., 2015). Thus, an integrated strategy going beyond technological innovation per se must be accorded highest priority so that effective mobility solutions can enhance the quality of city living.

Figure 9.1. Sustainable and smart mobility solutions.

Source: https://sustainablyforward.com/wp-content/uploads/DALL%C2%B7E-2024-02-09-17.12.36-A-horizontal-featured-image-that-exemplifies-sustainable-mobility-solutions-showcasing-a-harmonious-integration-of-electric-vehicles-bi-cycles-publi-900x515.webp.

To address the multi-faceted character of sustainable mobility, adopting a wiser mobility vision encompassing not only technological innovation but also the social, economic, and environmental dimensions of transport is essential. This perspective encourages collaborative and integrated methods in urban transportation planning, which are capable of responding to diverse needs of communities. When cities suffer from congestion, pollution, and equitable transportation accessibility, the weakness of technologically based approaches becomes apparent. By grasping the customer experience and adopting new governance patterns in the context of smart mobility, cities are able to create more equitable and accessible public transport systems. In this respect, the public institutions are to play a central part since their involvement in adopting innovative technologies can lead to inclusive city solutions ultimately resulting in successful, sustainable cities (Monte et al., 2017). Sustainable and smart mobility solutions require a readiness to venture into ongoing research and stakeholder collaboration. As cities in the world grapple with the challenges of climate change and the scarcity of resources, finding new technologies and innovative procurement approaches, such as the Public end Users Driven Technological Innovation (PDTI), can offer solutions to urban mobility issues. The approach not only spurs technological development but also aligns with public interests, making the solutions useful and effective. With the more visible interaction between community demand and technology, city planners are compelled to rethink how success is defined in mobility systems. By stepping back from purely consumerist models and choosing a more systemic approach, cities can begin the doorway to creating resilient, sustainable, and inclusive cities. Thus, the search for successful realization of sustainable and smart mobility solutions necessitates an extensive understanding of both technological innovations and their respective implementations in the urban environment (Figure 9.1).

Definition of Sustainable Mobility

Sustainable mobility is a holistic strategy to transform transport systems so that they bring forth the maximum well-being in society with a minimum footprint on the environment. It is beyond the standard thinking about mobility that is primarily focused on efficiency or speed. Sustainable mobility in essence seeks to bring together various modes of transport, including public transport, cycling, and pedestrian infrastructure, as an integrated web with a focus on equity and accessibility. The multinet character of sustainable mobility highlights policymakers' and planners' need to rethink infrastructure and service delivery with the ultimate goal of reducing carbon footprints. By adopting measures that boost shared mobility, electric vehicles, and other forms of personal car alternative options, cities can design environments favorable to sustainable behavior, harmonizing human transport needs with environmental preservation. The definition of sustainable mobility also places social and economic dimensions first, taking into account how investment in infrastructure can stimulate local economies while meeting the needs of diverse populations. By making all members of society have access to safe and efficient travel, sustainable mobility maintains social equity. The dependency on new technologies must not overwhelm the need for inclusive planning but instead support efforts to enhance the quality of life for the inhabitants of urban areas. As much as smart mobility solutions offer futuristic opportunities for urban transportation systems, the focus must be on the development of spaces that support communal interfaces and guardianship of the environment. Therefore, technological integration in transportation systems must be in line with wider sustainability and urban livability goals, rather than a limited emphasis on consumerism or efficiency that may undermine the spirit of sustainable mobility (Lauwers et al., 2015). In addition, encouraging sustainable mobility requires governance patterns that are collaborative in nature, involving input from diverse stakeholders such as technology specialists, urban planners, and residents themselves. Such a strategy is in support of a shift from techno-oriented or consumer-oriented models to a symbiotic conceptualization of mobility emphasizing social participation and interaction. New technologies have the capability to bring about profound changes in mobility patterns, specifically in small and medium-sized towns, where tailored solutions can strengthen regional public transport systems and enable sustainability (Melé et al., 2020). Therefore, the definition of sustainable mobility continues to be dynamic, capturing both technological advancement and the need for incorporating governance systems. By understanding the interplay between technology, governance, and sustainability as a complex phenomenon, cities can expect to design mobility solutions that are not only effective but also fair and eco-friendly.

Importance of Smart Mobility Solutions

The worth of smart mobility solutions lies in how they can transform city transport, making it more efficient with less environmental cost. The previous decade witnessed cities battling congestion, pollution, and rising demand for accessible transport services. Smart mobility brings together a mix of technologies like real-time data analysis and networked systems to enhance transit network management and facilitate informed choice making by users and providers alike. As cities grow, traditional transportation plans are found to be wanting, hence the necessity for cohesive systems that take advantage of technological innovations to achieve sustainable urban development. Through a focus on interconnected systems that are user-focused and green, smart mobility solutions can solve numerous problems at one time, building a stronger city environment. However, there are challenges to implementing smart mobility solutions in reality due to an overemphasis on technology or consumer-centric strategies. Criticism of such strategies reveals a deeply disconcerting disconnect between the theoretical possibility offered by technology and urban realities. For instance, an overly strong techno-focused strategy can overlook key social imperatives, such as fairness of access to mobility options. Similarly, consumerist views that reduce users to mere customers threaten to ignore broader social objectives, including the happiness of communities and environmental stewardship. With those limitations understood, parties can look for a wiser mobility paradigm, one centered around collective, inter-disciplinary strategies and solutions that transcend single, discrete technical advances. This approach involves that intelligent mobility needs to be envisioned within the context of a complex socio-environmental system, which needs a unified approach that realizes a variety of urban dynamics and includes civic engagement. Further, the necessity for innovative public procurement strategies also becomes important in mobilizing smart mobility solutions. Evidence indicates towards cities being ignorant of the transformational potential delivered through technology for solving their own problems. By applying models like Public end Users Driven Technological Innovation (PDTI), cities can identify new solutions that are more appropriate for public demand. In addition to pushing technological R&D, this also triggers cooperation among stakeholders in order to assure solutions are tailored to real needs. Through actively pursuing such practices, cities can turn themselves into laboratories of innovation, taking the lead and triggering peer-to-peer learning between cities. In the end, the convergence of innovative public procurement and intelligent mobility policies can have the power to create dynamic, sustainable cities that proactively respond to the evolving needs of their citizens, enhancing overall quality of life (Figure 9.2).

Figure 9.2. Importance of smart mobility solutions.

Source: https://pub.mdpi-res.com/sensors/sensors-21-02143/article_deploy/html/images/sensors-21-02143-g005.png?1620823287.

Overview of the Essay Structure

A well-structured essay is essential in presenting ideas and arguments effectively, especially for a subject as intricate as Sustainable and Smart Mobility Solutions. The first paragraph lays the foundation for the discourse by introducing the thesis statement, which states the essays focus and arguments. Following this, the body paragraphs form the cornerstone of the essay, wherein each paragraph must concentrate on a specific aspect of mobility solutions, for instance, technology integration, environmental effects, or socio-economic implications. This approach provides certainty and cogency, which enables the reader to glide smoothly through complex interdependencies among these factors. Thus, the structure of an organization within an essay not only allows information to flow but also ensures that the arguments presented work synergistically towards supporting the master thesis. Entering the body part of the essay, it is now imperative that a seamless connection between paragraphs be maintained. Each body paragraph should begin with a Topic Sentence that reflects the main argument being researched. For instance, when looking at the case of the contribution made by technology to smart mobility solutions, there is a need to explore how innovation drives transformation in urban transport infrastructure. This could entail an exploration of emerging technologies such as autonomous vehicles or intelligent traffic management systems. The subsequent explanation must contain evidence to support these arguments, thereby creating an effective narrative. Furthermore, the incorporation of relevant citations can support the argument more, as can be observed in arguments regarding shifts in teleworking behavior and its effects on mobility (Carter et al., 2008). Such connections form a structured analysis that connects theoretical concepts with practical usage, thus making the general argument more comprehensive. Lastly, the conclusion of the essay matters in the process of summing up the arguments provided in the body. It must not only restate the thesis but also respond to the broader implications of sustainable and intelligent mobility solutions. This entails thinking through how they could change city spaces and enhance social equity. Most importantly, thoughts regarding how open educational resources (OER) could make individuals access these ideas are crucial (Law et al., 2013). Thus, through such explicit section differentiation and intentional connection between them, the organization of the essay attains a coherent examination of the interrelated and intricate relationship of environment, technology, and society regarding mobility solutions. The entire scenario presents an enormous window of opportunity for ideas to be developed richly, and readers are challenged to think through the significance and prospect of intelligent mobility in existing realities.

9.1. THE CONCEPT OF SUSTAINABLE MOBILITY

Sustainable mobility is the inter-generational integration of transportation solutions that seek environmental sustainability, social justice, and economic sustainability. As urban populations grow, the role of traditional transport systems in air quality, energy use, and overall city livability grows. Sustainable mobility works to alleviate these impacts through practices that maximize public transport, cycling, and walking as alternatives to single-occupancy car use. Such measures not only reduce greenhouse gas emissions but also make cities healthier. By placing priority on low-cost modes of transport, cities can redirect traffic from saturated roads and enhance urban livability. The question is to find a way of effectively applying these measures to different urban contexts, and for this, one should have a deep understanding of the interconnected character of transport systems and their effects upon city living. Smart mobility brings a technological twist to sustainability action, but there is a necessity to approach it critically. Though smart mobility, characterized by the application of information technology and digital solutions, is likely to enhance operational performance, it simplifies the complexity of urban transport systems. The "smarter mobility" idea is a plea for more teamwork than the mere technological solution entails by amalgamating social necessity and town planning. As proposed under existing critiques, solving the limitations of both techno- and consumerist approaches is crucial for inclusive transport solutions. This interdisciplinarity not only improves the discourse of sustainable mobility but also aligns the objective of creating dynamic urban spaces catering to mixed demographic populations, ultimately bridging the gap between functionality and innovation (Lauwers et al., 2015). The Battery Electric Vehicle forms a focal point in discourse of sustainable mobility, particularly for emissions mitigation and urban pollution. As highlighted in recent research, the transition to BEVs is a potential avenue for mitigation of the environmental impact of conventional vehicles. With the BEV market expanding at a rate that has never been witnessed before, however, some issues still need to be addressed in order to facilitate their widespread adoption and integration into city transport systems. For instance, battery production concerns, charging infrastructure, and lifecycle emissions necessitate a comprehensive examination of BEV systems. Stakeholders, ranging from policymakers to producers, need to emphasize innovative solutions that answer such challenges, thereby helping to create a cleaner city environment. Embracing a comprehensive approach in advancing BEVs together with other sustainable transport means has the potential to improve mobility solutions to greater heights, towards a more resilient and sustainable city future (Dalton et al., 2013).

9.1.1. Principles of Sustainability in Transportation

Sustainability principles of transportation provide grounds for the development of efficient and sustainable solutions to mobility. Central to the principles is the quest for reducing environmental impact through cleaner technologies and practices being adopted. Urban areas, particularly, are adopting electric and hybrid cars while simultaneously enhancing public transportation infrastructure to reduce dependence on fossil fuels. This transformation is not merely a technical one but demands robust systems that combine efficiency and accessibility. The sustainable transport also values active mobility — cycling and walking — as means to decrease carbon emissions and promote public health. Therefore, handling transport sustainability calls for an overall strategy that includes the re-design of cities with an eye toward promoting multi-modal solutions, thereby a landscape of mobility that offers both environmental sustainability and quality of life. Added prosperity to transport sustainability principles is the concept of Smart Cities that harnesses data analytics and information and communication technologies (ICTs) to enhance mobility in the urban area.

The integration of intelligent technologies not only facilitates real-time monitoring of transport infrastructure but also enables planners to achieve efficient decisions with empirically derived information. The method is eco-friendly as routes are optimized, congestion is minimized, and pollutant emissions are reduced in the end. As has been noted, smart city projects are characterized by their utilization of data collection and analytical software that transforms raw data into usable intelligence (Stahl et al., 2017). These innovations are crucial in building responsive transport networks that have the ability to react to changing urban dynamics and improve the coordination between stakeholders in seeking sustainable solutions. Innovative public procurement is also required in bringing sustainability into transport, as demonstrated in up rushing frameworks with the potential to sustain the development of new technologies. The Public Driven Technological Innovation (PDTI) process is an illustration of the way cities can identify the specific problem and craft tailored robot solutions to cater to specific urban mobility needs. Such initiatives not only raise public acceptance levels of cutting-edge technologies but also encourage higher levels of public engagement from both the public administration and people (Monte et al., 2017). Through promoting cooperative frameworks and co-creation, cities are capable of releasing the innovation potential of different actors. Ultimately, sustainability and smart mobility can offer end-to-end solutions that appeal to the concerns of society and set the stage for future transport infrastructure development (Figure 9.3).

Figure 9.3. Principles of sustainability in transportation solutions.

Source: https://www.transport.gov.scot/media/48002/prioritising-sustainable-transport.jpg.

9.1.2. Environmental Impacts of Traditional Mobility

Traditional mobility systems have historically been associated with adverse environmental externalities, which continue to be challenging for policymakers and urban planners. Relying on fossil fuel-based vehicles has a massive contribution towards greenhouse gas emissions, which exacerbate climate change and reduce air quality. Moreover, increased traffic congestion, a result of traditional models of transport, results in higher emissions as cars operate inefficiently under stop-and-go conditions. This environmental expense is combined with land use schemes that encourage urban sprawl, necessitating lengthy journey times and the expenditure of yet more energy. Therefore, the imperative to seek out sustainable solutions has been recognized, as cities grapple with enhancing mobility but lessening environmental degradation. A shift from traditional mobility schemes demands new thinking and collaboration across sectors to ensure environmentally sound practice takes precedence. The environmental effects of traditional mobility extend beyond air pollution to other types of significant environmental degradation. Excessive vehicular traffic produces noise pollution that disrupts ecosystems and negatively affects human health. Traditional transportation infrastructure such as highways and parking lots contribute to habitat fragmentation, which prevents animals from inhabiting their natural environments. Hard surface runoff also causes water pollution, affecting aquatic habitats and ecosystems. Because these systems lack emphasis on sustainability in favor of efficiency and speed, the environmental cost keeps increasing beyond sustainable means. Addressing these issues is necessary for the creation of cityscapes that meet the needs of human and ecological health. Thus, by converting to sustainable transport solutions that limit dependency on traditional means, cities can work towards retaining their natural environment as they accommodate growing populations.

Figure 9.4. Environmental impacts of traditional mobility systems.

Source: https://bellengineering.net/wp-content/uploads/2023/10/environment-impact-of-gasoline-vs-electric-cars-1024x577-optimized.png.

To fulfill their role in countering the environmental impact of traditional mobility, new models and technologies must be embraced. Initiatives such as the Vehicle Design Summit (VDS) most epitomize this, involving the creation of eco-efficient vehicles and embracing green business models for the mobility sector. These types of initiatives encourage a paradigm shift from working on the creation of add-ons in existing vehicle technology towards system innovations in general. Through a redesign of car design, ownership, and operating schemes, stakeholders can make a more sustainable mobility system possible. With an emphasis on convergence of the green technologies as well as on driving convergence among leaders, dramatic differences are made for urban transportation. This redesign of mobility not only has the capacity to reduce environmental impacts holistically but also makes thriving city spaces with quality living possible (Lauwers et al., 2015; Ceschin et al., 2008) (Figure 9.4).

9.1.3. Benefits of Adopting Sustainable Practices

The adoption of green practices in mobility is instrumental in reducing the adverse effects on the environment, which is central to addressing climate change. By saving greenhouse gases and energy, these practices support cleaner air and improved environments. The integration of electric vehicles (EVs) into urban mobility systems is the ideal example of adopting these changes. EVs have no tailpipe emissions, which mean cleaner air in cities than traditional internal combustion technology. Secondly, the creation of effective public transport systems, such as electric buses and trains, reduces the consumption of fossil fuels, thereby accelerating the move towards a greener future. With the continued adoption of green mobility solutions by urban areas, the public's overall health is enhanced, curtailing a healthcare system burden and enhancing citizens' quality of life, ultimately validating the entire set of advantages in adopting green mobility solutions (Browne et al., 2014). Other than environmental advantages, adopting green mobility methods may incur substantial economic dividends. Green transport transition not only encourages technological development but also creates job opportunities in new industries. For instance, growing demand for electric vehicle manufacturing and infrastructure boosts economic growth and job prospects in the manufacturing, engineering, and installation sectors. In addition, energy-efficient transport solutions can lead to lower operating costs for both customers and transport operators. Upgrades of public transportation, such as pedestrian-centered infrastructure and bike-sharing programs, enhance users' convenience and accessibility to promote an integrated community. Finally, these economic benefits buttress the argument in favor of changing to sustainable mobility habits with the guarantee that economic development aligns with environmental stewardship (Carter et al., 2008). In addition, embracing sustainable mobility habits in the mobility system also enhances social fairness and improves the well-being of all citizens. Sustainable mobility initiatives focus on being inclusive and offer low-cost and accessible mobility to traditionally excluded groups by conventional systems. Through investment in public transport, cycling infrastructure, and pedestrian walkways, cities are in a position to develop connectivity and mobility for low-income citizens and thus end their reliance on private vehicles. The creation of shared mobility services, such as carpooling or ridesharing, also enhances socialization and communal engagement. Further, enhanced transport arrangements can make the roads less congested and release residents from some time to pursue leisure and family life. As more integrated greener practice is embraced in urban planning, they not only enhance operational efficiency but also sense of place and community, hence making the urban area livable and equitable (Browne et al., 2014).

9.2. SMART MOBILITY TECHNOLOGIES

The use of smart mobility technology is a milestone in transforming urban transport systems. The technologies employ data analysis, IoT sensors, and artificial intelligence to regulate traffic, optimize public transport systems, and facilitate the use of cleaner modes of transport. But the concept of "smart mobility" has been unjustly attacked for its exclusively technical concentration on uses of technology within mobility systems and how this can lead to a vision-reality gap. If concentrating on a "smarter mobility" approach, urban planners are able to break away from a technologically restricted point of view and take a more holistic path incorporating user needs and cross-disciplinary perspectives. This change is required for ensuring innovation in intelligent technologies is converted to tangible improvements in urban livability and mobility to address evolving challenges facing different communities in changing urban environments (Lauwers et al., 2015). Moreover, smart mobility technologies stimulate a platform on which innovative solutions can be developed to improve living conditions in cities. Urban areas are increasingly confronted with sophisticated problems like congestion, pollution, and accessibility, all of which require effective and responsive countermeasures. For example, governments can use emerging technologies, like robotics and automated systems, to make transportation services more efficient and interact with citizens in a better way. Introduction of Public end Users Driven Technological Innovation (PDTI) creates a platform for public institutions to utilize such innovations, outlining clear urban challenges that can drive the development of robotic solutions to address the needs of citizens. This policy is a major change towards more resilient and human-centered urban mobility solutions because cities can more readily integrate technological innovation directly with public demand and quality-of-life improvement (Monte et al., 2017). Moreover, the successful application of smart mobility technologies relies on a participative approach with multiple stakeholders' engagement and feedback. This requires an intersectoral partnership among technology innovators, urban planners, and local community groups to render mobility solutions equitable and accessible. Public education and outreach on the importance of smart technologies can assist in developing public acceptance and a culture of innovation in cities. As these technologies become more reliant upon cities, it is essential that not only the technological aspects are prioritized but also the social implications and relationships that they develop. That way, city mobility can indeed turn itself into a sustainable and just system that really addresses the dreams and requirements of everyone and fosters vibrant, resilient urban communities (Lauwers et al., 2015) (Figure 9.5).

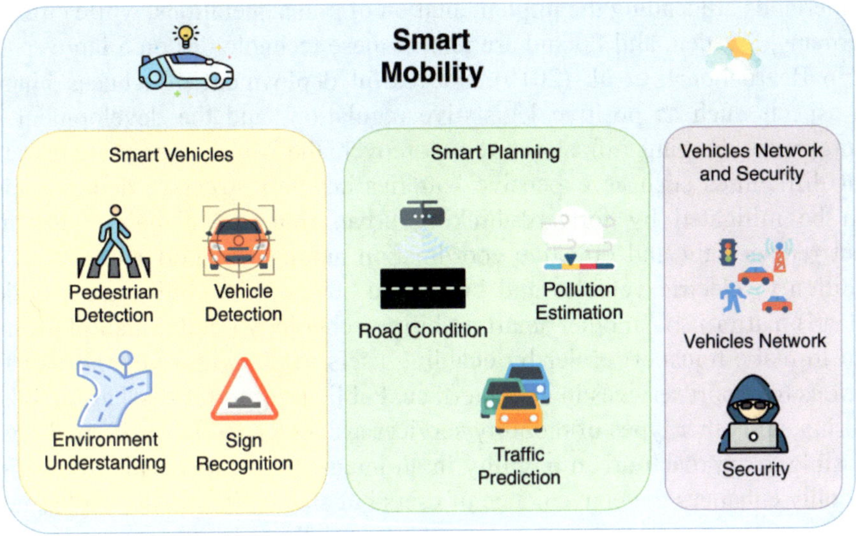

Figure 9.5. Smart mobility technologies.

Source: https://www.mdpi.com/electronics/electronics-14-01801/article_deploy/html/images/electronics-14-01801-g001.png.

9.2.1. OVERVIEW OF SMART MOBILITY TECHNOLOGIES

The mobility environment of the city is changing with smart mobility solutions emerging as actual answers to combat the drawbacks of traditional transport systems. Developments in battery electric vehicles (BEVs), self-driving cars, and integrated public transport systems are reshaping the way people travel through urban spaces. BEVs, in particular, are gaining popularity since they can offset urban air pollution and reduce fossil fuel dependence, factors that have been significant sources of environmental degradation, as Dalton et al. (2013) reports. Improved battery technology and charging systems, currently under development, are driving their uptake by customers, integrating them into sustainable transport in cities. As cities consider embracing such technologies, the integration of smart mobility solutions can enhance the livability of cities and prevent global warming. Another major trend in the era of smart mobility is the utilization of electric buses (e-buses) in urban cities across Europe. These buses are not only a cleaner alternative to conventional diesel buses but also demonstrate the capacity of local strategic planning to facilitate a transition to cleaner city transport. Analysis reveals that nations like the UK and the Netherlands are leading the implementation of e-bus operations, while cities in Germany, Sweden, and Poland are testing these technologies on a large scale, as in Bezruchonak et al. (2019). Successful deployment of e-buses hinges on aspects such as positive legislative regulations and the development of appropriate charging infrastructure. Moreover, the Minsk case demonstrates that difficulties such as expensive batteries and infrastructure dependencies can be mitigated by forward-thinking urban transport planning targeting fleet replacement and effective coordination among transport authorities. In addition to electric vehicles and buses, the advent of mobility as a service (MaaS) platforms is another smart mobility technology trend. MaaS platforms aim to make transport easier by enabling users to plan, reserve, and pay for various transport services in one interface. Public transport, ride-sharing, bike-sharing, and other types of mobility services are integrated in MaaS, making it possible to approach urban mobility in an integrated manner. This innovation not only enhances the convenience of users but also optimizes the deployment of resources across modes of transport, essentially reducing congestion and environmental impacts. Through the utilization of data analytics and real-time information, MaaS systems can also optimize the production of more responsive and efficient transport networks. While cities adapt to these innovations, it becomes increasingly apparent on a daily basis that smart mobility solutions are essential to designing sustainable and connected urban environments that can meet the needs of diverse populations (Figure 9.6).

Figure 9.6. Overview of smart mobility technologies transportation systems.

Source: https://www.intellistride.com/wp-content/uploads/2024/11/How-Singapore-Smart-Traffic-System-is-Redefining-Urban-Mobility.webp.

9.2.2. Role of Data Analytics in Transportation

The application of data analytics in transport is central to the attainment of sustainable and intelligent mobility solutions. Through the use of data from several sources—such as traffic flow, rider behavior, and vehicle efficiency—transport planners and policymakers can make data-driven decisions that optimize system performance and user satisfaction. Data analytics provides for very detailed insight into how transportation systems work, and it can highlight bottlenecks and inefficiencies that can be smoothed out through strategic interventions. Further, with cities increasingly emphasizing smart mobility solutions, integration of data analytics is essential to create responsive systems that react to actual-time realities, thereby optimizing routes and reducing congestion. Realignment of emphasis towards a harmonized practice with technology and urban planning can help cities leverage analytics, leading to an interlocked and sustainable transport system as envisioned in the debate surrounding smart mobility solutions (Lauwers et al., 2015). In addition to efficiency, data analytics fosters a more contextualized understanding of transportation users' needs as central to the services design that captures heterogeneous groups' appeal. The application of big data makes it possible for transit authorities to segment users by patterns of travel, attitudes, and demographic traits, translating into more personalized service provision. This consumerism-oriented approach shifts focus from technology advancement in isolation to an understanding of the participants in the transport system. However, this approach also demands caution because relying on user metrics alone may not take into account overall urban mobility goals. Instead, a holistic approach that combines socio-economic factors and urban planning is paramount in setting inclusive transport solutions. Thus, effective use of data analytics must combine user data with broader systemic enhancements in order to guarantee that transport initiatives are incorporated into sustainable urban planning principles (Stahl et al., 2017). In the smart city environment, the combination of data analytics and transportation management is essentially intended to collectively solve urban issues. Applications that emphasize information and communication technologies (ICTs) enable cities to harvest and aggregate data efficiently, which is then converted into actionable insights for transportation systems. Furthermore, building organizational structures that encourage cooperation among a broad set of stakeholders, from government agencies to community groups and private business, is needed to drive this data into new solutions. The dynamic between governance and data that comes about encourages not only the efficacy of operations but also the development of vibrant urban spaces that enhance quality of life. This synergetic approach to mobilization of data analytics is central to the effort for smarter and sustainable cities, a pointer to the need for an umbrella framework rather than discrete technological application (Lauwers et al., 2015; Stahl et al., 2017) (Figure 9.7).

Figure 9.7. Role of data analytics in transportation.

Source: https://thefamilyoffice.ch/wp-content/uploads/2024/06/1717881463.png.

9.2.3. Integration of Internet of Things (IoT) in mobility

The application of the Internet of Things (IoT) to mobility is a paradigm shift towards smart and sustainable transportation. At its core, IoT enhances connectivity between modes of transport and the environment around them, facilitating smarter and more informed decisions. For example, smart traffic lights can tune their patterns based on up-to-date data from networked traffic and pedestrians, reducing congestion and emissions by an enormous amount. In addition, IoT devices can monitor car performance and ambient conditions, providing predictive maintenance and best practice operation of cars. This connectivity enables more active management of mobility, with cities using data analytics to optimize traffic flow and resource use, ultimately resulting in reduced urban carbon emissions and enhanced livability of urban lifestyles. As more cities implement smart mobility solutions, the use of IoT extends beyond cars to entire transport networks. Integrating IoT into public transportation systems may enhance user experience and improve operational efficiency. Connected buses, for instance, can provide real-time information to passengers regarding arrival and potential delays, therefore ensuring overall service dependability. Furthermore, integration of IoT technologies within the public transportation aids in effective fleet management, which reduces fuel usage and optimizes routes.

Figure 9.8. Integration of internet of things (IoT) in mobility.

Source: https://media.monolithicpower.com/wysiwyg/Educational/ADC_Chapter_9_Fig1-_960_x_813.png.

This is extremely crucial with regards to sustainability since it encourages increased utilization of public transport and, by extension, less reliance on individual vehicles. The shift towards IoT-based public transit systems strongly continues in tandem with global sustainability goals, paving the way for more smooth and clean urban mobility solutions. But while the benefits of IoT integration to mobility are immense, several challenges have to be overcome if its full potential is to be optimized. Issues related to data privacy, security, and interoperability between systems pose enormous barriers. For example, as automobiles and infrastructures increasingly become interconnected, the risk of cyberattacks becomes greater, necessitating strict security systems and protocols. Also, successful implementation of IoT solutions requires broad collaboration among stakeholders such as city governments, traffic authorities, and tech corporations. Such collaboration can lead to new models of governance with a focus on sustainability and adaptability in urban planning. The learning that can be gained through IoT integration can influence future research and development, ultimately shaping best practices to develop sustainable urban mobility systems across the world (Mercadé Melé et al., 2020; Copeland et al., 2019) (Figure 9.8).

9.3. PUBLIC TRANSPORTATION INNOVATIONS

Transportation innovations are a key focus in the development of sustainable mobility systems for the growing urban populations. The introduction of technology-driven services such as real-time tracking, mobile tickets, and integrated public transit apps has transformed urban operations. Such features facilitate seamless connectivity, making the citizens optimize their trips and reduce waiting time, thus improving overall satisfaction with public transportation. Besides, in order to meet the needs of sagacious planning in cities, efforts are also being put on automating transportation systems. Combination of autonomous transport and intelligent public procurement means can result in better efficiency of transit networks. This is in line with the requirement of innovative solutions to urban problems, as evidenced by the continuous innovations in technology that serve the needs of the public within various smart cities (Monte et al., 2017). With technological advancement also comes the need to re-examine the basic frameworks for mobility to enable effective public transport innovations. The notion of "smart mobility," while gaining traction, has been vilified as being too technology-driven at the expense of holistic planning and user experience. It is worth mentioning that urban mobility improvement is not simply a matter of unleashing new technology, but rather requires a sophisticated understanding of the social and environmental contexts in which such technologies operate. Therefore, adopting a policy of "smarter mobility" that seeks inter-disciplinary coordination and gives precedence to user-centric solutions will guarantee significant enhancement in the quality of city life. With the injection of knowledge from the fields of urban planning and environmental science, public transport systems can become increasingly responsive to the needs and aspirations of city dwellers, as reflected in the emerging paradigms of urban mobility (Lauwers et al., 2015). Lastly, the future of sustainable and smart public transportation innovations hinges on the willingness of stakeholders to embrace overall strategies that address urban mobility concerns at a bigger scale. This involves not only installing advanced technologies but also ensuring these solutions are accessible and beneficial to different populations of people. People participate in implementing the process, leading to ownership and pushing people to have more trust in transit systems. Additionally, as urban areas pilot new mobility schemes, constant monitoring of their social and environmental effects will be needed in order to scale up and fine-tune successful schemes. This collaborative effort emphasized by adaptive policymaking will propel the creation of productive, efficient, and sustainable urban transport systems that can serve as a blueprint for additional development within public transport (Monte et al., 2017; Lauwers et al., 2015) (Figure 9.9).

Figure 9.9. Public transportation innovations.

Source: https://www.centreforcities.org/wp-content/uploads/2023/11/Cambridge-feature.png.

9.3.1. Electrification of Public Transport Systems

Electrification of public transport systems is a crucial step towards the creation of eco-friendly and smart mobility. Moving away from traditional fossil fuel-based cars to electric buses and trains entirely eliminates the release of greenhouse gases, resulting in cleaner air and reduced city pollution. More importantly, cities are gradually realizing the importance of electrification not just from an environmental point of view but also in rendering public transport more effective. The advent of electric vehicles (EVs) offers an opportunity to utilize cutting-edge technologies, such as smart grids and renewable energy sources, hence promoting an inter-connected, robust transit system. As urban populations increase, there is an increased demand for sustainable transport, necessitating rapid action and investment in electric public transport systems. Moreover, the unique characteristics of various locations, especially in small island developing states (SIDS), present special opportunities for initiatives in electrification. Locations such as Fiji, Mauritius, and Barbados provide examples of how small geographical presence and a concentrated tourism sector allow for swift transitions of public transport to electrical modes without going through traditional carbon lock-in.

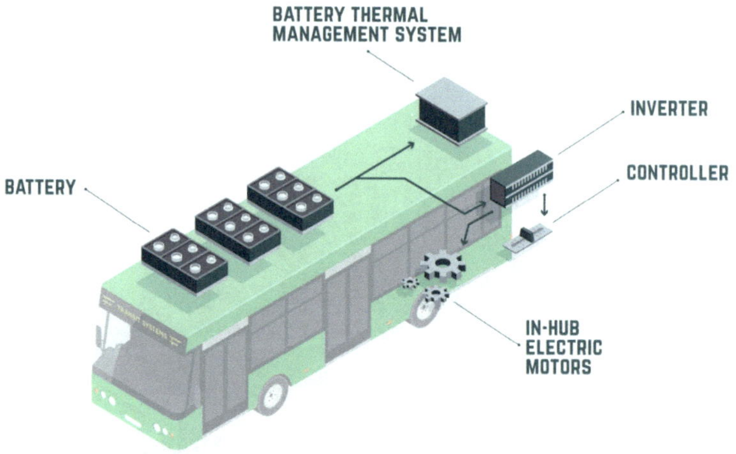

Figure 9.10. Electrification of public transport systems.

Source: https://images.squarespace-cdn.com/content/v1/5a668f1080bd5e 34d18a 7e76/1562049452197-F6CDV67PWQ8TSOMSH5PL/ke17ZwdGBToddI8pDm 48 kAfDp1G1nRpUemC3XWslcjR7gQa3H78H3Y0txjaiv_0fDoOvxcdMmMKkD syUqMSsMWxHk725yiiHCCLfrh8O1z4YTzHvnKhyp6Da-NYroOW3ZGjoBK y3azqku80C789l0mlM0or4nqX7jrn5yWu0hA0z07G9XBJxpAeGC-t-PxtKeuiN-PW4pm3yvNB-Jie_QnQ/TS_How_Electric_Bus_Works_InfoGraphic_02.jpg.

The tourism sector also serves as a midwife to innovation, providing local decision-makers with space to experiment with shared mobility programs that both educate locals and visitors. Through the implementation of electric fleets in the public transport industry, these regions are not only able to minimize the impact of climate change but become low-carbon leaders as well. This is a message to the rest of the globe that is empowering in regards to whether decarbonizing public transport networks is feasible by 2040, proving that sustainability can be achieved in various contexts (Blechinger et al., 2020). As cities grapple with the increasing demand for electric vehicles, the necessity to roll out massive EV infrastructure becomes ever more apparent. Urban planners, city officials, and utilities must collaborate to meet this challenge, making the placement of charging stations and facilitating technologies top priority, which offer the setting for adoption of electric public transport. The joint effort to transform mobility systems to encompass electrification, therefore, requires a paradigm shift in investment, community, and policy. Research has proven that cities leading the electrification process are those that embrace new trends and possess robust structures for the roll-out of EV infrastructure (Fox-Penner et al., 2019). The extensive incorporation of electrified public transport is not only a transport innovation but part of building up sustainable cities that ultimately result in smarter, cleaner cities (Figure 9.10).

9.3.2. Implementation of Smart Ticketing Solutions

Smart ticketing solutions' adoption is an important factor in promoting smart and sustainable mobility solutions in cities. Advances in these technologies make it easier to collect fares, allowing seamless integration of transport modes. Integration of mobile apps, contactless cards, and digital wallets allows passengers to utilize multiple services without the hassle of physical tickets. As mobility providers and transport authorities strive to create a seamless and integrated mobility network, smart ticketing is an essential element to deliver evolving mobility demands. By enhancing user experience via real-time information and convenient fare systems, smart ticketing enriches passengers while reducing operational sophistication for transport operators. Thus, this shift toward digitalization is an extension of a bigger movement toward holistic mobility solutions to develop public transport as the spine of urban mobility, as suggested in Ambrosino et al. (2016). Notably, the successful implementation of smart ticketing systems in a variety of contexts is a pointer toward their central role in public transport. For instance, the cities of Latin America have effectively integrated smart payment systems into their busway systems to facilitate greater user convenience and operational efficiency. The innovations in Curitiba and Bogota demonstrate the potential of streamlined fare collection not just to boost passenger satisfaction but also to reduce operational expenditure and improve higher ridership. With the use of technology, cities are capable of dynamically adapting to transport needs and maintaining the flexibility inherent in busway operations. The integration of smart ticketing with capacity corridors signifies the possibility to create flexible transport solutions that diverge from car-centric traditional paradigms, as in Wright et al. (2001). Hence, smart ticketing is a primary nexus between user experience and operational efficiency that has implications for sustainable urban transport initiatives. Moreover, the use of smart ticketing solutions has the potential to contribute significantly to environmental and social objectives. By promoting public transport usage through simplified payment procedures, cities are able to minimize congestion and carbon emissions associated with single-car use. Data also that the intelligent ticketing systems collect facilitate route planning and optimum services, thereby providing enhanced sustainability outcomes. In line with the functions of integrated mobility platforms, smart ticketing allows for the overall planning of urban transport systems, as indicated in Ambrosino et al. (2016). The growing number of cities embracing such new technologies can result in a transition towards shared mobility programs, which will enhance sustainable urban growth, increase civic engagement, and raise the overall quality of life. The growing adoption of smart ticketing as an integral infrastructure element thus makes it a significant driver of transformation of the urban transport environments (Figure 9.11).

Figure 9.11. Implementation of smart ticketing solutions.

Source: https://railway-news.com/wp-content/uploads/2018/04/FIME-Smart-ticketing-Quality-Assurance.jpg.

9.3.3. Enhancements in Accessibility and User Experience

Enhancements in user experience and accessibility are fundamental aspects of designing sustainable and smart mobility solutions. With urbanization, the necessity to be able to integrate sophisticated technology and user-centric design becomes a prerequisite. Good mobility solutions must be responsive to all types of users' diverse needs, including people with disabilities, and integrate them into shared mobility services. This is evident in attempts to fashion accessible public transport using multimedia materials, calling for effective information system designs that cater to varied user needs. By utilizing assistive technologies, cities can remove barriers to accessing transport, thereby ensuring inclusivity. Such advancements not only enhance people with disabilities but also promote a more equitable and inclusive urban mobility system, underscoring the importance of accessibility in the wider sustainability agenda. Furthermore, user experience is enhanced also through the use of technology to provide real-time information and tailored services. Smarter mobility solutions employ data analysis and customer feedback to optimize transit systems and improve overall travel conditions. This approach takes advantage of the limitations of traditional mobility means, shortening journey times and streamlining them. A case in point is smartphone applications that integrate various transport modes, enabling seamless transfers between services, reducing waiting time, and improving connectivity. The transition to a more intelligent mobility model targets an integrative approach that balances user experience with technological innovation. Critically, it is necessary to transcend the exclusive techno-centric understanding of mobility and embrace a model that prioritizes user experience and involvement in transport system design. The emphasis should continue to be on improving quality of life through reflective innovations that are sensitive to the needs of heterogeneous groups. Additionally, an interdisciplinary and collaborative environment is necessary for balancing technological innovation with demands for accessibility. Compared to the conventional consumerist and technocentric models that have come under criticism in contemporary literature (Lauwers et al., 2015), an integrated solution would necessitate engagement by stakeholders from diverse sectors, including urban planners, transport authorities, and disability groups. This leads to mutual comprehension of mobility challenges faced by marginalized groups as well as pushing for innovation and equity in solutions. The intersection of these perspectives can create multifunctional urban spaces emphasizing accessibility, security, and convenience. As such, as cities strive towards sustainable and smart mobility plans, an emphasis on inclusiveness and user experience will be the driving force for innovation, ensuring that the benefits of cutting-edge technology are accessible to all sections of society (Gregor et al., 2006).

9.4. ACTIVE TRANSPORTATION SOLUTIONS

Active transport solutions, which encompass modes such as walking, cycling, and public transport, provide a multi-faceted approach towards improving urban mobility at the least environmental expense. The switch to these solutions is founded on the argument that motorized transport is the primary source of urban congestion, air pollution, and drain on natural resources. Through their use of non-motorized transport, cities are able to induce healthier lifestyles among their populations at minimal environmental cost. The uptake of measures to encourage active travel—like pedestrian facilities and bike-sharing systems—has the capacity to transform urban landscapes, making cities more human-friendly, user-friendly cities. Increased reliance on active transport not only enhances the quality of urban life but also ensures that the city is in even closer alignment with the principles of sustainable development by lessening reliance on emissions-generating vehicles.

Figure 9.12. Active transportation solutions.

Source: https://s3-us-west-1.amazonaws.com/ehq-production-us-california/b5 b0 cd6b11239f16497a576e798cffdd8ed87027/original/1688083188/fa86cf69a74e-8ba0fa27679d0ca4cac7_IMG_0004_brighter.jpeg?1688083188.

Moreover, the "smart mobility" concept must intersect with active transportation options to form an integrated urban transit plan. As aforementioned, conventional thinking on smart mobility overemphasizes technology and consumerism, a strategy which may overlook the essential societal needs and urban planning priorities (Lauwers et al., 2015). Instead, a smarter understanding of smart mobility needs to include an integrated view of active transportation, wherein technology augments but does not dominate urban mobility planning. This entails integrating transport systems to be holistic and inclusive, along with enhancing users' experience and promoting lower carbon emissions. By moving beyond solely technical fixes, cities have the capacity to develop fair places where active travel is not just feasible but also appealing as a mainstream mode of mobility. Moreover, the interconnection of active travel solutions with innovation on battery electric vehicles (BEVs) increases the process of moving toward more sustainable urban transport systems. While traditional forms of transport have had a negative impact on urban air quality and energy consumption, BEVs hold the promise to provide cleaner transportation options (Dalton et al., 2013). Still, these technologies are not meant to usurp the essential position of active transport but rather be regarded as complementing elements of an inclusive strategy that is based on walking, cycling, and public transport. This balanced approach will optimize urban mobility to prevent new technologies from aggravating existing traffic congestion and urban sprawl issues. Ultimately, the harmonized system that integrates active transportation with advanced BEV systems will allow cities to embrace a future of sustainable mobility decision-making that is equitable, efficient, and environmentally sound (Figure 9.12).

9.4.1. Promotion of Cycling and Walking Infrastructure

Encouragement of cycling and walking infrastructure is a core element in forming sustainable and intelligent mobility plans. As urban areas face increasing issues of traffic congestion, air pollution, and global warming, the development of non-motorized travel methods becomes increasingly important. Making available dedicated bicycle lanes, pedestrian-only streets, and other enabling infrastructure will assist in enabling cities to foster a positive environment that encourages cycling and walking. This not only decreases car dependency but also promotes healthy living among citizens. Research has indicated that cities with well-designed cycling facilities witness an increase in cycling and, accordingly, decreases in traffic emissions as well as improved public health outcomes. Therefore, taking these green mobility plans together is a precautionary step that serves both environmental causes and the welfare of the public too, as per the general aims of sustainable urban development. Additionally, effective rollout of walking and cycling infrastructure necessitates an overall approach that considers the different needs of city inhabitants. Understanding the local mobility patterns and demographics allows for tailored strategies that are relevant to residents. This need-based strategy is exhibited in studies conducted in Berlin, where there was an attempt to uncover the specific mobility needs of city center residents (Hoor et al., 2019). Rather than clinging to traditional transport planning strategies, this research concentrated on stakeholder engagement and qualitative data collection, hence contributing to a higher ability to deliver effective transport solutions. Putting the voice of the community first, planners will be able to design infrastructure that accommodates not only cyclists and pedestrians but also integrates fully into current public transport systems, hence enhancing the mobility context. Moreover, the supply of well-planned walking and cycling infrastructures has significant benefits that further advantage particular communities but also have broader societal effects. The contribution of sustainable mobility strategies towards reducing environmental degradation and supporting social justice cannot be overstated. As outlined in a number of papers, inclusive sustainable mobility indicators provide a platform on which these infrastructures can be assessed across various contexts (Gillis et al., 2015). This enables policymakers to apply best practices and monitor progress towards aspirational levels for sustainability. Through the promotion of cycling and walking as alternatives to car travel, cities can establish equitable spaces that put emphasis on access while spurring economic growth and environmental management. Overall, these activities form a vital part of the conversation regarding sustainable and smart mobility solutions, making resilient cities that benefit current and future residents.

9.4.2. Development of Pedestrian-Friendly Urban Spaces

The progressive development of walkable urban spaces is pivotal to the development of sustainable and smart mobility solutions. With the city getting increasingly crowded, there is a need for improving walks, which is a paramount aspect in urban planning. Walkable spaces not only enhance the quality of life of individuals by promoting physical exercise and socialization but also assist in reducing traffic from cars, and thus the production of greenhouse gases. Cities designed with walkers in mind are likely to have wide sidewalks, parks, and well-designed crossings, encouraging individuals to make walking a convenient means of mobility. In this way, good design can play a vital part in changing cities to meet the pressing challenges of those who live in them, making them healthier, more vibrant places. By prioritizing pedestrian infrastructure, cities can simultaneously address environmental sustainability and civic engagement. Transportation planning and housing development are heavily influenced by the pedestrian-oriented environment design. For instance, as is illustrated by the case of San Mateo County, a housing shortage leads to a multiplier effect on transportation issues with increasing commuting times and environmental degradation. The jobs-housing imbalance in these regions demands the master planning of urban development that integrates affordable housing and transit. By creating pedestrian-friendly cities that connect housing to central amenities and employment hubs, planners are able to decrease congestion and create a more sustainable urban model. This integrated approach can be utilized to provide transport solutions that sustain housing schemes, effectively addressing the housing and mobility needs of high-density areas (Alexander et al., 2019). Moreover, the Smart Cities idea is inherently linked with walking-friendly strategies, as technology can facilitate improved mobility in cities.

Cities are able to respond more effectively to their city challenges through the application of innovative public procurement methods and technology. As demonstrated through various pilot schemes across European cities, the use of smart technologies can enhance pedestrian experiences by optimizing urban infrastructures, such as smart traffic light signals and real-time information systems that guide pedestrians. Consequently, the technologies have the capability to make pedestrian networks not only efficient but also user-sensitive. By embracing a holistic policy that synthesizes intelligent technological systems with people-oriented urban growth, cities are able to align themselves towards a more sustainable future where traffic issues are solved by innovative, people-oriented methods (Monte et al., 2017).

9.4.3. Benefits of Active Transportation on Public Health

Active transport, such as walking and cycling, is of crucial importance to the public's health as it encourages individual and community well-being. Active transport helps to create regular physical activity, which helps to prevent long-term diseases such as obesity, diabetes, and coronary heart disease. Integrating active transport into people's lifestyles makes the daily recommended levels of physical activity more easily achievable, resulting in general well-being. Moreover, when individuals value active transport infrastructure such as cycling lanes and sidewalks highly, they establish environments where healthy choices are present and facilitated, which then further supports the social shift towards more sustainable mobility patterns. The resulting increase in physical activity benefits individual health outcomes as well as diminishes healthcare costs, thereby positively impacting public health expenditure. Furthermore, active transport greatly affects mental health and overall social well-being. Research indicates that regular physical exercise can reduce symptoms of depression and anxiety, which leads to a better mental state. Walking or cycling to the workplace is encouraged as an opportunity for individuals to meet other people, an aspect critical in building community bonds and a sense of belonging. With the neighborhood being accessible by bicycle or walking, inhabitants are more likely to engage in socialization with one another, hence increased community cohesion and social support systems. This social component of active transport reinforces the argument that such policies are more about creating healthier, more cohesive populations, and less about mobility. With these benefits in view, policymakers and urban developers must make choices that would serve to enhance the encouragement of active transport as a flagship policy in public health strategies. Lastly, the environmental benefits of active transport automatically support public health initiatives. Through the mitigation of motor vehicle reliance, populations can effectively eliminate traffic-related air pollution and greenhouse gas emissions, widely established public health hazards. Cleaner air immediately means healthier respiratory systems and fewer instances of related disease. Active transportation also is a proactive step to address the pressing threats posed by climate change because it supports sustainable development compatible with health objectives. Acknowledgment of such environmental-public health relationships is the starting point for stakeholders seeking to create more sustainable cities. Transitioning toward integrated transport planning, focusing on active transport, is a promising route in reconciling public health and environment sustainability goals, creating cities that are smarter yet healthier for future generations (Hoor et al., 2019; Dalton et al., 2013) (Figure 9.13).

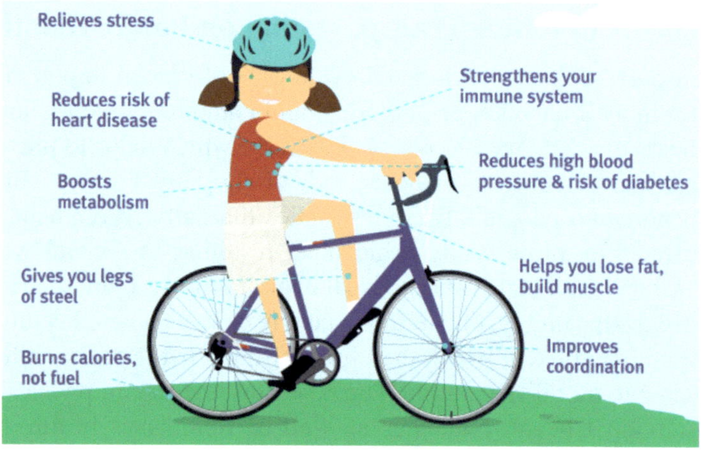

Figure 9.13. Benefits of active transportation on public health.

Source: https://ridetowork.bike/wp-content/uploads/2018/01/Health-Benefits-of-Cycling-no-header.png.

9.5. POLICY AND REGULATORY FRAMEWORK

The policy and regulation environment of sustainable and smart transportation choices is most essential in planning effective transitions within the transport industry. These frameworks act to neutralize the negative environmental consequences of traditional modes of transport, particularly those of routine or habitual passenger cars that are largely accountable for air pollution and energy use (Dalton et al., 2013). Governments and regulators must enact adaptive policies that promote integration of new mobility solutions. These include the establishment of incentives for electric vehicles, investment in charging networks, and promotion of research and development of alternative transport technology. Through the provision of a favorable regulatory environment, stakeholders can promote market development, encouraging manufacturers and consumers to take up sustainable practices while addressing the pressing pressures of climate change. Aside from this, clarity in regulatory policy is the key to executing successful smart mobility programs. Policies must not only target the construction of infrastructure but also harmonize with safety and environmental regulations to ensure free movement of electric and autonomous cars. While the market is poised to make a shift toward battery electric vehicles (BEVs), it is important that regulation is forward-thinking and looks ahead to see what might happen in terms of future developments and challenges (Dalton et al., 2013). It is important to coordinate efforts between different levels of governments, private interests, and community in order to support measures for sustainable transportation solutions. The policy and regulation design that will set the tone for the deployment and use of smart mobility technologies will significantly influence public acceptance and trust, and thus the success or failure of smart mobility policies. Lastly, the convergence of telecommuting trends and smart mobility policy accentuates the evolving nature of work and commuting behaviors in the modern economy. The shift towards telework, with an accelerated growth in the number of teleworkers, carries a prospect of reducing the demand for conventional transportation infrastructure. The shift calls for associated modifications of mobility policy to meet new citizen commuting needs in the direction of an environment-conscious approach to transport. Regulatory frameworks should thus encourage coordination between employers, urban planners, and tech developers towards the development of cutting-edge solutions such as variable transport, which can be applied directly to fight congestion and pollution. Through understanding and application of these dynamics, policymakers will enable a more efficient and resilient transportation system, which also fosters an economic and environmental resilience space.

9.5.1. Government Initiatives for Sustainable Mobility

Government initiatives are a key element in creating sustainable mobility as cities strive to build effective and eco-friendly transport systems. An example is through the creation of smart city plans that are made possible by innovative public procurement practices. Through mechanisms such as the Public end Users Driven Technological Innovation (PDTI), governments can target urban issues related to mobility head-on. This pre-emptive strategy encourages the design of solutions through interaction with technology developers, culminating in new areas such as robotic urban transit. The new merging of such technologies aims at enhancing public transport services while reducing the associated carbon footprint for traditional modes of transport, consequently serving to enhance sustainable urban mobility programs and serving as a model for other cities to follow (Monte et al., 2017). The use of intelligent technologies also complements government efforts towards sustainable mobility. Information and communication technology (ICT) programs establish the foundation for evidence-informed decision-making that addresses metropolitan transport issues. Data creation and aggregation regarding traffic patterns, public transport need, and customer behavior enable governments to inform strategic planning that aligns with sustainability goals. Analysis software that interprets this data assists the authorities in identifying inefficiencies and developing specialized solutions. In addition, organizational structures founded upon stakeholder collaboration—government agencies, private sectors, and civil society—fuel innovation, such that smart mobility solutions are practical and efficient in alleviating congestion and environmental impacts (Stahl et al., 2017). In addition to enhancing public transport, government sustainable mobility initiatives typically involve alternative transport-friendly schemes such as cycling and walking. With investment in the development of safe, accessible, and networked multi-modal transport systems, public authorities can encourage their citizens to transition from reliance on fossil fuel-based vehicles to more environmentally sustainable alternatives. Pedestrian-friendly landscapes and dedicated cycle lanes in urban design facilitate this shift. In addition to that, electric vehicle subsidies and car-sharing schemes can mitigate barriers to adopting sustainable mobility patterns. These kinds of integrative measures offer examples of the roles of governments in facilitating an environment in which sustainable mobility is a viable and desirable option for metropolitan residents, ultimately making urban locations all over the globe more sustainable and livable (Figure 9.14).

Figure 9.14. Government initiatives for sustainable mobility.

Source: https://jugnoo.io/wp-content/uploads/2022/03/Govt-initiatives-future-of-mobility-1-1024x576.png.

9.5.2. Role of Urban Planning in Mobility Solutions

Urban planning plays an important role in the creation of successful mobility solutions in the setting of smart and sustainable cities. Through strategic land use and infrastructure planning, urban planners can enhance accessibility, reduce congestion, and encourage sustainable modes of transport. An integral component of this is the integration of Information and Communication Technologies (ICTs), which aid in the collection and analysis of transport information, thereby supporting decision-making processes. By enhancing public participation and collaboration among various actors, urban planners create multimodal transportation systems that cater to the needs of various users. In addition to promoting efficient movement, such an integrated system is conducive to enhanced public health and urban livability, thus upholding broader sustainability goals as outlined in several smart city proposals (Stahl et al., 2017). In addition, the advent of electric cars presents an excellent chance for urban planners to revolutionize mobility solutions. With cities yet to deal with increasing demand for eco-friendly transport, it is crucial to introduce robust EV infrastructure. Urban planners must address the nuances of integrating charging points into existing cityscapes to ensure mass adoption of electric vehicles. These involve public-private collaborations and collaboration between municipal administrations and private developers to provide the necessary infrastructure. Significantly, as highlighted by recent research, effective EV deployment strategies must consider equity, accessibility, and sustainability in addition to addressing the needs of the environment, so that all members of a given population gain from cleaner transport (Fox-Penner et al., 2019). Finally, the role of urban planning extends beyond short-term issues of infrastructure; it encompasses visions for long-term, sustainable urban growth. Sound urban planning directs the development of walkable neighborhoods, the facilitation of public transit systems, bike trails, and pedestrian paths that encourage non-motorized transportation. In designing inhabitable environments that promote sustainable approaches, planners can significantly lower reliance on single-occupancy vehicles, ultimately reducing carbon emissions and the city's air quality. In addition, the joint nature of urban planning allows for the application of new technologies and data-based strategies, making society adaptive and resilient in the face of shifting mobility challenges. This integrated planning strategy not only promotes environmental goals but also enhances the quality of urban residents' life, within the overall schemes of smart mobility solutions (Stahl et al., 2017).

9.5.3. Incentives for Adopting Smart Technologies

The transition to sustainable mobility solutions is expedited by the deployment of smart technologies, with various incentives that bring stakeholders from across industries. Some of the most promising reasons for embracing such innovations include the drastic reduction in environmental impacts, particularly urban air quality and energy consumption. Under the imperative of rampant air pollution caused mainly by conventional transport systems, alternative options such as battery electric vehicles (BEVs) emerge as viable options that reverse these effects. It is noted in research that the transport industry remains the prime cause of environmental degradation, positioning smart technologies as central agents of initiating sustainable practices (Dalton et al., 2013).

Figure 9.15. Incentives for adopting smart technologies.

Source: https://geospatialmedia.s3.amazonaws.com/wp-content/uploads/2024/01/Nov-Dec18.jpg.

Thus, government and industry incentives for tax rebates, grants to purchase electric vehicles, or incentives to build efficient charging networks are essential to make consumers and businesses invest in these technologies. Economic advantages of adopting smart technologies also substantiate the need for their widespread adoption. Initial investment in electric vehicles and smart transport systems can be recovered through extended fuel and servicing savings, effectively in the form of lower total cost of ownership to consumers. Furthermore, as urban areas fight to contain increasing traffic jams and emission limitations, smart systems have innovative solutions to enhance operational efficiency to the utmost. For instance, smart grids of electric vehicles and real-time traffic management systems can yield considerable return on investment to private and public investors. Not only are these incentives attractive to individual buyers but also they induce collective investment in infrastructure development, which renders cities better placed to absorb the anticipated demand for electric vehicles and associated services (Fox-Penner et al., 2019). Moreover, having cross-public-private partnerships is crucial in order to successfully implement smart technologies in cities. By collaborating with stakeholders like government, technology developers, and urban planners, cities are able to create a concerted effort towards sustainable mobility. Shared resources and expertise can be utilized through partnerships to enhance the efficiency of incentive programs to embrace smart technologies. Furthermore, cities that promote the development of infrastructure for electric vehicles will also be more likely to witness economic development and employment opportunities, highlighting the several benefits involved in adopting smart mobility solutions. Incentives to facilitate such collective endeavors would not only stimulate the adoption of sustainable technology but also stimulate more widespread change in urban infrastructure towards resilient and intelligent future mobility systems (Dalton et al., 2013) (Figure 9.15).

9.6. CHALLENGES AND BARRIERS

Adoption and implementation of sustainable and intelligent mobility solutions are confronted with numerous challenges and hurdles that undermine their effectiveness and adoption. The lack of comprehensive infrastructure capable of supporting cutting-edge transportation technologies represents one of the most imperative challenges. Urban centers are typically plagued by outdated systems that cannot support evolved mobility solutions, leading to inefficiencies and added congestion. In addition, public institutions usually lack awareness of the technological potential that exists to correct these urban mobility problems and consequently generate resistance to investing in needed evolution. These infrastructural limitations hamper the successful application of sustainable solutions, creating a gap between conceptual innovations and practical deployment in urban centers. Moreover, funding constraints seriously impede the design of sustainable mobility schemes. Urban areas typically possess budget restrictions that restrict their ability to invest in required technologies and infrastructural development. The funding of mobility projects is often fragmented, with other rival parties seeking limited funds. Competition in such situations can result in the establishment of short-term requirements in place of long-term sustainability goals, exacerbating transport challenges in urban areas. The creativity of Innovative Public Procurement instruments, particularly those for the promotion of technological innovation, represents an opportunity to reduce some of these economic limits. Unless appropriate frameworks exist that facilitate the involvement of public institutions and private sector innovators, the leverage for effective solutions remains largely unexploited (Browne et al., 2014). Last but not least, the social and cultural dimensions of mobility in the city cannot be dispensed with when addressing the obstacles to sustainable solutions. Public opinion towards new technologies such as robotics often involves suspicion and resistance to change, possibly inhibiting their adoption and usage in normal life. The concept of the Smart City is also reliant to a large degree on citizens embracing smart mobility systems, but if stakeholders believe that they are intrusive or unwanted, they will be opposed to embracing them. Addressing these problems involves a comprehensive public outreach policy informing citizens of the benefits of sustainable mobility and how technology can be used to enhance urban life. Ensuring awareness and appreciation of new solutions will enable cities to overcome opposition and build receptiveness to embracing change that is fundamental to achieving sustainable and smart mobility outcomes (Monte et al., 2017).

9.6.1. Financial Constraints in Implementing Solutions

The implementation of eco-friendly and intelligent mobility solutions is generally threatened by various budget constraints, which have the potential to greatly impede progress toward the deployment of clean transport systems. The biggest among these hindrances is the high cost of investing in new technology and infrastructure. For instance, it takes massive investment not only on the vehicles but also on infrastructure such as charging points and garage facilities to switch to electric or alternative fuel vehicles. This expense could be daunting for small towns or developing regions with already minimal budget planning. In such a case, scarce financial resources can push policymakers to prefer immediate infrastructural needs to environmentally friendly long-term goals, thereby reinforcing a cycle of dependency on conventional mobility alternatives that are not ecologically prudent. Furthermore, rural-urban differences in financial capacity also play a role in the speeding up of the uptake of sustainable mobility alternatives. Cities also receive more investment and enjoy economies of scale, thus it is simpler to raise capital for large projects. Their rural counterparts, on the other hand, have limited access to funds because their reduced population densities and lower tax bases limit the amount of necessary innovations they can undertake. For small island developing states (SIDS), where there are even fewer resources, the task of finding sustainable mobility solutions is that much harder. SIDS' unique financial characteristics and exposure to climate change highlight the necessity of strategic investments in order to facilitate zero-carbon mobility systems by 2040, as recent literature has called for innovative funding mechanisms to overcome these budget constraints (Blechinger et al., 2020). The institutional arrangements under which these financial decisions are made can also further impede the adoption of sustainable mobility solutions. Poor coordination between the public and private sectors has the tendency to lead to disconnected initiatives that cannot attract investment. Small and medium-sized towns in various regions, for instance, will not be able to use the innovative potential necessary to improve their mobility systems due to bureaucracy and absence of overall policy environments. It has been revealed through evidence that governance facilitation, adoption of technology, and sustainability practices are essential to ensure innovation in such environments (Melé et al., 2020). Hence, transcending the budgetary constraints by effective governance and collaborative cooperation can result in smarter implementation of smart mobility solutions with an ability to transition towards more sustainable transport systems complementing both local and global aspirations (Figure 9.16).

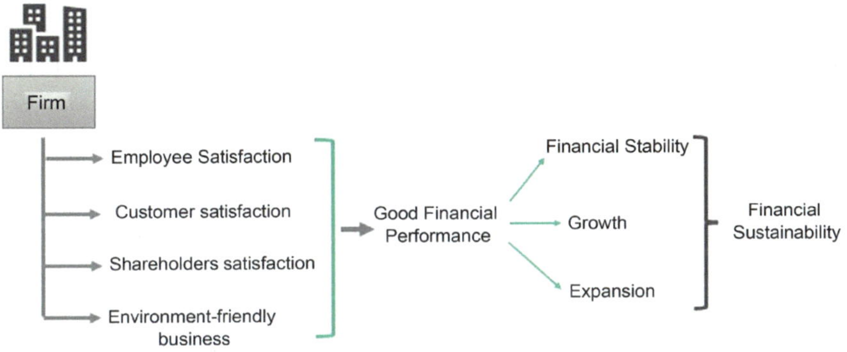

Figure 9.16. Financial constraints in implementing solutions of sustainable.

Source: https://wallstreetmojo-files.s3.ap-south-1.amazonaws.com/2023/01/Financial-Sustainability.png.

9.6.2. Public Resistance to Change

Society resistance to change is a major barrier to the embracement of sustainable and intelligent mobility solutions. People are generally against alterations to tested-and-proven habits and living patterns, since they are primarily based on uncertainty. The current transport facilities are a level of familiarity, therefore causing fear regarding new developments. Public opinion tends to favor the status quo as seen in many studies and takes time and effective communication strategies to alter this. Engaging people in conversations about new mobility solutions will make them feel they are the owners of the idea, ultimately assisting in silencing resistance. Moreover, by transparently making visible the benefits and imperative for change, there is potential for capturing more public support, hence facilitating the transition towards more sustainable transport modes and urbanization. Secondly, the complexity of mobility policy can multiply suspicion among the public. In jurisdictions such as Flanders, for instance, matters are exacerbated by decentralized governance structures, under which local governments possess a lot of power but are inclined to act in rigid formulas. Such rigidity has the effect of creating the perception of ineffectiveness and inefficacy in addressing mobility problems. To present stakeholders with bureaucratic inertia leads them to feel powerless and question leadership. The disunited nature of neighborhood and more central transportation plans can create doubt about the benefits of change plans, solidifying existing opposition among groups. Intriguingly, building and promoting new bottom-up proposals can demonstrate the potential for effective solutions, which can abate some of the difficulties presented by entrenched systems and trigger more action. Lastly, the public opposition dynamic must be framed in terms of local conditions and capabilities. Some regions, such as small island developing states (SIDS), might possess unique opportunities for change due to their island geographies and specialized economies. As emphasized in recent literature, these places can leverage their environments to advance the rapid electrification of transport through targeted investments. But this is most often lost due to a lack of public awareness and understanding of the sustainable alternatives that they are able to avail. Educating the community regarding efficient models, such as shared mobility systems, has the ability to spur enthusiasm and participation. By establishing the success of such projects and correlating them with regional economic benefits, policymakers can alter the mindset of the people, thereby lessening resistance to change and building a coordinated movement towards low-carbon mobility options (Blechinger et al., 2020; Boelens et al., 2015).

9.6.3. Technological Limitations and Infrastructure Needs

As metropolitan cities strive for sustainable mobility solutions, issues of extreme technological limitations always slow them down. Rapid urbanization has placed immense pressure on existing infrastructures, thereby necessitating creative out-of-the-box solutions. The solutions, however, are often stumped due to legacy technology and investment in key systems. For instance, the majority of metropolitan cities remain reliant on legacy traffic systems that do not directly integrate with new smart technologies. It is not just a question of rolling out new technologies but ensuring that the underlying infrastructure will be able to sustain such developments. The importance of integrated technological frameworks becomes necessary in order to meet the specific mobility issues of city dwellers. Thus, cities must focus on a policy of double investment: the technology upgrade of already existing technologies and the development of new infrastructure that can accommodate smart solutions, ultimately optimizing urban transport networks (Melé et al., 2020). In addition, the integration of smart mobility solutions necessitates profound understanding of the specific infrastructure needs of different urban contexts. Not all cities are capable of deploying advanced technological systems, and this understandably generates inequalities in the quality of mobility as well as accessibility. These inequalities necessitate strategies that respect local characteristics and constraints. The focus on the deployment of technology to alleviate mobility challenges has culminated in promoting smart cities; however, this concept leans towards overlooking the central role that existing infrastructure plays in making it a possibility. Urban planners must link technological innovation with functional infrastructure expansion to be in a position to address growing urban population needs effectively. Emphasizing local context not only facilitates the utilization of innovation but also reinforces collective engagement, thereby translating into inclusive and sustainable mobility solutions (Melé et al., 2020).

Figure 9.17. Technological limitations and infrastructure needs.

Source: https://novatech.net/wp-content/uploads/2023/10/Top-5-Challenges-in-IT-Infrastructure-and-How-Infrastructure-Management-Can-Help.png.

Finally, addressing the intersection of technological limitations and infrastructure needs is essential in facilitating effective deployment of sustainable mobility solutions. When cities go more towards smart technologies to facilitate efficient public transport and reduce congestion, they must overcome the biggest challenges posed by outdated systems. Deploying new technologies such as autonomous cars or real-time data analytics requires existing infrastructure to be heavily upgraded. This paradigm shift calls for collaboration among various players like policymakers, urban planners, and business stakeholders to make sure that the development of infrastructure is strategically aligned with technology goals. Through developing a framework that fosters innovation while, at the same time, addressing the infrastructure deficit, cities can open up channels towards sustainable, resilient, and smart transportation systems. It is only through these well-rounded strategies that city centers can truly address the mobility needs of their inhabitants and welcome the future of transport (Adderley et al., 2015) (Figure 9.17).

SUMMARY

In sum, the investigation of sustainable and smart mobility solutions reaffirms the imperative of innovative strategies to improve urban transport systems. As cities expand, traffic congestion, air pollution, and energy use issues become more and more critical. The case study evidence, for example, ASAYISH Township study, indicates the potential of intelligent transport systems, electric cars, and shared mobility services to transform the urban landscape. The use of these smart transport solutions has a dramatic impact on increasing the efficiency and sustainability of small metropolitan transport, thereby influencing policy priorities that respond to environmental wellness and city resilience (Aslamy, 2023). Lastly, there is a need for collective action of stakeholders, including the government, transport planners, and people, in leading the way to the future of cleaner and smarter mobility systems. The transition towards battery electric vehicles (BEVs) is one important element of sustainable transport schemes. Although the BEV system holds potential in reducing environmental impacts, there are still challenges that need ongoing research and development. Importantly, issues surrounding battery production, charging infrastructure, and lifecycle impacts of electric vehicles necessitate shared attention from manufacturers and policymakers. Technological innovation of BEV technology takes a historical trajectory with varied stakeholder concerns and priorities at various scales, as highlighted in the literature. The Focus-Attention-Scale-FAS perspective helps us to deduce a more profound insight into market evolutionary dynamics in enabling sustainable transport choices (Dalton et al., 2013). Therefore, promoting cooperation between automakers, researchers, and governments is imperative in preparing for challenges and achieving mass adoption of BEVs. Simply put, balancing smart and sustainable mobility solutions is not an ideal vision but a compelling course towards more human-friendly cities. The complex interlink between technology, urban planning, and public policy presents a strange fact where development can be realized, only when there is a clear framework for its realization. Development of intelligent transport systems needs to be prioritized by policymakers to ensure accessibility and equity among all citizens of the city are made available. Cities can in fact bank on the prospects of new paradigms of mobility that favor the environment as well as public good by adopting cooperative frameworks. The mobility future in cities relies on our capacity for innovation and learning; hence, addressing the complex issues that lie ahead will be a prerequisite to realizing a sustainable mobility vision that addresses the needs of generations to come (Aslamy, 2023; Dalton et al., 2013).

While investigating the present situation of sustainable and smart mobility solutions, there is no doubt that urban electric vehicle (EV) deployment is a part of sustainable mobility. With the global shift towards electric mobility, the majority of cities are bearing the burden of rising EV demand. This necessitates an overall approach to enhance EV infrastructure in urban areas. Urban cities can also play a fundamental role through collaboration with private developers and utility companies to offer the necessary support to charge EVs. This is important in propelling increased adoption of electric mobility, which is vital to reduce urban carbon levels as well as improve environmental sustainability (Fox-Penner et al., 2019). Moreover, cities are being more and more set up as climate action leaders in an environment of hamstrung national efforts within America. Numerous cities, from Los Angeles to Boston, are going further to reduce greenhouse gas emissions through innovative mobility measures. The collective effect of city actions demonstrates a proactive response to climate change and exemplifies the capacity of city governments to implement meaningful policies. When cities invest in projects that enhance their mobility solutions, they are contributing to an overall climate action plan that emphasizes the role of city-level leadership in global sustainability (Cleveland et al., 2019). These projects not only help minimize emissions but also help drive economic development through the development of green technology. Lastly, the argument for sustainable and smart mobility choices highlights cities making bold action in ratcheting up their climate mitigation ambition. The evidence brings forward that city mobility plans have to be remodeled and reimagined, integrating new technologies and green approaches to deliver real emissions reductions. Moreover, ongoing investment in infrastructure is essential in facilitating a transition toward more sustainable transport systems. By prioritizing projects grounded in electric mobility and climate-resilient city planning, the cities have the capacity to influence large-scale national and global trends of sustainability and thereby transform their roles as passive players in climate action into front-line players in a quest for a more sustainable future (Cleveland et al., 2019).

REFERENCES

1. Adderley, S., Brindley, C., Larson, M. J., Lichy, J., et al. (2015). 'Smart cities' –Dynamic sustainability issues and challenges for 'Old World' economies: A case from the United Kingdom. https://core.ac.uk/download/42137828.pdf (accessed on 21st October 2025).
2. Alexander, S., Appleyard, B., Garrett, M., Kott, J., et al. (2019). Housing and mobility toolkit for San Mateo County. https://core.ac.uk/download/215422910.pdf (accessed on 21st October 2025).
3. Ambrosino, G., Gini, S., & Nelson, J. D. (2016). The pivotal role of public transport in designing the integration of mobility services and in operating MaaS offer: The concept of Shared Mobility Centre and the experience of Arezzo. https://core.ac.uk/download/55284735.pdf (accessed on 21st October 2025).
4. Aslamy, Y. (2023). Smart mobility solutions for efficient and sustainable transportation in small metropolitan areas: A case study of ASAYISH township. https://core.ac.uk/download/588493759.pdf (accessed on 21st October 2025).
5. Bezruchonak, A. (2019). Geographic features of zero-emissions urban mobility: The case of electric buses in Europe and Belarus. https://core.ac.uk/download/225150005.pdf (accessed on 21st October 2025).
6. Blechinger, P., Creutzig, F., & Soomauroo, Z. (2020). Unique opportunities of island states to transition to a low-carbon mobility system. https://core.ac.uk/download/322792155.pdf (accessed on 21st October 2025).
7. Boelens, L., Lauwers, D., & Van Brussel, S. (2015). Evolution of mobility governance in Flanders: Opening up for bottom-up initiatives or suffering from lock-in? https://core.ac.uk/download/55784060.pdf (accessed on 21st October 2025).
8. Bolea Monte, Y., Casanovas Garcia, J., Grau Saldes, A., Puig-Pey Clavería, J., et al. (2017). Public entities driven robotic innovation in urban areas. https://core.ac.uk/download/224975022.pdf (accessed on 21st October 2025).
9. Browne, M., Leonardi, J., Allen, J., et al. (2014). Internal report cluster 1: Urban freight innovations and solutions for sustainable deliveries (3/4). https://core.ac.uk/download/322492286.pdf (accessed on 21st October 2025).
10. Carter, S., Mason, C., & Tagg, S. (2008). Business begins at home. https://core.ac.uk/download/9559147.pdf (accessed on 21st October 2025).
11. Ceschin, F., & Vezzoli, C. (2008). Product service systems in the automotive industry: An alternative business model for a sustainable satisfaction system. https://core.ac.uk/download/9633146.pdf (accessed on 21st October 2025).

12. Cleveland, J., Fox-Penner, P., Hatch, J., Silano, M., et al. (2019). Cities and climate change: Strategic options for philanthropic support. https://open.bu.edu/bitstream/2144/39182/1/CitiesStrategicOptions_Hatch_Cleveland-et-al-_Final_3.12.19b.pdf (accessed on 21st October 2025).
13. Copeland, R., Crespi, N., Rehmani, M. H., Saleem, Y., et al. (2019). Internet of Things-aided smart grid: Technologies, architectures, applications, prototypes, and future research directions. http://arxiv.org/abs/1704.08977 (accessed on 21st October 2025).
14. Dalton, R., ElBanhawy, E., & Thompson, E. M. (2013). Eras of electric vehicles: Electric mobility on the verge. Focus Attention Scale. https://core.ac.uk/download/17299059.pdf (accessed on 21st October 2025).
15. Fox-Penner, P., Hao, J. (R.), Hatch, J., Helveston, J. P., et al. (2019). The critical role of public charging infrastructure. https://open.bu.edu/bitstream/2144/39112/1/MeltingIceBook_Web_FINAL.pdf (accessed on 21st October 2025).
16. Gillis, D., Lauwers, D., & Semanjski, I. (2015). How to monitor sustainable mobility in cities? Literature review in the frame of creating a set of sustainable mobility indicators. https://core.ac.uk/download/55765085.pdf (accessed on 21st October 2025).
17. Gregor, P., Sloan, D., & Stratford, J. (2006). Using multimedia to enhance the accessibility of the learning environment for disabled students: Reflections from the Skills for Access project. https://core.ac.uk/download/14392.pdf (accessed on 21st October 2025).
18. Hoor, M., & Schwedes, O. (2019). Integrated transport planning: From supply- to demand-oriented planning. Considering the benefits. https://core.ac.uk/download/286083113.pdf (accessed on 21st October 2025).
19. Lauwers, D., & Papa, E. (2015). Smart mobility: Opportunity or threat to innovate places and cities. https://core.ac.uk/download/55823526.pdf (accessed on 21st October 2025).
20. Law, A., Law, P., & Perryman, L. A. (2013). Developing sustainable business models for institutions' provision of open educational resources: Learning from Open Learn users' motivations and experiences. https://core.ac.uk/download/18306493.pdf (accessed on 21st October 2025).
21. Mercadé Melé, P., Molina Gómez, J., & Sousa, M. J. (2020). Technology, governance, and a sustainability model for small and medium-sized towns in Europe. https://core.ac.uk/download/322918717.pdf (accessed on 21st October 2025).
22. Stahl, E., & Dupuis, N. (2017). Trends in smart city development. https://core.ac.uk/download/83111937.pdf (accessed on 21st October 2025).

CHAPTER 10

POLICY, ETHICS, AND FUTURE TRENDS

Contents

INTRODUCTION

Transportation engineering stands at a decisive juncture marked by rapid advancements, evolving policies, and ethical considerations that need innovative solutions. With urbanization expanding and populations becoming increasingly mobile, demands for efficient and environmentally-friendly transportation systems become more urgent. Policymakers and engineers must navigate through an interconnected web of issues from climate change to infrastructural safety and technology integration. The confluence of these factors mandates examination of existing paradigms and the potential of pending trends. As such, an initial foray into the practice of transportation engineering must address the effect of new technologies, regulatory reactions, and ethical issues that define this dynamic field, with the interfacing points of pragmatism and innovation in engineering solutions.

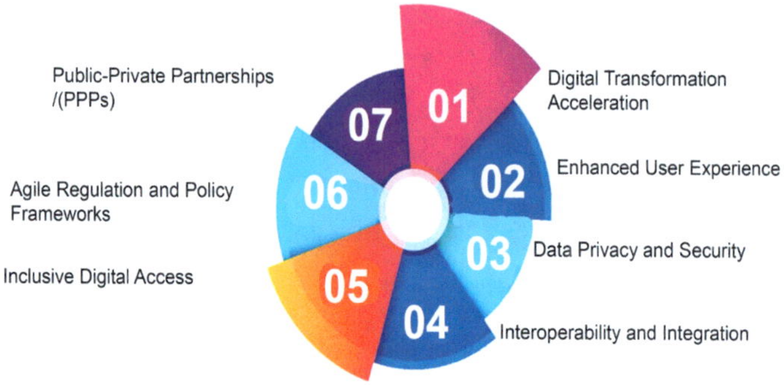

Figure 10.1. Policy, ethics, and future trends.

Source: https://fastercapital.co/i/E-government-policy-and-regulation-Unlocking-Business-Opportunities--E-Government-Regulations-Demystified--Future-Trends-and-Recommendations.webp.

The complexities of transportation extend beyond the mere logistics; they involve a diverse range of social and environmental concerns. As put forward in the recent conferences, for instance, the Fourth Annual World Class Supply Chain Summit, firms must not just adhere to normal supply chain fundamentals but also be adaptive in responding to changing market demands and environmental imperatives (Haughton et al., 2019). This principle is appealing to the landscape of transport engineering, where the mantra is flexibility. Whether through the adoption of intelligent technology to foresee traffic volumes or the design of greener forms of transport, engineers are asked to balance efficiency with sustainability. As the transport sector grapples with such pressing concerns, there is a call to consider policy paradigms that not just promote technological innovation but also emphasize the public interest and ethical ideals of engineering practice. Besides, world changes, such as the effects of Arctic change, further underscore visionary transport policies (Ashcraft et al., 2018). These developments impact not only local dynamics but also extend to have wide-ranging implications for trade, environmental balance, and socio-economic order. With shifts in shipping lanes due to climate change, transport engineers must counter such development with innovative solutions in infrastructure that consider these far-reaching impacts. The incorporation of new paradigms in transport engineering therefore has to include interconnectivity; there must be multidisciplinary cooperation between engineers that transcends traditional disciplines. Addressing these multifaceted issues requires a new strategy integrating ethical aspects, sustainability, and technological advancements, opening the doorway to a sustainable transport system capable of meeting future demands (Figure 10.1).

Definition of Transportation Engineering

Transportation engineering is a very diverse discipline that is primarily focused on planning, designing, building, and operating the transport infrastructure. It includes roads, bridges, railroads, airports, and public and private transport systems, all aimed at the safe and convenient movement of goods and people. The key role played by transportation engineering is to satisfy the growing need for sustainable transport modes, particularly in the context of increasing urbanization and environmental concerns. Through the inclusion of policy and ethics within its design, transportation engineering can create not only quick streams of traffic and improved infrastructure but also safety and quality of life among citizens. This integrative approach enhances more inclusive objectives, such as reducing carbon emission and congestion, thereby affecting global sustainability projects. Furthermore, transportation engineering plays a fundamental role in the design and implementation of new technologies that enhance transport systems. As issues such as congestion, safety, and the environment become more pressing, engineers are increasingly being asked to integrate advanced solutions such as intelligent transportation systems (ITS), which leverage real-time data to optimize traffic flow. As well, novel technologies such as delivery drones and air taxis are reshaping personal mobility expectations, playing in the low-altitude space, and changing dominant transportation paradigms. The intersection of such technologies with traditional engineering processes warrants precise examination of their social and environmental impacts, and therefore it is essential to adopt a comprehensive approach to policy-making that considers both expected advantages and potential hindrances to implementation (Biehle et al., 2020). In addition, the ethical dimension of transportation engineering cannot be overemphasized because decisions in this field have direct consequences on public health and the environment. Equity of access to transportation, special population effects, and environmental stewardship are issues that must be carefully balanced. As engineers strive to design and implement transportation systems, they are confronted with unraveling complex interactions among technology and ethics. Green supply chain management (GSCM) practices are an instance of how transport engineers can help bring about a greener future through adopting green technologies and practices. This alignment of engineering practice with social and environmental ethics not only contributes to firm performance but also responsible decision-making in transportation (Saenchaiyathon et al., 2019). These integrative methods are crucial in the development of transportation engineering to be innovative and ethically oriented (Figure 10.2).

Figure 10.2. Transportation engineering.

Source: https://www.statiker.com.au/wp-content/uploads/2021/07/shutterstock_575630101.jpg.

Importance of Policy and Ethics in Transportation

Policies and ethics spearhead the future of transport through the management of the challenges of implementation, sustainability, and equity. In a rapidly changing industry with background advances in technology, such policy frameworks direct the transportation system to be not just effective but equitable and green. For instance, increased openness of new sea routes due to climate change necessitates robust policies responding to these changes while retaining ecological integrity. The University of New Hampshire event indicated a central imperative for expertise in responding to the economic and environmental implications of Arctic change, illustrating how strong policy can minimize negative effects and enhance adaptive strategy that enhances community resilience (Ashcraft et al., 2018). It is thus imperative that stakeholders in transportation engineering set great priority on policy formulation that ensures the principles of ethics, thereby resulting in future development that is sustainable. Ethics in transport goes beyond regulation compliance; it involves the role of professionals to champion technologies and practices that are of greater benefit to the public in general. For example, when transportation engineers grapple with the integration of emerging technologies such as driverless cars, ethical dilemmas with regard to safety, privacy, and equity arise. There is a need for a framework that facilitates dialogue among experts, policymakers, and citizens to effectively address these challenges. Fourth Annual World Class Supply Chain Summit emphasized the need for diverse viewpoints in handling new concepts, particularly when dealing with developing adaptive, data-driven strategies that prioritize societal needs as well as operational efficiency (Haughton et al., 2019). With the increasing technological advancement in transport systems, experts must remain committed to those ethical values that are sustainable in accordance with the global values of the societies they are in. Also, the interface between policymaking and ethics carries a profound influence on public perception and trust within the transportation system. Whether a transport project is valid may be based on openness in policymaking and ethical virtues embedded in it. With greater stakeholder engagement being more crucial, maintaining an enabling environment may raise awareness and accommodation by the public. This evolution manifests in ongoing changes in supply chain paradigms, where customer needs responsiveness and environmental awareness are of primary concern. Transportation engineers therefore need to navigate these dynamics cautiously, so that policies and ethical guidelines are not only robust but responsive to the changing environment as well. By prioritizing these areas, the transportation sector can guarantee that it not only meets present needs but also looks forward to the future with honesty and vision to impending challenges.

Overview of Future Trends in the Field

One of the most critical trends in transportation engineering with the advancement of transportation engineering is at the intersection of technology and sustainability. Development in autonomous vehicles, smart infrastructure, and advanced traffic management systems is revolutionizing how cities manage transportation. These technologies have the ability to make transport more efficient, reduce congestion, and enhance safety, all critical considerations in the context of urbanization and ecological concerns. Moreover, the utilization of renewable energy in transport systems can help minimize carbon footprints while enhancing energy resilience. Recent workshops suggest that stakeholders have to navigate sophisticated regulatory regimes and public policy regimes in an effort to exploit these technological advances and ensure that they are harmonized with ethical concerns and social equity (Ashcraft et al., 2018). Hence, the future of transportation engineering must walk a fine line between innovation and good governance so as to accrue benefit for all communities. Another trend emerging in the domain of transportation engineering is the increasing emphasis on multimodal transport networks. Multimodal transport networks facilitate harmonious interaction of various modes of transport—like public transport, cycling, walking, and driverless vehicles—allowing an efficient and friendly mobility experience. Cities are embracing the necessity of reducing single-occupancy vehicle use, which requires new infrastructure designs and policy interventions that encourage alternative modes of transportation. In addition, increasing the use of electric vehicles and enhancing cycling infrastructure can help offset adverse environmental impacts as climate change remains an issue. Academic papers have referred to the success of urban areas where multimodal networks were prioritized, demonstrating how they facilitate not only improved transport flow, but improved equity and accessibility (N/A, 2017). By promoting a wide range of transport modes, transportation engineering can contribute significantly towards supporting sustainability goals as well as social equity. Furthermore, the availability of big data analytics will transform the practice of transportation planning and management. As sensors and networked devices have spread at an ever-faster rate, engineers possess more data than ever before on traffic flow, user behavior, and infrastructure condition. This information can be leveraged to inform more informed decisions on transportation policy, infrastructure investment, and long-term planning initiatives. Predictive analytics allows stakeholders to deploy resources in a more optimal manner and react to evolving user demands more effectively. Convergence workshops also point out that utilization of big data can provide vital insights into how adaptive measures towards climatic changes and emerging city challenges can be ensured (Ashcraft et al., 2018). Thus, the inclusion of data-driven strategies in transportation engineering not only enhances operational efficiencies but also guarantees the application of proactive measures in accordance with broader ethical and policy objectives.

10.1. HISTORICAL CONTEXT OF TRANSPORTATION ENGINEERING

The evolution of transport engineering is inextricably interwoven with the course of history, instigated by society's needs and technological advancement. From the earliest rudimentary routes through to sophisticated infrastructure supporting modern transport networks, each phase reflects an accommodation for the complication in human mobility. Major innovations such as the discovery of the steam engine and construction of railroads not only increased travel speed but also enhanced business and communication. As cities began to expand, they became preoccupied with finding efficient metropolitan transit systems, and hence innovations such as streetcars and subways emerged. This history describes how transportation engineering has continually adapt to accommodate population growth, geographic conditions, and economic objectives, laying the groundwork for the practices in place today and what will be needed in the future. Throughout the 20th century, as the world moved forward, several significant issues were created that would require ethics in transportation engineering. The shift toward greater mechanization and automobile dependency brought challenges like urban sprawl, environmental degradation, and social injustices. The car culture and rapid urbanization increased traffic pollution and congestion, becoming sustainability and social responsibility issues. These were the ethical dilemmas that compelled the transportation engineering profession to steer towards mainstreaming more sustainable policies and practices into its agenda. Current advancements, such as mass transit systems and the emphasis on active transport modes of cycling and walking, echo the lessons of past errors. Thus, a thorough understanding of these historical ethical considerations is required for upcoming transportation policy-making aimed at balancing expansion with environmental wellbeing, mirroring a need for wise engineering practice. In the past few years, the intersection of transportation engineering and emerging technologies has characterized an inspiring milieu that promises to revolutionize the mobility of the future. The integration of Big Data within smart city initiatives exemplifies how modern analytics can enhance transportation systems efficiency and effectiveness. By utilizing data-driven approaches, engineers can optimize traffic flow, reduce congestion, and improve safety measures. The development of autonomous vehicles further revolutionizes this field, with potential implications for policy, public safety, and ethical standards surrounding liability and privacy. Besides, as seen in projects like the one described in Ashcraft et al.(2018), real-time data collection and analysis are key roles in preparation for and action against environmental change and socio-economic forces. As the technologies continue to advance, they are sure to shape not only the engineering practice but also the policy frameworks for transport infrastructure, which will make it a future-oriented means of addressing the challenges as indicated in Doctor et al. (2020).

10.1.1. Evolution of Transportation Systems

Transport system development is very much influenced by the convergence of moral analysis, policy regulations, and technological innovation. Historically, transport has evolved from horse-drawn wagons and primitive routes to advanced railroads, motor roads, and aerial transport systems. With each innovation, there were revolutionary transformations in city planning, human interaction, and economic development. Advancements in technology during the 20th and

21st centuries have prompted the creation of intelligent transportation systems (ITS), which improve traffic management and increase safety through real-time data analysis and communication technologies. On the basis of research at the Fourth Annual World Class Supply Chain Summit, adaptation to shifting consumer requirements and competitive landscapes has underscored the call for responsive supply chain approaches in transport, presaging constant paradigm shift within the field (Haughton et al., 2019).

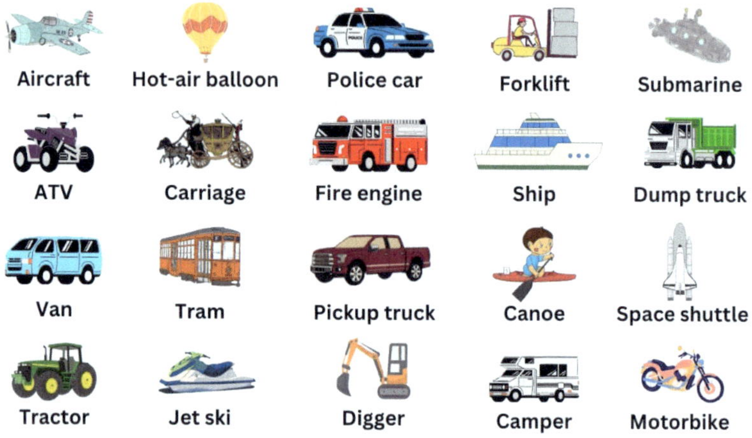

Figure 10.3. Evolution of transportation systems frameworks.

Source: https://i.ytimg.com/vi/1U69NiVJ6V8/maxresdefault.jpg.

The influence of data analysis also reforms the paradigm frames of operating in transport systems, particularly in the evolving concept of smart cities. These modern-day cityscapes leverage Big Data to optimize transport networks, aiming not just for efficiency, but for sustainability and accessibility as well. Leveraging advanced computational intelligence techniques, networks are being designed that can predict traffic patterns, avoid congestion, and optimize public transit. A case study of smart transport discovered that the implementation of a Hierarchical Spatial-Temporal State Machine (HSTSM) can greatly optimize city mobility by creating prediction models for transport demand (Doctor et al., 2020). This discovery marks a significant shift towards evidence-informed decision-making policies that are centered on sustainable outcomes, and the implications on policies ensuring ethical usage of such information. Despite the advantages presented by modern transportation technologies, these ethical concerns of equity, privacy, and environment sustainability continue to persist. As cities become more and more advanced with smart technologies, inequalities are set to become more pronounced, dis-enfranchising low-income groups with the least access to technological ability. Policymakers must address such disparities through equitable and inclusive transportation systems. Also, transport Big Data raises fundamental concerns regarding data privacy and the ethical use of information for commercial and public purposes. The key challenge today is how to create regulatory frameworks that balance innovation and ethics, creating a transport ecosystem in the future that is considerate of human rights while fostering social mobility (Haughton et al., 2019; Doctor et al., 2020). The history of transport is thus inextricably linked with the ethical dilemmas and policy decisions that shape its path (Figure 10.3).

10.1.2. Key Legislative Milestones

The path of transportation engineering has been profoundly influenced by pivotal legislative milestones which represent the changing policy context regulating the industry. Flagship legislation typically arises due to urgent social necessities, especially with regard to safety, sustainability, and infrastructure growth. For instance, the formation of the National Highway Safety Act of 1966 was a turning point in the regulation of road safety at the federal level, initiating a series of steps that saw states need to meet equivalent safety standards. This act provided the groundwork for further developments in auto safety regulation, showing the manner in which congressional efforts can stimulate improvements in technology and engineering practices. The impact of such legislative strides can be realized in the reduction of accident rates and in safer advanced technologies that have continued to define transportation engineering to this day. Apart from safety, environmental concerns have become increasingly important as an area of focus under transportation legislation. The 1990s amendments to the Clean Air Act required the states to develop plans to reduce vehicle emissions, revolutionizing the landscape of urban transportation planning. This shift towards focusing on environmental sustainability is a moral obligation to maintain public health and natural resources while encouraging infrastructure. These guidelines have compelled transport engineering to advance in electric vehicle systems and alternative fuels. The inclusion of such aspects not only complies with legislation but also remedies moral problems on the long-term impacts of transport systems on global warming. By weaving ethical considerations into policy, policymakers have made sure that future transportation engineering advances broader agendas of sustainability. In the coming years, Smart City is a prominent legislative direction poised to revolutionize transportation engineering within cities. As cities experience congestion and environmental degradation, weaving technology and data-driven strategies is becoming ever-more vital to planning more efficient transport networks. Smart City-supportive regulations trigger the adoption of intelligent transportation systems, micromobility sharing services, and eco-friendly city-planning methods. These need to evolve in response to create responsive urban spaces that can manage the requirements of their denizens (Cohen et al., 2019). Second, the embedding of metrics and evaluations into pilot projects enables sustained improvement and scaling-up of effective policies, thereby allowing the valuable lessons derived from such projects to be translated for the benefit of communities as a whole. As policy approaches come to embrace these innovations, transportation engineers will be uniquely positioned to lead the transition towards more intelligent, more sustainable urban mobility solutions.

10.1.3. Impact of Historical Policies on Current Practices

The legacy of previous transport policies on current practice is significant, informing the development of infrastructure and policy-making. The shift from a thought style focused on the widening of road space to one that includes the use of public transport and active travel represents significant policy change. This transformation, contained in Good wins contemplation of transport policy shift in Great Britain, marks the necessity for demand management to counter congestion and damage to the environment. With urban areas continuing to battle against the effects of increasing automobile use, the previous policies are observed to reflect the failure of the previous measures and incriminate the necessity of sustainable options in transport engineering. From the experience of previous governments that failed to innovate, policymakers now can proceed with initiatives that promote multi-modal transport plans, therefore promoting flexibility and resilience in the face of evolving urban trends. Previous policies that emphasized road infrastructure over other transportation modes not only have contributed to growing congestion but also facilitated sustenance of economic and social disparities. The use of car-based urban planning was inclined to displace people who rely on public transport, leading to further disparities. Past project evidence from the past confirms that policies that omit investment in public transport result in increased social consequences, i.e., lower access to basic services by marginalized groups. This reality demands a rethinking of transport systems, emphasizing inclusivity and accessibility. The insights gained from workshops like those held at the University of New Hampshire on the implications of shifts in the Arctic also demonstrate how behaviors in transportation must adapt to address greater economic and environmental concerns in other areas, like New England as it adjusts to altering climate conditions. The intersection of transportation policy and environmental considerations invites space for ethical decision-making in future transportation engineering practice. As history has already shown in the past scenarios, failure to adopt sustainable transport policies has been far-reaching, with implications of accountability and the ethical credibility of policymakers' choices. The emphasis on sustainability now is a response to previous shortcomings, requiring transparent, equitable policies that focus on environmental health and all citizens. Besides, empirical evidence-based projects and stakeholder engagement have the potential to generate innovative solutions for aligning modern transportation practice with evolving ethical expectations. By convergence of past knowledge and forward-looking strategies, transportation engineering can embrace a future where there is social responsibility and sustainability, ultimately reengineering the environment where mobility takes place and delivering a balance between technical advancement and environmental conservation.

10.2. CURRENT POLICIES IN TRANSPORTATION ENGINEERING

New developments in transport engineering have spawned calls for revision of current policies in order to meet emerging trends within logistics and supply chain management. The Fourth Annual World Class Supply Chain Summit emphasized flexibility and rapid response to an increasingly complex environment. Industry experts and researchers claimed that while firms must possess traditional supply chain fundamentals such as building trust-based relationships and robust IT underpinnings, they must now emphasize responsiveness to quickly changing conditions generated by customers' imperatives and competitive pressures. This new focus requires creativity-enhancing policies and flexible transport networks that can adjust fast to fluctuations in market forces without neglecting environmental and social concerns. For this purpose, current policies have to evolve by adding a trade-off between economic performance and environmental sustainability with regard to how these two intersect in transport engineering (Haughton et al., 2019). Not only do current transportation policies have to address operational flexibility, but also the socio-economic implications of global warming, particularly for regions affected by altering climatic conditions. The University of New Hampshires Arctic change workshop placed the broad implications for Northern regions and lower latitude coastal areas such as New England in the spotlight. Disappearing Arctic ice and newly accessible maritime trade routes require transportation policy that leads the way in terms of trade impact, conservation goals, and community adjustment strategies. Policymakers are urged to seek convergence research initiatives that integrate both scientific inquiry and actual implementation. This coming together will aid in crafting policies that, apart from promoting economic growth, build resilience against climate-related disruptions. Proper transportation engineering policies must then demonstrate a systemic understanding of socio-economic and environmental systems to thrive in a rapidly changing world (Ashcraft et al., 2018). Finally, the overall impacts of current transportation policies need an ethical foundation on the principles of equity and sustainability. With growing urbanization and population expansion, transport systems are the key to regulating the access to basic services and opportunities. Policymakers owe it to themselves to bridge investment gaps in infrastructure so that poor communities are not left behind in the move towards more advanced transport systems. Furthermore, the drive for sustainability should neither be narrowed down to seeking technological improvements but also embrace social responsibility and equality. By involving stakeholders and the public in the process of policy-making, transport engineering can more effectively integrate the values and wishes of people. Finally, the making of transport policy will require an attempt to combine ethical ends with effective solutions to deliver lasting benefit on the fronts of society, economics, and the environment (Figure 10.4).

Figure 10.4. Current policies in transportation engineering.

Source: https://engineeringfactsz.com/wp-content/uploads/2023/06/Transportation-Engineering.jpg.

10.2.1. Federal Transportation Policies

Federal transport policies have a critical function in determining the infrastructure and regulatory environment that directs the transportation of commodities and individuals across the nation. They are employed to guarantee safety, efficiency, and fairness in the transport industry, considering the diverse interests of urban and rural nodes. Following the pace of growing technology and increased attention to sustainability, federal regulations have come to encompass programs that create smart transport systems. Such programs encourage the adoption of intelligent transport systems (ITS) that allow for real-time monitoring and management of traffic, greatly increasing the agility of transport organizations. Also, the federal funding alignment for strategic objectives that aim to reduce emissions also reflects a growing appreciation for the role of transportation in addressing climate change. These developments point to the importance of policy frameworks that not only lobby for infrastructure investments but also demand ethical considerations in transportation planning and implementation. With transportation concerns growing more complex, federal legislation must also adapt to the complexity of animal-vehicle crashes, impacting both human safety and wildlife. Examination of these collisions reveals significant implications that extend far beyond immediate fatalities, having implications for ecological balance and economic stability. As Seiler et al. (2003) suggests, auto-wildlife collisions are set against broader conservation practices and demand a holistic approach involving wildlife management, road safety, and education of the public. Federal policy must therefore adopt measures to check such collisions, such as wildlife corridors and adaptive road planning, towards making human transport a synergy with wildlife movement. By prioritizing interdisciplinary actions, policymakers can counterbalance the ethics of transportation impacts on environments while promoting the safety of drivers and animals, displaying the intricate interdependence of transportation engineering and environmental ethics. Furthermore, the effects of changing climates, particularly in the Arctic, are precipitating a re-evaluation of federal transportation policies concerning emerging shipping routes and consequent risks. As highlighted in Ashcraft et al. (2018), emerging sea routes through melting ice offer possibilities and implications for trading, marine ecosystems, and coast-based societies. Policymakers are faced with the dilemma of balancing economic interests with environmental concern, particularly since increased deeper penetration by shipping in these areas may lead to increased ecological degradation and population alteration. To navigate these complexities, federal policy needs to be visionary and agile, learning from convergence research that emphasizes collaboration on multiple dimensions, including economic, environmental, and social models. Multidisciplinary will play a crucial role in shaping predictions for the changing transportation landscape so that policy-making not only addresses current needs but also gazes into the future looking for transportation engineering trends and challenges.

10.2.2. State and Local Regulations

State and local regulations play a vital role in transportation engineering, with the regulatory environment shaping the framework under which infrastructure projects are conceived and undertaken. They dictate from zoning laws and environmental impact studies to security measures and government subsidies. Their significance is particularly acute as cities cope with the consequences of rapid urbanization and climate change. For instance, local planning measures must consider the unique geographical, economic, and cultural characteristics of every locality such that infrastructure growth is geared towards community needs and sustainability goals. In addition, with the emergence of new technology, such as autonomous vehicles and smart traffic management systems, regulators must refine existing policy to both promote innovation and safeguard public interests, helping to highlight the delicate balance regulators must maintain between advancement and prudence. The effectiveness of government regulations at the state and local levels in transport engineering is generally tested by economic factors such as competitiveness and productivity.

Figure 10.5. State and local regulations in transportation engineering.

Source: https://www.maclogistics.co.uk/wp-content/uploads/2025/01/DALL % C2 % B7E-2025-01-10-16.50.22-A-professional-and-detailed-illustration- depicting-smooth-international-shipping-while-navigating-customs-regulations.-The-scene-features-a-shipping-c-1080x675.webp.

There is a state of mutual interdependence between regulatory systems and the environment for doing business such that excessive regulation will slow down economic growth. Reduced competitiveness in Canada, for example, has been due in part to institutional inefficiency as well as overly prescriptive regulations discouraging business innovation as well as productivity (Mayman et al., 2025). As they seek to attract investment and foster development, local administrations are forced to enact policies that spur rather than impede advancement. This requires a re-examination of existing policies to ensure that not only does regulation protect the public good, but also enables the free movement of transportation systems that can keep up with emerging fads and innovation. In the future, transportation engineering will likely be determined significantly by the shifting dynamics of state and local law. As societies look forward to threats related to environmental change, such as those that have been prevalent in the debates about the impact of the Arctic (Ashcraft et al., 2018), it is crucial that regulatory bodies remain responsive and adaptable. This would involve promoting cooperation among engineers, policymakers, and stakeholders to develop holistic plans addressing not just immediate transportation requirements but also long-term sustainability. Regulations in the future need to integrate sophisticated data analytics and modeling technologies to anticipate and counter potential issues, enhancing the resilience of transport systems. As innovation in transport is accelerating, the regulative authority must catch up with it while ensuring that ethical dimensions of new technology are properly thought through, ultimately facilitating sustainable development in this vital industry (Figure 10.5).

10.2.3. Role of International Agreements

International agreements play a crucial role in establishing a framework for cooperation among nations, particularly in the rapidly evolving field of transportation engineering. These agreements serve not only as a conduit for sharing best practices and technological innovations but also as a platform for addressing complex challenges like safety, sustainability, and environmental impact. Through collaboration, transnational treaties enable countries to establish mutual strategies that enhance efficiency in transportation systems while reducing adverse effects on communities and ecosystems. Treaties that can govern maritime navigation are examples which can be used in shipping route management in regard to adjustments caused by climate change, hence the necessity for adaptable frameworks that remain significant as situations change across the world (Ashcraft et al., 2018). Including ethical concerns in global treaties makes them stronger and enforceable. The ethical considerations of transportation engineering involve protecting the public, the environment, and social justice. Including ethical concerns in international policy allows nations to ensure together fair labor practices, equal access to transportation facilities, and equitable disaster relief schemes. Not only are these guidelines used for making policy decisions, but they also encourage stakeholder involvement in decisions that affect local residents. Moreover, as transportation systems converge with technologies such as autonomous vehicles and smart infrastructure, ethical issues are of utmost relevance to consider while analyzing the role of these technologies on social norms and values and in opening the doors to visionary and ethical engineering practices (Franek et al., 2012). The shifting dynamics of international agreements are most likely to dictate the pace for future trends in transportation engineering. With more pervasive global issues, such as climate change and urbanization, comes increased demand for multi-national policy frameworks which will adapt to these shifting circumstances. This adaptability must incorporate agencies across multiple borders in such a way that practice is shared and implemented for maximum gain. Moreover, international agreements can facilitate research and development projects that contribute to solving emergent transportation issues, such as the move towards alternative fuel and incorporation of new technology. Through the cultivation of a global cooperation and information exchange commitment, these agreements will ultimately serve as the basis for sustainable transportation systems capable of thriving in an uncertain future, proving significance to the ongoing transportation engineering policy and ethics discussion (Ashcraft et al., 2018).

10.3. ETHICAL CONSIDERATIONS IN TRANSPORTATION ENGINEERING

In the branch of transportation engineering, ethical considerations play a considerable part in shaping the planning and implementation of transportation infrastructures. Interference of the transportation systems with wildlife populations is one of the problems, where human activities lead to animal-vehicle collisions. Not only are such incidents economically costly, but also their occurrence endangers the existence of some species. For instance, studies indicate that road mortality has a lethal impact on wildlife populations, particularly on sensitive species like badgers, whose high rates of vehicle collisions can ultimately outstrip their reproductive capacities. Engineers should thus embrace a more holistic approach to transportation planning, such as wildlife corridors and showing concern for environmental stewardship, to ensure thus that infrastructure expansion does not occur at the expense of ecosystem balance (Seiler et al., 2003). In addition, the emerging technologies of air taxis and delivery drones raise challenging ethical questions that will have to be addressed by transportation engineers. Because these technologies can potentially revolutionize city mobility through low-altitude flight, they also come with their own set of societal and environmental issues that will need to be evaluated judiciously. Arguments against drones propose that while they hold out the potential for greater efficiency, they could also perpetuate prevailing social injustices and unleash unforeseen environmental factors (Biehle et al., 2020). Discussion surrounding drones involves weighing the detriments and advantages, highlighting the need for scientific evidence to support arguments made about their effectiveness (Biehle et al., 2020). Transportation engineers thus have an ethical duty to ensure that new technology encourages not just technical effectiveness but equity of access and sustainability within the urban environment. Finally, ethical considerations in transportation engineering extend beyond questions of environmental impact and technology integration to larger societal norms, including public participation and accountability. Because transportation projects typically usher in significant impacts on communities, engineers must involve local communities in planning. Community involvement assists in fostering openness and addressing public concerns, ultimately leading to better project outcomes. Community involvement also ensures that the ethical ramifications of transport decisions are considered, enabling a just transition to sustainable mobility. The designers must reconcile the technical aspects of the design with the ethical obligation to serve the greater good of the public, demonstrating how transport is designed as a form of enhancing people's well-being while restricting damage to humans and the environment (Seiler et al., 2003; Biehle et al., 2020) (Figure 10.6).

Figure 10.6. Ethical considerations in transportation engineering.

Source: https://americanprofessionguide.com/wp-content/uploads/2024/08/Ethical-Issues-in-Transportation-Engineering.jpeg.

10.3.1. Safety and Public Welfare

The convergence of public welfare and safety in transport engineering necessitates the blending of man's behavior and environment. With escalated road traffic volumes, there is a growing menace of animal-vehicle accidents whose consequences on wildlife populations as well as human lives can be devastating. The analysis of these accidents displays not only their immediate effects on human health, i.e., death and injuries, but also the broader implication on the environment and species preservation. For instance, research indicates that factors like motor vehicle speed play a significant role in the frequency of wildlife accidents, and this calls for public education campaigns and road development prioritizing safe driving behavior. Hence, public welfare promotion is beyond regulatory measures; it involves safety culture development through education, policy reform, and technological innovation. Beyond the remedies to animal-vehicle conflict risks, the transportation engineering future also needs to face the ethical challenge of technology advancement. The introduction of autonomous vehicles brings possibilities as well as challenges regarding safety and public welfare. For instance, in as much as they claim to reduce accidents caused by human error, they also raise ethical concerns regarding liability and decision-making in certain extreme circumstances. Engineers and policymakers will be faced with setting standards to guarantee such systems prioritize passenger safety above all else and minimize risk to other road users, including pedestrians and cyclists. Furthermore, the application of these technologies should be done cautiously, with all having equal access and potential for socioeconomic disparities in transportation safety taken into account. This equitable perspective is necessary for maintaining public confidence and achieving a sustainable, safe transportation system. Finally, the common theme of safety and public benefit through transportation engineering requires commitment to long-term sustainability within the process of research and development. Recent trends are such that short-term fixes tend to dominate the underlying problems in transport systems, potentially resulting in unintended effects to safety and environmental integrity. Transcending is the need for integrating all-around research designs that take into account social dynamics, economic considerations, as well as environmental aspects to design effective transport policies. By taking a systems approach to transportation, engineers are able to come up with solutions that minimize hazards associated with traffic accidents and create ecological balance. These measures are in line with the goals outlined in the TP Organics Vision Research Agenda, concerning long-term priorities in addressing the imbalances between economy, ecology, and common benefit among modes of transport, and, finally, ensuring a safer world for all using roads and the environment in general (Seiler et al., 2003; Halberg et al., 2008).

10.3.2. Environmental Impact and Sustainability

With the international community battling the effects of climate change and environmental degradation, the transport sector is at crossroads. Evolution and implementation of green measures are crucial to rescuing the environment. Electric vehicles, public transport systems, and better rail systems are great milestones towards sustainability. But all these need to be supplemented with integrated policies that encourage environmentally friendly modes of transportation and address reducing emissions. These measures have to consider regional impacts, such as those described in adaptations required due to Arctic transformations, that potentially shift shipping routes and affect ecosystems much farther south than the northernmost regions. Addressing these related challenges, transportation engineering can evolve to integrate sustainability while accommodating additional growth and economic growth, thus making environmental concerns top of mind when preparing for the future (Ashcraft et al., 2018).

Figure 10.7. Environmental impact and sustainability the transportation.

Source: https://media.licdn.com/dms/image/v2/D4E12AQGK1heVLJZIJQ/article-cover_image-shrink_720_1280/article-cover_image-shrink_720_1280/0/1702298221592?e=2147483647&v=beta&t=CHvK-Deg--a8400mHhd4ZYS19UTHPq7-qwtfmakwGLM.

In addition, the incorporation of green supply chain management (GSCM) in transportation engineering activities plays a significant role in reconciling environmental and operational sustainability. Institutional isomorphism has been shown to significantly influence the application of GSCM practices whereby external forces from customers and the government can push corporations towards greener alternatives (Saenchaiyathon et al., 2019). Furthermore, the transportation entities have to comprehend the importance of cooperation with industry partners in order to optimize sustainability activities. Through sharing pooled resources and expertise, the transport sector is able to effectively reduce its ecological footprint while, simultaneously, acquiring a competitive advantage. This cooperation not only assists in remitting instant environmental concerns but also helps realize overall economic goals, thereby facilitating a positive feedback loop of sustainability and profitability within the industry. Future transportation engineering trends ought to be informed by the ethical values of environmental stewardship. Engineers and policy makers need to concentrate on social justice and inclusivity while designing the infrastructure of transportation so that sustainable models do not increase disparities. In this context, the ethical implications of technological advancements should be considered because uninhibited implementation can lead to unintended consequences, particularly for marginalized groups who are most affected by environmental degradation. Effective governance frameworks that involve ethical, social, and environmental factors are essential in realizing holistic transportation sustainability. By embracing a holistic model that involves ethical decision-making, the transport sector can discover its path towards sustainability and therefore create sustainable communities that will be in a position to thrive against the challenges caused by climate change (Saenchaiyathon et al., 2019) (Figure 10.7).

10.3.3. Equity and Accessibility in Transportation

Access and equity in transport are inherent aspects of contemporary urban planning that determine the mobility capacity of heterogeneous populations within and across neighborhoods. Transport equity is not only equal access to means of transport but also differential impact consideration across segments of the population, particularly marginalized groups. When cities are growing, the integration of intelligent technologies into transport infrastructure can enhance accessibility but, at the same time, can compound inequalities unless it is well managed. Through offering frameworks that emphasize inclusivity, planners in cities can ensure that interventions benefit all residents, particularly the ones who previously had their access to transport services hindered (Cohen et al., 2019). This commitment to fairness is consistent with broader moral attitudes that promote justice and justice in policy-making, which calls for solutions that acknowledge the unique needs of marginalized populations. The emphasis on sustainable transport solutions necessitates rethinking how policies think about equity and access. As much as new technologies like autonomous cars and shared micromobility schemes promise to revolutionize urban travel, they are likely to overlook the needs of elderly or disabled individuals, inadvertently perpetuating urban disadvantages. This is checked by inclusive policy solutions that not only incorporate technological innovation but also prioritize public consultation and involvement (Cohen et al., 2019). Transportation engineers thus must advocate for policies that ensure equal access to benefits by all users, as long as the benefits of technological innovations are equitably distributed. Then, transportation policies must be designed with the diverse needs of different socio-economic groups in mind, resulting in a fair urban infrastructure. Further, the intersection of climate change and socio-economic disparities additional muddles the conversation around accessibility and equity in transport. Climate change disproportionately affects poor and vulnerable groups of people, who do not always possess the ability to cope with environmental changes or transport disruption brought about by adverse weather phenomena. Employing equity-weighted indicators of the impacts of transport, policy-makers can better establish how differences in transport infrastructure may worsen harm to, or benefit, different communities. These equity-weighted solutions are essential in building sustainable transportation policy that does not expand but instead attempts to correct current inequalities. Pondering the broader implications of transportation planning within the context of climate stress, transportation engineers can assist in building resilient systems that prioritize accessibility for all members of society and prevent any from being left behind as society makes its way toward a sustainable future.

10.4. TECHNOLOGICAL INNOVATIONS IN TRANSPORTATION

The rapid growth of technology has changed the face of transport in a revolutionary way, leading to unprecedented innovations with impacts on mobility, efficiency, and sustainability. One of the most notable advances is the emergence of autonomous cars, which was welcomed with both excitement and controversies on safety, responsibility, and impacts on society. While proponents argue that such vehicles would greatly reduce car accidents, opponents indicate the impact of widespread employment loss and the challenges of developing secure models of operation. With application in mass transit, policymakers must deal with issues of regulation, making sure legislations and safety codes stay abreast of technology. Understanding how these innovations could enhance urban space and address ethical concerns is valuable in directing future transport policy. Air taxis and delivery drones are a good example of another new field of transport technology that can revolutionize urban mobility by utilizing airspace at low altitudes. Nevertheless, research indicates that the adoption of the technologies is faced with significant regulatory and technical challenges, which delay their adoption and application (Biehle et al., 2020). The potential benefits of reduced travel times and improved delivery services are a clear call to investment and expansion. Yet the aerial mobility debate must be inclusive, holding regard for public opinion as well as ecologic considerations. Critical evaluation suggests that economic potential for these innovations is promising, although stringent empirical evidence must be provided to substantiate the environmental and social consequences of these innovations, thereby suggesting the importance of encouraging public engagement in the regulation of future transportation technology. At the same time, green supply chain management (GSCM) represents a pivotal point for technology and environmental responsibility in transportation engineering (Saenchaiyathon et al., 2019). Through the focus on sustainability, organizations are increasingly forced to adopt GSCM practices that, in addition to being performance-improving, are also sustainable. The isomorphic institutional pressures specified in the existing literature suggest that producers ought to involve their customers in collaboration as a major ingredient of successful practice adoption. Moreover, government interventions that seek to create awareness and assist in the adoption of GSCM can assist in better industry standards. This dual-pronged approach not only promotes environmental stewardship but is also consistent with technological breakthroughs which act towards creating more efficient and sustainable transportation networks, thus making the future a place where economic desire equals ecological well-being (Figure 10.8).

Figure 10.8. Technological innovations in transportation.

Source: https://safarimagazines.com/wp-content/uploads/2024/06/Technological-Innovations-in-Modern-Transportation-like-GPS-navigation-and-electric-cars2.webp.

10.4.1. Autonomous Vehicles

The advent of autonomous vehicles (AVs) is a transport engineering revolution that has posed various policy and ethical concerns. Among these are the assurance and safety needs, as outlined in the Technical Topic Notes of the Towards Identifying and Closing Gaps in Assurance of Autonomous Road Vehicles (Tigars, 2025) project. With resilience and safety requirements being the mainstay, public trust and regulatory recognition are guaranteed. Policy suggestions need to consider exhaustive regulations and standards that ensure the safe deployment of AVs, particularly as they engage with human drivers and other road users. As AV technology advances, it is important to establish a strong foundation for formal verification and static analysis of machine learning systems behind their operation. This will not only ensure technical competence but also ethical compatibility with social requirements and norms, absolutely necessary for a seamless regulatory clearance process (Bloomfield et al., 2020). AVs' ethical concerns also emerge when pertaining to liability and decision-making algorithms. For instance, in case of an inevitable accident, who is to be blamed—the manufacturer, software developer, or owner? These are questions that demand a strong ethical framework to answer not only legal liability but also moral obligation. The transparency of decision-making algorithms is necessary in order to gain public acceptance and trust in autonomous vehicles. With the growing prevalence of AVs, embedding ethical values into their programming and operating procedures will become indispensable. This creates the issue of how one should most rightly program such vehicles to make snap decisions which are value-based for society. There is a profound need for inter-disciplinary engagement in the creation of standards of what is ethical for such technologies, that they concur with broad social values and promote the common good. Looking ahead, the governance of self-driving cars will also need to change to encompass the rapid developments in technologies associated with them, such as drones for use in the delivery industry and city air mobility. The systematic review of new technology literature indicates that, while there are enormous anticipated benefits such as reduced congestion and easing of the environment, there are equally complex problems of regulatory as well as technical challenges. With transportation systems evolving to integrate vertical mobility solutions, an integrated policy response will be crucial in addressing these challenges and realizing the full potential of both AVs and air taxis. Cross-sector stakeholder consultations are required to design transportation systems that not only innovate but also consider societal and environmental consequences in an integrated manner. The intersection of policy, ethics, and technology will all ultimately determine the future trajectory of transportation engineering in the next few years (Biehle et al., 2020).

10.4.2. Smart Infrastructure

The creation of smart infrastructure is increasingly important in handling the complexity of modern transportation systems. With cities growing and the need for efficient mobility on the rise, integration of technology into infrastructure presents a significant opportunity to enhance connectivity and enhance operational efficiency. Smart infrastructure utilizes cutting-edge communication technologies, sensors, and data analytics to provide end users with real-time data to enhance traffic flow, reduce congestion, and lower emissions. These technologies drive economic growth while addressing heightened environmental challenges. However, effective use of smart infrastructure demands collusion from the stakeholders that includes policymakers, engineers, and society as a whole, along with a solid framework that allows for ethical issues to be made part of the decision-making process behind such developments. Moreover, integration of smart infrastructure in transport engineering brings about advanced ethical issues that must be addressed with care. The use of such technologies as autonomous cars, for instance, is problematic in terms of liability, privacy, and the potential for socio-economic disparity among different groups of people. Equal sharing of technology's benefits is essential, to avoid further marginalization of already excluded groups. Researchers point out that involving the public in conversation around smart infrastructure projects is essential in a bid to integrate diverse views in policy and implementation systems. Such a commitment to public participation has the potential to build trust and help to create a sense of mutual ownership of the advantages delivered by smart transport technology, preventing the risk of perpetuating pre-existing inequalities. Addressing these ethics issues is crucial in creating a sustainable transport future that is inclusive and equitable. Emerging technologies such as drones and air taxis hold tremendous potential to reshape smart infrastructure, and by extension, the transport engineering sector. As pointed out by recent studies, such innovations hold the promise of injecting a new dimension into mobility, minimizing city blockages and improving accessibility (Biehle et al., 2020). However, incorporating them into existing frameworks is fraught with significant challenges, ranging from regulatory obstacles to public security issues. Effective policy must be crafted in a way that addresses these challenges without compromising the social benefits of such technologies. Furthermore, empirical data are critically necessary to substantiate assumptions regarding the advantages of such new systems. Overcoming anticipated challenges and participatory public discussion enable stakeholders to pave the way for a transportation evolution that optimizes efficiency, sustainability, and ethics and ensures that future transportation engineering trends will be innovative as well as responsible (Figure 10.9).

Figure 10.9. Smart infrastructure modern transportation systems.

Source: https://scwcontent.affino.com/AcuCustom/Sitename/DAM/028/STEng_image_traffic_tech_smart_cities_rt.jpg.

10.4.3. Data Analytics and Transportation Systems

The use of data analytics in the transportation industry has been a game-changing force in the evolution of infrastructure and urban planning within present-day cities. With cities becoming increasingly complex, utilizing big data analytics makes decision-making more holistic across various modes of transport. This analytical technique employs vast amounts of data obtained from sensors, GPS, and user input to optimize traffic flow and reduce congestion. Concepts that have been explored in the past in cities represent an intrinsic move towards smart city development, where transportation networks are linked with real-time data analysis. It is not cheap to change, as issues of data privacy, security, and ethical use emerge. Debates about such problems underscore the importance of formulating comprehensive policies that provide direction for the application of data in the transport sector, whereby stakeholders understand how to negotiate the space between innovation and ethical responsibility in their utilization (Doctor et al., 2020). Moreover, transport systems are viewed from supply chain management lenses, emphasizing the aspect of reactive adaptation to dynamic environments. Current events, such as the Fourth Annual World Class Supply Chain Summit, indicate industry professionals' need to embrace real-time data analytics if they are to remain competitive. Seminars by industry experts and scholars displayed a mutual examination of the ways in which advanced analytics can be employed to inform decision-making that impacts not only the effectiveness of transportation systems but also their adaptability to a shifting market. Firms currently are challenged to redesign distribution networks, requiring them to make their organizations more agile and adaptable. The lessons learned from such discourses demonstrate the pivotal convergence of data analytics and logistics systems, which is becoming increasingly important to tackle the sophistication of modern logistical problems (Haughton et al., 2019). The future of transport engineering relies on proper inclusion of data analytics and the creation of current and responsive policies. When the transportation scenario evolves, policymakers will have to make sure that policies suitably respond to shifting analytics while addressing ethical concerns of data privacy and tracking. The smart city plans will require the combined initiative of government bodies, tech firms, and civil organizations to create inclusive analysis-driven objectives. These frameworks must provide equitable access to transportation assets so that big data analytics solutions do not worsen inequalities. Ultimately, the marriage of policy, ethics, and data analytics will not only fuel greater transportation efficiency but provide the doorway to sustainable practices that will benefit generations to come. This integrated framework will establish the future of transportation engineering, focusing on collective responsibility to innovate responsibly and equitably (Doctor et al., 2020) (Figure 10.10).

Figure 10.10. Data analytics and transportation systems.

Source: https://external-content.duckduckgo.com/iu/?u=https%3A%2F%2Fwww.sansi.com%2Fupload%2Fimages%2F20220630105552493.jpg&f=1&nofb=1&ipt=f632c51a056af333526573ea70fc39921421e24ae1572425d94c2117f37cec45.

10.5. FUTURE TRENDS IN TRANSPORTATION ENGINEERING

As transport engineering advances, the use of advanced technologies is at the forefront of future trends. Automation and artificial intelligence are transforming urban transit networks, optimizing efficiency and security. Autonomous vehicles are not just a concept but an ever-present reality, which has the potential to reduce human error and establish new paradigms in urban planning, such as less parking space and reconfigured road layouts. Further, smart infrastructure deployment via sensors and real-time data analytics facilitates better traffic management and better utilization of resources. This new path toward smart systems requires policymakers to consider ethical implications like data privacy and fair access to new mobility technologies. As mentioned at the Fourth Annual World Class Supply Chain Summit, the necessity of being agile to adapt to rapid technological changes is becoming more necessary for all the stakeholders in the transportation sector (Haughton et al., 2019). Future transportation engineering trends also have environmental sustainability as a central aspect that is pushing innovation. As global concerns around climate action intensify, transport systems must reform themselves to minimize carbon footprints. This is through the application of electric and hybrid vehicles and integration of renewable sources of energy into public transportation networks. In addition, infrastructure development will see restructuring material and means to ensure they harmonize with environmental principles. In workshops assessing the Arctics shifting shipping routes, emphasis was placed on the aspect that economic and environmental impacts of transportation systems extend beyond their immediate environment (Ashcraft et al., 2018). Thus, policymakers and engineers must collaborate to implement solutions that address these intricate matters, incorporating design philosophies that emphasize putting sustainability and climate resilience at the core. As the infrastructure changes, so too must the regulatory frameworks that manage those policies. Further, the expansion of integrated transportation systems emphasizes the importance of cooperation among industries. Future transportation engineering will involve greater interconnection between land, sea, and air transport systems, making them more interdependent and efficient. Building multimodal systems, which efficiently integrate multiple modes of transport, will require sound planning, innovative technology, and effective stakeholders' engagement. Efficient transfer between transport modalities will allow cities to enhance mobility while reducing congestion and pollution. But these developments also create important ethical and policy concerns, mainly related to funding arrangements and the equitable allocation of resources. The players need to address these issues to provide equal access to sophisticated transportation modes, most importantly for disadvantaged groups. An integrated approach, evidenced in action from summits and workshops on future trends, is necessary for creating transport systems that are capable of consistently responding to future demand while promoting social equality (Haughton et al., 2019) (Figure 10.11).

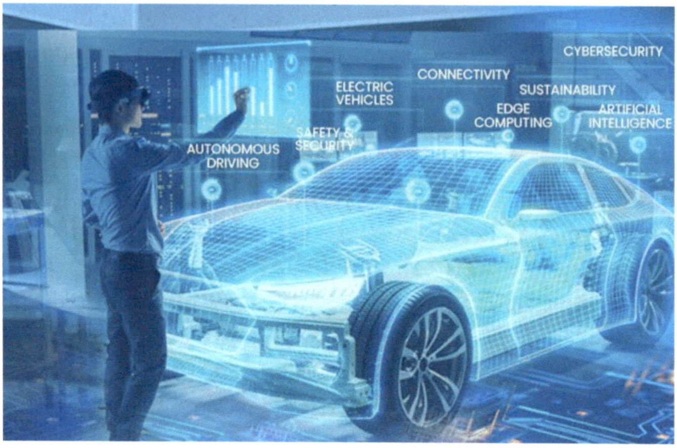

Figure 10.11. Future trends in transportation engineering.

Source: https://www.taaltech.com/wp-content/uploads/2023/08/8-current-future-trends-disrupting-automotive-embedded-systems-web-banner.jpg.

10.5.1. Electrification of Transportation

The electrification of transport is a watershed in addressing the interrelated challenges of energy consumption, greenhouse gas emissions, and city air pollution. While societies across the globe are struggling with the payoffs of pollution and climate change, renewable energy (RE) sources present a viable pathway to decouple economic development from dependence on fossil fuels. This shift is significant, especially in the context of SD policy aimed at enhancing energy security while promoting social equity. Technological advances in electric vehicle (EV) technology, together with conducive government policies and incentives, enable this shift. By employing RE for charging infrastructure, there is potential for significantly decreasing the greenhouse gas emissions sourced from transportation in accordance with worldwide goals for climate goals. These revolutionary developments do not simply demand technological improvements but also robust policy frameworks emphasizing sustainability, accessibility, and morality in transport engineering (Angerer et al., 2011). Vehicle electrification in the transportation system is not a mere technical step; it also poses ethical issues that need to be thoroughly taken into account. Policy frameworks that regulate the process of electrification must be in a way that access to clean transportation technologies is made equitable, particularly to marginalized communities that are disproportionately impacted by pollution. Additionally, manufacturing and discarding electric vehicles pose certain issues with respect to resource extraction and environmental justice. For instance, the mining for rare earth elements for batteries has harmful environmental and social impacts. Therefore, although electrification of transport is capable of appeasing some of society's problems, it is paramount to examine its richness from a multi-dimensional point of view. EV component production and recycling with sustainable approaches must be employed to minimize harmful effects and permit the advantage of electrification to be shared equitably. Hence, policymakers must integrate ethics in their approaches to incorporating technology (Abdul-Rahman et al., 2013). Ahead, transport engineering future developments indicate a convergence of electrification, urban planning, and integration of smart technologies. The appearance of autonomous electric cars and the development of world smart grid infrastructure have the potential to reshape mobility in cities in a dynamic manner. This evolution is dependent on imaginative urban designs made possible by policies that encourage electrified transport systems as a way of enhancing connectivity with reduced congestion and environmental degradation. Furthermore, public-private partnership will play a crucial role in mobilizing the necessary investment for infrastructure development. Transitioning to electrification in transportation also calls for extensive education campaigns in order to make people aware of the environmental and social impacts of these technologies. Thus, the construction of an educated and active citizenry through this transition process will be essential to the successful fulfillment of electrification alternatives in transport (Angerer et al., 2011).

10.5.2. Integration of Multimodal Transportation Systems

Multimodal transport systems' integration is key to ensuring the increase in the efficiency and sustainability of modern transport networks. Through integration, multimodal transport systems combine various modes of transport—rail, road, air, and sea—to enable easy transfers from one mode to another. Policymakers can, in turn, address increasing mobility demands of urban residents. Further, as urbanization gains momentum, calls for an integrated transport system are hastened to reduce congestion, reduce travel time, and enhance service delivery. Understanding the complexity in modern transport systems demands an analytical paradigm that not only relies on technological advancement but also on socio-economic and environmental implications. The need of ports, as a case in point, to evolve and diversify in terms of addressing various users' demands illustrates how complexity seems operationally. According to (Karalis et al., 2025), new ports must accommodate several service providers to remain competitive in the growing dynamic global economy. In addition, the ethical implications of consolidating multimodal transportation systems cannot be overlooked, especially as policymakers attempt to reconcile equity and efficiency. The transition towards more consolidated systems creates concerns regarding access, affordability, and environmental equity. Policy-making here has to be inclusive and take into account the interests of various population groups, and most importantly those who have been traditionally marginalized from the transportation planning process. The moral issues include the provision of making sure all citizens in the community are benefited by improved transport services while mitigating adverse effects on local environments. The call for strategic, visionary solutions that fit these moral paradigms is highlighted by issues of global warming, resource constraints, and public health crises. These are essential to creating a transport environment that not only addresses current needs but also anticipates future requirements of society, as indicated in Cández et al. (2014).

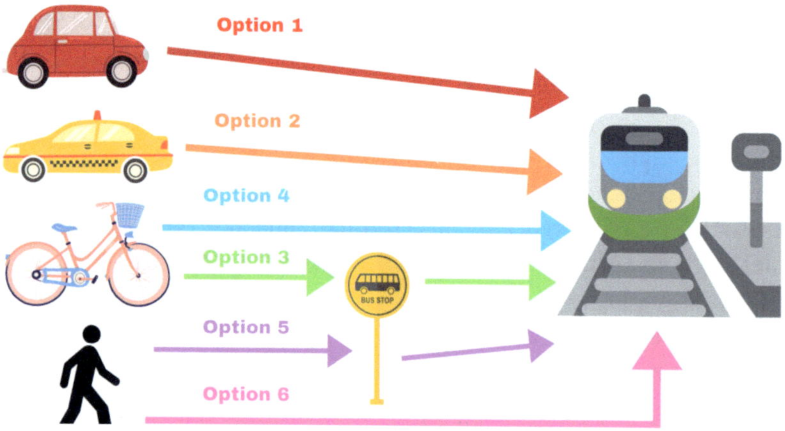

Figure 10.12. Integration of multimodal transportation systems.

Source: https://www.iunera.com/wp-content/uploads/Multimodal-transport-options-1024x576.png?v=1617956471.

The fate of integrated multimodal transport systems in the future will likely hinge on technological advances and learning governance. Decision-makers must undertake end-to-end planning that embraces forthcoming megatrends, such as moving towards green energy and changing urban mobility patterns. Combining data analytics, real-time information exchange, and automated networks can maximize the scheduling and efficiency of multimodal networks. Besides, incorporating these systems must be considered a revolutionary exercise that redefines traditional transport modes rather than substituting fossil-fuel-based modes with alternative fuels. This visionary approach requires collaboration among governments, the private sector, and communities in developing flexible frameworks that will be capable of accommodating technological advances and changing user demand. Initiating such a receptive, visionary approach will be paramount to establishing resilient transport systems that will be capable of thriving in a changing world environment as noted in Cández et al. (2014) (Figure 10.12).

10.5.3. Resilience and Adaptation to Climate Change

The paradigm of adaptation and resilience for climate change shifts increasingly to the center of transportation engineering in response to new global dangers. Because the character of transportation systems by necessity is subjected to climate-related disruptions like sea-level rise and extreme weather, engineers must focus on how infrastructure capacity for resilience can be enhanced. More recent attempts, such as those outlined in the idea of crossing the New Arctic, show that areas other than those normally associated with these changes, i.e., New England, will also be affected. Economic, ecological, and societal processes will be modified by new shipping routes and coastal fauna generated as a result of Arctic change, demanding a forward-looking response from engineering disciplines. This response needs to incorporate short-term infrastructural modifications and long-term strategic planning in order to mitigate the far-reaching effects of climate change on transportation systems (Ashcraft et al., 2018). Building resilience within transportation systems means not only technical modifications, but emphasizes ethical aspects related to community susceptibility and equity. The Global Risks Report identifies the immediacy of climate warming impacts, where millions stand to be displaced by environmental degradation. Policymakers and engineers thus have a moral responsibility to prioritize vulnerable populations in transport planning. By making equal access to effective and secure transport options available, they can insulate the most vulnerable groups from climate impacts. This demands a cooperative environment that mobilizes local wisdom and stakeholder perspectives to make feasible responses that are not only effective, but socially responsible. The development of such environments is critical to constructing resilience in the situation of increasing global threats and environmental imbalances documented in the Report (N/A, 2016). Finally, transportation engineering adaptation and resilience action will be ideologically determined by new technological advances and policy institutions. Technologies such as clean energy and smart infrastructure can be pivotal to the establishment of responsive capacities that are a counter-mitigation to climate change. For instance, the integration of real-time climate data and intelligent transportation systems enhances the decision process during weather-related disasters. Also, policy interventions must encourage investment in projects that have resilience as a priority while ensuring sustainability and reducing carbon emissions. As new global challenges create a new normal, transportation engineering must keep pace with these innovations, so that adaptive measures are forward-looking and grounded in ethical analysis. This synergy of technological development with forward policy will be essential to making transportation systems prepared to thrive under uncertainty created by changing climate (N/A, 2016).

10.6. CHALLENGES FACING TRANSPORTATION ENGINEERING

The transport engineering profession is presently confronted with various challenges spurred by the rapid rate of technological transformation and evolving socio-economic conditions. Perhaps one of the most daunting issues is how to incorporate smart technology into existing infrastructure, which demands considerable finance and technical know-how. As networked transportation systems and autonomous vehicles grow in prominence, engineers will be forced to deal with the issue of creating roads that communicate with cars and the other pieces of the transportation infrastructure. Further, a green shift in practice requires that transportation engineers go looking for green materials and practices, as well as adapt to new policies that promote environmental responsibility. This transformation necessitates not just an awareness of state-of-the-art technologies but an allegiance to uphold mature engineering axioms while striving toward visionary and adaptive systems capable of solving the demands of future mobility (Haughton et al., 2019). A further challenge is the interdisciplinary nature of transportation engineering, which more and more requires collaboration between various stakeholders such as government agencies, private companies, and research centers. The divergent interests and agendas of such organizations are able to complicate project development, as every stakeholder may have varying goals and expectations. Harmonization of these interests while maintaining safety, efficiency, and environmental sustainability is not merely a logistics undertaking but a relevant ethical issue. Transportation engineers must navigate these complex dynamics so that the projects are equitable and for the common benefit of the society. This collaboration with diverse perspectives can enhance solution quality but also creates conflicts that must be properly managed to allow for effective project outcomes (Ashcraft et al., 2018). Finally, transportation engineers face the daunting challenge of transitioning infrastructure to accommodate climate change effects, which significantly endanger existing transportation infrastructure. With more frequent and intense extreme weather occurrences, engineers must include resilience in their designs. This includes predicting flooding, erosion, and other environmental changes that impair transport infrastructure. Increasingly sophisticated design techniques are necessary to withstand such forces as well as the use of sustainable techniques in maintenance and reconstruction activities. Besides, engineers must advocate for policies that support investment in infrastructure and good environmental planning within transport systems. Challenging these issues requires not just technical expertise but commitment to positive, forward-looking policies to render transport systems resilient, safe, and efficient in anticipation of imminent environmental change (Haughton et al., 2019) (Figure 10.13).

Figure 10.13. Challenges facing transportation engineering.

Source: https://jugnoo.io/wp-content/uploads/2021/08/multimodal.png.

10.6.1. Funding and Budget Constraints

Budget constraints have become a watchword in the world of transportation engineering in its very nature, determining scopes of projects and making possible the justification of funding allocations. As governments struggle to manage limited resources, with it comes a sensitive evaluation of what transportation projects will yield high economic and social returns. With increasing costs of infrastructure creation and upkeep, particularly for green transport modes, engineering professionals must navigate the complex interplay of finance levels and project feasibility. Above all, public safety issues and environmental considerations now compete for funding, with the result that there is a high-stakes balancing act to be achieved within transport policy processes. And as highlighted in recent studies, the imperative to renew must be answered with financial measures that are not only effective but also equitable, and where investments are aligned with broader social goals and moral imperatives guiding the development of transport infrastructure. Additionally, advances in transport technology added a new dimension to budget constraints, particularly with the integration of sophisticated systems like smart traffic management and autonomous vehicles. New technologies do come with enormous initial investment costs, and therefore there are controversies surrounding how the available budgets are channeled into innovation compared to traditional infrastructural construction. The budgetary issue is also aggravated by institutional structures that are not dynamic enough to accommodate emerging technological developments. In publicly sponsored transport projects, for instance, policymakers and engineers are responsible for enhancing fiscal models of sponsorship that can accommodate innovation but with regard to budget needs. A collaborative funding strategy, like public-private partnerships, can be a solution for addressing the funding gap for innovative projects in the field of transportation engineering, paving the way for sustainable growth and development. Finally, addressing issues of funding and budget in transportation engineering demands a participatory strategy involving stakeholders from industry as well as the community. Cooperative strategies can render the funding more open and accountable, with funds allocated based on public interests as well as ethical values. Calls for additional public engagement emphasized in recent studies speak to the need for adaptive regulatory frameworks that are able to respond to the mounting complexities of urban mobility and ecological impacts. By the integration of citizens' contribution into the decision-making process, transport engineers can more effectively allocate priorities to projects that reflect the aspirations and values of the people. Not only does this maximize the effective use of limited resources, but it also adds more strength to the ethical underpinning upon which transport policies are created, the building block of a more sustainable and equitable transport engineering future.

10.6.2. Public Resistance to Change

Public opposition to change is a key factor in framing policy choices within the domain of transportation engineering. As novel technologies and infrastructural improvements appear, public trepidation tends to arise as an impediment to change. Such resistance may be generated by a range of factors, such as fear of the new, skepticism regarding governmental intentions, and deeply established societal values. Parallels drawn from history demonstrate that high levels of public anxiety can shape regulatory policies. For instance, social protest against agri-genetic modification led to a total reconsideration of GM crop policy, reflecting a broader requirement for safety and transparency (Waele et al., 2006). In transport, the same kind of public sentiment can halt advancements in technologies such as autonomous vehicles, in which safety, privacy, and ethical issues are foremost in the minds of the public. As transportation engineering evolves with technologies such as electric vehicles, intelligent infrastructure, and shared mobility programs, stakeholders must address the complex universe of public opinion. It can minimize opposition by engaging the public in genuine discourse through embracing a more participatory process of change. For that, public participation in the policy-making process must be prioritized to the greatest extent so that the public voice concerns and engage in public discussions regarding prospective transport projects. As emphasized in the Global Risks Report, emerging trends such as climate change and technological dependency necessitate a transdisciplinary strategy to embrace new technologies and manage public phobias (N/A, 2016). Proactively engaging stakeholders in the conversation helps transportation policymakers and engineers enhance social trust and facilitate the path towards more sustainable solutions. Besides, understanding the ethical side of public opposition is essential in enabling the embrace of change in transport projects. The precautionary principle, as elsewhere, is a valuable framework that enhances the review of potential risks prior to the adoption of new technologies. In transportation engineering, this involves considering environmental impacts, social impacts, and ethical concerns of equity and access. Public resistance is likely to reveal deeper social injustices and can serve as a catalyst for more equitable policy approaches. The ongoing public debate with regulatory norms reflects a need for sustained ethical consideration of transportation solutions (Waele et al., 2006). Addressing these ethical dilemmas head-on, engineers and policymakers can move toward transforming opposition into a unifying force for a more sustainable, technology-driven transportation future.

10.6.3. Balancing Innovation with Regulation

The rapid evolution of transport technology brings opportunities and challenges that must be addressed by effective regulation. Among them are technologies such as electric vehicles, autonomous transport, and unmanned aerial vehicle delivery systems, which have the potential to greatly improve efficiency, safety, and environmental performance. But they also pose complicated regulatory issues that policymakers must work through to capture their benefits while ensuring the public interest. For instance, autonomous car deployment means rewriting current traffic law, insurance infrastructures, and liability standards, to ensure that the law keeps up with evolving capabilities. The question is how to design regulation that is powerful enough to manage risks but not suffocating to innovation, and thus the idea of adaptive regulatory regimes that are able to grow alongside the technologies they aim to regulate. In this case, the balancing acts between the promotion of innovation and safety will be the solution to a sustainable transportation future. In recent years, green supply chain management (GSCM) has been in the center of the spotlight as a primary area where regulatory needs must converge with innovation. Institutional isomorphism increasingly affects GSCM practices through which firms respond to pressures from regulatory bodies, their competitors, and the environment-related concerns of customers. There is evidence that compliance with such pressures can lead to improved firm performance, where there is evidence of the synergistic relationship between regulation and innovation (Saenchaiyathon et al., 2019). The transport sector is one where regulatory needs prompt investment in clean technology, thus compelling organizations to innovate. Through this manner, effective regulation is able to spur innovation by providing clear parameters that allow for sustainable practices along with giving businesses the autonomy to come up with future-oriented solutions that comply with or exceed the stipulated requirements. New technology such as air taxis and drones also creates new levels of complexity for the transportation engineering regulatory landscape. Although these technologies hold the promise of improved mobility and reduced congestion, they also raise significant regulatory concerns regarding safety, privacy, and environmental impact. A literature review of civil drones identifies numerous anticipated barriers to their integration with existing transport infrastructures, emphasizing the need for open regulatory systems to address such issues. The socio-technical debate around the use of drones emphasizes the importance of regulators taking equal heed not only of the economic benefits but also social and environmental issues that result from such technologies (Biehle et al., 2020). Policymakers must thus balance the debate with stakeholders in order to come up with regulations that foster technological advancement without compromising public safety and ecological balance.

SUMMARY

Briefly, policy making for transportation engineering has a direct connection with ethical concerns and the new trends forced by the recent difficulties of the world. With the demand for eco-friendly transport solutions increasing, addressing the mounting issues of mobility in cities becomes unavoidable. Implementing master policies that move towards sustainability necessitates, beyond innovative engineering mechanisms, structures that enable stakeholders to participate actively. This approach is particularly relevant to greenhouse gas emissions management, as shown by studies on the challenges of estimating campus-emissions quantification. Universities, for instance, must create a proper accounting process of baseline emissions as outlined in (Kawahara et al., 2019). This necessity indicates wider significance for transport policy since universities are increasingly recognizing their contributions to limiting climate change impacts. Furthermore, the ethical dimensions of transport engineering reveal a strong need for equitable policy making. Attempts to expand road capacity or to invest merely in more vehicle capacity can only perpetuate inequality and environmental degradation. In contrast, a shift in paradigm to demand management, investment in public transport, and alternatives such as cycling and walking is the only way toward sustainable development. The principles advocated in Goodwins research to examine transport policy ring a collective consensus that has over time been developed and demands the immediate integration of sustainable practices into engineering designs. As noted in Anable et al. (2011), the original concepts regarding sustainable transport still dominate; however, actual change is often undermined by entrenched political forces. Progressive governance models thus are key to dismantling barriers that limit significant progress in transportation policy. The future of transportation engineering, as such, relies on the capacity to embrace shifting societal values and technological advancements. For sustainability to be realized, it not only warrants reconsidering existing infrastructure but also embracing new modalities reflecting the transforming transportation landscape. The ethical implications of embracing technology in transport imply that policymakers and engineers have to exercise caution in ensuring they keep safety, efficiency, and equity at a high priority. The research findings assessing campus travel behavior unveil the importance of applying evidence-based decision-making when investigating the efficacy of transport policy. As transportation engineering progresses, the incorporation of interdisciplinary strategies will be essential in responding to interdependent problems of mobility, environmental accountability, and social equity, steering the field toward a sustainable future.

Transportation engineering increasingly intersects with policy and ethics, particularly as we project into future industry trends. Key lessons from recent controversies are that flexibility will be crucial for practitioners operating in environments of evolving technology, environmental concerns, and legislative regulations. The Fourth World Class Supply Chain Summit illustrated this imperative, with focus on practitioners having robust IT infrastructures in place and being responsive to customers and competitors' requirements. Participants engaged in discussions regarding the necessity of agility in traditional supply chain models, which is also so essential in transportation networks. This adaptability shapes contemporary policy mentality, where policymakers must take account of what approach best serves evolving conditions affecting transport systems (Haughton et al., 2019). Moreover, ethics in transport engineering emerge in the environmental sustainability and social responsibility debate. The search for green practices, as substantiated by studies on green supply chain management

(GSCM), shows how institutional pressures are driving businesses toward environmentally responsible practices. Evidence indicates not just that the practice of GSCM is positively associated with business performance, but also that it constitutes an ethical necessity to hold in check environmental impacts. As firms face pressures from customers and regulatory compliance, the ethical aspects of transport engineering come to the forefront. Addressing these ethical issues involves referring to different stakeholders in order to have policies that accommodate ecological preservation and financial viability (Saenchaiyathon et al., 2019). In the coming years, transportation engineering will be shaped by the advances in technology that will enhance operational efficiency and support sustainable operations. With new technology being incorporated into transportation systems, it is necessary to understand the policy implications of such technological development. Autonomous vehicles and smart city technologies threaten existing regulations and necessitate the formulation of visionary policies that address safety, privacy, and environmental concerns. Enabling interdisciplinary collaboration among engineers, policymakers, and ethicists will be essential in creating holistic solutions that not only embrace technological advancement but also prioritize public good and environmental stewardship. Such alignment will be important for ensuring transportation engineering evolves responsibly to cater to future generations (Haughton et al., 2019).

REFERENCES

1. Abdo, L., Coupland, G., Griffin, S., Kemp, R., et al. (2019). Biodiversity offsets can be a valuable tool in achieving sustainable development: Developing a holistic model for biodiversity offsets that incorporates environmental, social and economic aspects of sustainable development. https://core.ac.uk/download/232131035.pdf (accessed on 21st October 2025).
2. Abdul-Rahman, H., Anumba, C., Arango, S., Assaf, S., Batlle, A., Bautista, M., Bhide, A., et al. (2013). The impact of sustainable construction and knowledge management on sustainability goals: A review of the Venezuelan renewable energy sector. https://core.ac.uk/download/148667100.pdf (accessed on 21st October 2025).
3. Anable, J., Banister, D., Banks, N., Begg, D., et al. (2011). The transformation of transport policy in Great Britain? 'New Realism' and New Labor's decade of displacement activity. https://core.ac.uk/download/110265.pdf (accessed on 21st October 2025).
4. Angerer, G., Bauer, C., Bazilian, M., Brecha, R., et al. (2011). Renewable energy in the context of sustainable development. https://core.ac.uk/download/13750495.pdf (accessed on 21st October 2025).
5. Ashcraft, C. M., Brewer, J. F., Burakowski, E., Coffin, A., et al. (2018). Preparing for a Northwest Passage: A workshop on the role of New England in navigating the new Arctic. https://core.ac.uk/download/215537367.pdf (accessed on 21st October 2025).
6. Biehle, T., Fischer, L., & Kellermann, R. (2020). Drones for parcel and passenger transportation: A literature review. https://core.ac.uk/download/323212148.pdf (accessed on 21st October 2025).
7. Bloomfield, R. E., Fletcher, G., Imai, K., Khlaaf, H., et al. (2020). Towards identifying and closing gaps in assurance of autonomous road vehicles–A collection of technical notes Part 1. https://core.ac.uk/download/305120054.pdf (accessed on 21st October 2025).
8. Castro Fernández, A., Christ, A., Hoppe, M., Seppänen, S., et al. (2014). Transformation in transportation? https://core.ac.uk/download/162589185.pdf (accessed on 21st October 2025).
9. Cohen, A., Davis, R., Dowd, M. K., Shaheen, S., et al. (2019). A framework for integrating transportation into smart cities. https://core.ac.uk/download/289097213.pdf (accessed on 21st October 2025).
10. Davies, J. C. (2008). Nanotechnology Oversight: An Agenda for the New Administration. https://core.ac.uk/download/71353341.pdf (accessed on 21st October 2025).

11. De Waele, D., Devos, Y., Reheul, D., Van Speybroeck, L., et al. (2006). The interplay between societal concerns and the regulatory frame on GM crops in the European Union. https://core.ac.uk/download/55749078.pdf (accessed on 21st October 2025).
12. Doctor, F., Iqbal, R., Mahmud, S., More, A., et al. (2020). Big data analytics: Computational intelligence techniques and application areas. https://core.ac.uk/download/228156656.pdf (accessed on 21st October 2025).
13. Franek, A., Loucks, D. P., Koncagul, E., Unver, O., Connor, R., Grego, S., Abete, V., et al. (2012). Managing Water Under Uncertainty and Risk: The United Nations World Water Development Report 4. https://core.ac.uk/download/71360160.pdf (accessed on 21st October 2025).
14. Halberg, N., Niggli, U., Schlüter, M., Schmid, O., et al. (2008). Vision research agenda to 2025. https://core.ac.uk/download/10929353.pdf (accessed on 21st October 2025).
15. Haughton, M. (2019). World class supply chain 2019: Next generation ideas. https://core.ac.uk/download/225556480.pdf (accessed on 21st October 2025).
16. Hepburn, C., Anthoff, D., & Tol, R. S. J. (2025). Equity weighting and the marginal damage costs of climate change. https://core.ac.uk/download/pdf/6461183.pdf (accessed on 21st October 2025).
17. Karalis, A. V., Pallis, A. A., & Chlomoudis, C. I. (2025). Port Reorganization and the Worlds of Production Theory. https://core.ac.uk/download/pdf/9314684.pdf (accessed on 21st October 2025).
18. Kawahara, T., Liu, B., Pande, A., Thigpen, C., et al. (2019). Moving from walkability? Evaluating traditional and emerging data sources for evaluating changes in campus-generated greenhouse gas emissions. https://core.ac.uk/download/289097246.pdf (accessed on 21st October 2025).
19. Mayman, K., & Lefebvre, R. (2025). Fading Productivity–Making Sense of Canada's Productivity Challenge. https://core.ac.uk/download/pdf/6922690.pdf (accessed on 21st October 2025).
20. N/A. (2016). The Global Risks Report 2016, 11th Edition. https://core.ac.uk/download/71338494.pdf (accessed on 21st October 2025).
21. N/A. (2017). SciTech News, Volume 71, No. 1 (2017). https://core.ac.uk/download/83049835.pdf (accessed on 21st October 2025).
22. N/A. (2025). Report of the Advisory Committee on the Future of the US Space Program. https://core.ac.uk/download/pdf/42818545.pdf (accessed on 21st October 2025).

23. Saenchaiyathon, K., & Wongthongchai, J. (2019). The key role of institution pressure on green supply chain practice and the firm's performance. https://core.ac.uk/download/288626090.pdf (accessed on 21st October 2025).
24. Seiler, A. (2003). The toll of the automobile. https://core.ac.uk/download/11694660.pdf (accessed on 21st October 2025).

INDEX

M

N

O

P

T

U

V

W